The Discrete Mathematical Charms of Paul Erdős

Paul Erdős published more papers during his lifetime than any other mathematician, especially in discrete mathematics. He had a nose for beautiful, simply stated problems with solutions that have far-reaching consequences across mathematics. This captivating book, written for students, provides an easy-to-understand introduction to discrete mathematics by presenting questions that intrigued Erdős, along with his brilliant ways of working toward their answers. It includes young Erdős's proof of Bertrand's postulate, the Erdős–Szekeres Happy End Theorem, De Bruijn–Erdős theorem, Erdős–Rado delta-systems, Erdős–Ko–Rado theorem, Erdős–Stone theorem, the Erdős–Rényi-Sós Friendship Theorem, Erdős–Rényi random graphs, the Chvátal–Erdős theorem on Hamilton cycles, and other results of Erdős, as well as results related to his work, such as Ramsey's theorem or Deza's theorem on weak delta-systems. Its appendix covers topics normally missing from introductory courses. Filled with personal anecdotes about Erdős, this book offers a behind-the-scenes look at interactions with the legendary collaborator.

VAŠEK CHVÁTAL is Professor Emeritus of Concordia University, where he served as Canada Research Chair in Combinatorial Optimization (2004–11) and Canada Research Chair in Discrete Mathematics from 2011 until his retirement in 2014. He is the author of *Linear Programming* (1983) and co-author of *The Traveling Salesman Problem: A Computational Study* (2007). In the 1970s, he wrote three joint papers with Paul Erdős. He is a recipient of the CSGSS Award for Excellence in Teaching, Rutgers University (1992, 1993, 2001) and co-recipient of the Beale–Orchard–Hays Prize (2000), Frederick W. Lanchester Prize (2007), and John von Neumann Theory Prize (2015).

"Vašek Chvátal was born to write this one-of-a-kind book. Readers cannot help but be captivated by the evident love with which every page has been written. The human side of mathematics is intertwined beautifully with first-rate exposition of first-rate results."

— Donald Knuth, Stanford University

"This book is a treasure trove from so many viewpoints. It is a wonderful introduction and an alluring invitation to discrete mathematics – now a central field of mathematics identified mostly with the hero of this book. With lucid, carefully planned chapters on different topics, it demonstrates the unique way in which Paul Erdős, one of the most prolific and influential mathematicians of the 20th century, invented and approached problems. Sprinkled with historical and personal anecdotes and pictures, it opens a window to the unique personality of "Uncle Paul". And implicitly, it reveals the charming and candid way in which Vašek Chvátal, an authority in the field and a lifelong friend and collaborator of Erdős, likes to combine teaching and story-telling."

— Avi Wigderson, IAS, Princeton

"Paul Erdős is one of the founding fathers of modern combinatorics, whose ability to pose beautiful problems greatly determined the development of this field and influenced many other areas of mathematics. This book uses some basic questions, which intrigued Paul Erdős, to give a nice introduction to many topics in discrete mathematics. It contains a collection of beautiful results, covering such diverse subjects as discrete geometry, Ramsey theory, graph colorings, extremal problems for graphs and set systems and some others. It presents many elegant proofs and exposes the reader to various powerful combinatorial techniques."

— Benjamin Sudakov, ETH Zurich

"This is a brilliant book. It manages in one fell swoop to survey and develop a large part of combinatorial mathematics while at the same time chronicling the work of Paul Erdős. His contributions to different areas of mathematics are seen here to be part of a coherent whole. Chvátal's presentation is particularly appealing and accessible. The wonderful personal recollections add to the mathematical content to provide a portrait of Erdős' mind recognizable to those who knew him."

— Bruce Rothschild, University of California, Los Angeles

"Vašek Chvátal's book is a gem. Paul Erdős' favorite problems and best work are beautifully laid out. Readers unfamiliar with Erdős' work cannot fail to appreciate its power and elegance, and those who have seen bits and pieces will have the pleasure of seeing it thoughtfully and lovingly presented by a master. It's hard to imagine now, but there was a time when combinatorics was thought to be a jumble of results without depth or coherence. "Uncle Paul" understood its heart and soul, and nowhere is this more evident than in Chvátal's wonderful compendium. This volume belongs on every math-lover's night-table!"

– Peter Winkler, Dartmouth College

"Beautiful mathematics is presented with great care and clarity in Vašek Chvátal's book, complemented with well-written anecdotes and personal reminiscences about Paul Erdős. This combination makes the book a very enjoyable reading and a lively tribute to the memory of one of the most prolific mathematicians of all time. Studying discrete mathematics from this book is likely to give a great experience to students and established researchers alike."

– Gábor Simonyi, Hungarian Academy of Sciences

"A fantastic blend of math, history and personal anecdotes; a true mathematician's perspective on the legacy of a legend."

– Maria Chudnovsky, Princeton University

Vašek Chvátal, 1972. Photo©Adrian Bondy

The Discrete Mathematical Charms of Paul Erdős

A Simple Introduction

VAŠEK CHVÁTAL

Concordia University

Shaftesbury Road, Cambridge CB2 8EA, United Kingdom

One Liberty Plaza, 20th Floor, New York, NY 10006, USA

477 Williamstown Road, Port Melbourne, VIC 3207, Australia

314–321, 3rd Floor, Plot 3, Splendor Forum, Jasola District Centre, New Delhi – 110025, India

103 Penang Road, #05–06/07, Visioncrest Commercial, Singapore 238467

Cambridge University Press is part of Cambridge University Press & Assessment,
a department of the University of Cambridge.

We share the University's mission to contribute to society through the pursuit of
education, learning and research at the highest international levels of excellence.

www.cambridge.org
Information on this title: www.cambridge.org/9781108831833

DOI: 10.1017/9781108912181

First published 2021

A catalogue record for this publication is available from the British Library

Library of Congress Cataloging-in-Publication data
Names: Chvátal, Vašek, 1946– author.
Title: The discrete mathematical charms of Paul Erdős : a simple introduction / Vašek Chvátal.
Description: Cambridge ; New York, NY : Cambridge University Press, [2021] | Includes
bibliographical references and index.
Identifiers: LCCN 2021000630 (print) | LCCN 2021000631 (ebook) | ISBN 9781108831833
(hardback) | ISBN 9781108927406 (paperback) | ISBN 9781108912181 (ebook)
Subjects: LCSH: Erdős, Paul, 1913–1996. | Discrete mathematics. |
Mathematicians – Hungary – Biography.
Classification: LCC QA297.4 .C48 2021 (print) | LCC QA297.4 (ebook) | DDC 511/.1–dc23
LC record available at https://lccn.loc.gov/2021000030
LC ebook record available at https://lccn.loc.gov/2021000630

ISBN   978-1-108-83183-3   Hardback
ISBN   978-1-108-92740-6   Paperback

To Markéta

Another roof, another proof.

Paul Erdős

There will be plenty of time to rest in the grave.

Paul Erdős

Nostalgia isn't what it used to be.

Simone Signoret

Contents

Foreword

There will be an answer, let it be...

The Beatles

It is every scientist's dream to publish a work which is complete and perfect, which presents the best solution to a major problem or the ultimate survey of an important field of research. A work which will require no updates for a very long time. Unfortunately, this dream is almost never realized, for it would contradict the laws of the evolution of science. Sometimes the failure is dramatic. After completing his magnum opus, the *Grundgesetze,* in which he proposed a solid foundation for arithmetic based on mathematical logic, Gottlob Frege received a letter from Bertrand Russell, describing his famous paradox. Frege had to append an epilogue to his book, in which he wrote: "A scientific writer can hardly encounter anything more undesirable than that after completing his work, its foundations are shattered. I was put into this position by a letter from Mr. Bertrand Russell."[a]

Most cases, however, are more mundane. There are almost two hundred thousand mathematical papers published every year. Even if none of them shakes the foundation of a particular monograph in preparation, many may be relevant to the subject. A perfectionist author might want to reference at least the most significant ones. This is a never-ending struggle, and prolonging the process may actually harm the project. I have seen many beautiful mathematical manuscripts which, by the time they were published, doubled in size and became much less appealing. When Vašek sent me the first version of his book *The Discrete Mathematical Charms of Paul Erdős,* it struck me that this was not a draft. It was a piece of art, essentially complete and finished. I replied to him right away: "By adding more references, details, pointers, and clarifying paragraphs here and there, you would only blur this miraculous gem. Just correct the misprints and obvious errors and sit back! *Let it be!*" Vašek wrote back immediately. He was puzzled: How could I possibly have learned about the cover he proposed to his editor? I had no idea what he was talking about. Then I opened the attachment to his email, and it was my turn to be baffled: I saw a replica of the cover of the famous Beatles album, *Let It Be,* with the portraits of the four band members

---

[a] Einem wissenschaftlichen Schriftsteller kann kaum etwas Unerwünschteres begegnen, als dass ihm nach Vollendung einer Arbeit eine der Grundlagen seines Baues erschüttert wird. In diese Lage wurde ich durch einen Brief des Herrn Bertrand Russell versetzt, als der Druck dieses Bandes sich seinem Ende näherte [6].

replaced by four photos of Paul Erdős! We concluded that this must have been just another manifestation of Jung's synchronicity.

Most of the songs of *Let It Be* were conceived during the tumultuous year of 1968, which brought huge anti-war and civil rights protests in the United States, general strikes and student riots in France, as well as the Prague Spring, a major attempt to liberalize Soviet-type communism. Suddenly everything seemed possible: Yale College started admitting female students, the first manned spacecraft entered orbit around the moon, Pierre Trudeau became prime minister of Canada ... and the 22-year-old Vašek Chvátal left Czechoslovakia to discover and conquer the world. He was about the same age as Paul Erdős at the time Erdős left Hungary for England to escape growing anti-Semitism and chauvinism in his country. In 1934, the political atmosphere appeared much grimmer than in 1968. Nevertheless, it is hard not to notice that the quest for personal freedom played a similar role in the lives of Erdős and Chvátal.

"Von Haus aus,"[b] Erdős spoke very good German and English, albeit with a heavy Hungarian accent. When George Csicsery made a documentary [5] about him, Erdős's English words were consistently subtitled in English. He did not pay much attention to polishing his pronunciation or style, but his systematic use of a special, Erdős-esque vocabulary added a unique and humorous flavor to his speech.[c] Vašek learned English and French in record time. Four years after "making landfall" in Canada,[d] his short story [3] was included in Martha Foley's list of Best American Short Stories of the year.

By offering a new "unorthodox" combinatorics course at Concordia University, Chvátal set out on a very ambitious project: to give a concise introduction to some of the basic concepts and results of discrete mathematics through the work of Paul Erdős, one of the founding fathers of the subject. The present book grew out of this project. What is the best way to present this material? Vašek is a master expositor. At the beginning of his classic monograph [4], he illustrated his approach by quoting Ralph P. Boas [2]: "Suppose you want to teach the 'cat' concept to a very young child. Do you explain that a cat is a relatively small, primarily carnivorous mammal with retractile claws, a distinct sonic output, etc.? I'll bet not. You probably show the kid a lot of different cats saying 'kitty' each time until it gets the idea." This strategy is in perfect agreement with the spirit of Erdős's mathematics. He did not like to formulate big theories and metatheorems. He had a nose for seductively beautiful, simply stated mathematical problems with solutions that have far-reaching consequences in several areas across mathematics. Each section of this book presents one or more such questions, together with Erdős's brilliant attempts to answer them. Each answer, each theorem, leads to a multitude of further exciting problems.

---

[b] Straight from home.
[c] "Slave" for husband, "boss" for wife, "epsilons" for children, "Sam" for the United States, "Joe" for the Soviet Union, "poison" for alcohol, etc.
[d] Adrian Bondy's expression from Section 1 of [1].

I must warn you: If you are attracted to combinatorics and the chemistry is right, you are likely to fall in love with this book. In any case, quoting from [3], "either you get the telepathic shock, the unmistakable click, and then you just KNOW it – the only thing you can be absolutely sure about, outside the cogito-ergo-sum syllogisms world. Or you don't and then there is no way to put it into words for you."

János Pach
Rényi Institute (Budapest) and EPFL (Lausanne)

References

[1] D. Avis, A. Bondy, W. Cook, and B. Reed. Vašek Chvátal: A very short introduction, *Graphs and Combinatorics* **23** (2007), 1–25.

[2] R. P. Boas. Can we make mathematics more intelligible?, *American Mathematical Monthly* **88** (1981), 727–731.

[3] V. Chvátal. Déjà Vu, *PRISM international,* **11** 3 (1972), 27–40.

[4] V. Chvátal. *Linear Programming,* W. H. Freeman and Company, New York, 1983.

[5] G. P. Csicsery (director). *N Is a Number: A Portrait of Paul Erdős.* Documentary about the life of mathematician Paul Erdős, ZALA films, 1993.

[6] F. G. Frege. *Grundgesetze der Arithmetik – Begriffsschriftlich abgeleitet. Band I und II: In moderne Formelnotation transkribiert und mit einem ausführlichen Sachregister versehen*, edited by T. Müller, B. Schröder, and R. Stuhlmann-Laeisz, Paderborn, Mentis Verlag, 2009.

Preface

Paul Erdős (26 March 1913 – 20 September 1996) was an outstanding, prolific, influential, legendary mathematician.

Three times between January 2007 and December 2009, I taught at Concordia a one-term course of my own design entitled *Discrete Mathematics of Paul Erdős*. This was a graduate course, but it was open to undergraduates as well. Colleagues from other Montreal universities frequently sat in the audience, too, and this delighted me very much. The lecture rooms assigned to us were not always adequate. The sight of people huddled in the space overheated by their bodies and straining their ears in the doorway made me feel like a participant in clandestine gatherings of the early Christians. The present book is based on my lecture notes for that course. From time to time, I strayed away from the syllabus and talked about my own interactions with Erdős. A few of these recollections are recorded here, too.

Others were much closer to Erdős than I was. Others are better qualified to provide a tribute to him than I am (and some did). Still, large mosaics are made out of small fragments such as the reminiscences of mine that are sprinkled throughout the following text. I have been one of the blind men holding onto an elephant, and these vignettes form my report on what I felt.

The objective of my course was to survey results of Erdős and others that laid the foundations of discrete mathematics before it matured into the rich and vibrant discipline of today. Revisiting them after several decades brought back memories of pulling down heavy volumes from library shelves, leafing through their yellowing pages, and wondering at the treasures that they revealed. Memories like faded sepia photographs. Memories of a bygone era.

Paul Erdős had done a lot for me and enriched my life very much. Teaching that course and writing this book brought me to his presence again and sharpened my focus on him. They entailed my repeated silent thanks to him.

Acknowledgments

In 2015, Gösta Grahne told me about the website containing a declassified FBI file on Paul Erdős, and a few years later, Jermey Matthews told me about the book containing a part of this file. (Section C.5 of the Appendix refers to both.)

When I began turning my lecture notes into this book, I benefited from consultations with and help from Geoffrey Exoo, Nikolai Ivanov, Brendan McKay, Sergei Malyugin, and Staszek Radziszowski. A year later, Yogen Chaubey, Bipin Desai, and Tariq Srivastava joined me in tackling the mystery of the Majindar/Majumdar duality noted in Chapter 2.

Markéta Vyskočilová read early versions of the vignettes set here in sans serif, and her comments helped me very much in improving them. I am blessed to be married to such a thoughtful, discerning, brilliant editor.

Peter Renz guided me through the realm of publishing, and his advice has been invaluable to me.

Noga Alon, Laci Babai, Thomas Bloom, Xiaomin Chen, Javier De Loera, Mark Goldsmith, Peter Grogono, Steve Hedetniemi, Petra Hoffmannová, Svante Janson, Ida Kantor, Mark Kayll, Don Knuth, János Komlós, Laci Lovász, Tomasz Łuczak, János Pach, Balázs Patkós, Staszek Radziszowski, Bruce Reed, Charles Reiss, Peter Renz, Vojta Rödl, Bruce Rothschild, Lex Schrijver, Gábor Simonyi, Benny Sudakov, Michael Tarsi, Zsolt Tuza, Avi Wigderson, Pete Winkler, and Yori Zwols sent me comments on preliminary versions of the manuscript, which helped me in improving them.

The enthusiasm that Katie Leach expressed for an early version of the manuscript has meant much to me. In her capable care as Cambridge Editor, I felt nurtured and protected.

I am grateful to all of these people.

Introduction

The lecture notes that form the basis of this book have been distributed to graduate students, but the model readers I had in mind when writing them were contestants in a mathematical olympiad. As I did not want to intimidate such youngsters, I chose to include prerequisites that they might not be familiar with even if these prerequisites could be taken for granted when addressing graduates. (They are collected in Appendix A.) The term 'simple' in the subtitle of the book is the operative word: I aimed to make the material accessible to as wide an audience as possible.

I used the lecture notes in a course that consisted of twelve 90-minute lectures and a screening of George Csicsery's brilliant documentary "N is a number" [92]. In each of its editions, I covered at most nine of the following eleven chapters (and once also a large part of Appendix A) at a leisurely pace.

Here is how I arrived at the order of the chapters.

1. Erdős's first important achievement, his 1932 paper proving Bertrand's postulate, seemed a logical choice for the first chapter.
2. His next widely acclaimed result, published in a 1935 paper co-authored with George Szekeres, was the Happy Ending Theorem. Erdős's proof of it, chronologically second and quantitatively far superior to the first, is the starting point of Chapter 2. Its geometric nature suggests continuing with another early geometric interest of Paul Erdős, his conjecture that was confirmed by Tibor Gallai (né Grünwald) and became known as the Sylvester–Gallai theorem. As pointed out by Erdős in 1943, this theorem has a pretty corollary involving points and lines in the plane. In a 1948 paper, Erdős and Nicolaas de Bruijn proved a combinatorial theorem that subsumes this corollary and extends it far beyond the reaches of geometry. This De Bruijn–Erdős theorem and its several proofs round up Chapter 2.
3. Not to leave the reader in suspense for too long, we then backtrack to the Happy Ending Theorem and present its proof by Szekeres. This is done in Chapter 3, whose main theme is Ramsey's theorem. At the end of this chapter, I indulge myself by discussing my second joint paper with Erdős.
4. Another instance of such self-indulgence comes in Chapter 4, where I point out how a qualitative version of the Erdős–Rado theorem on Δ-systems can be viewed as a corollary of Ramsey's theorem. This observation is linked to a

conjecture of Erdős and Lovász on weak and strong Δ-systems, whose beautiful proof by Michel Deza concludes the chapter.

5. The Erdős–Rado theorem on Δ-systems opens the gates of extremal set theory, which is the subject of Chapter 5. One of the two results closing this chapter is Erdős's lower bound on the number of hyperedges in a k-uniform hypergraph of chromatic number greater than s.

6. This bound subsumes not only Erdős's lower bound on diagonal Ramsey numbers but also a lower bound on van der Waerden numbers, and so van der Waerden's theorem on arithmetic progressions is treated in Chapter 6.

7. In Chapter 7, we return to extremal set theory and survey its rich autonomous branch, extremal graph theory.

8. Chapter 8 stands out by having no links to other chapters. It begins with the Friendship Theorem of Erdős, Alfréd Rényi, and Vera Sós. Its proof by Herbert Wilf connects it to strongly regular graphs and the dazzling theorem on Moore graphs of diameter two by Alan Hoffman and R. R. Singleton.

9. After the detour, the next chapter begins with a reference to the Erdős–Stone–Simonovits formula of Chapter 7, which features the chromatic number of a graph. This invariant is the sole subject of Chapter 9. Several proof techniques used there are early instances of what has become known as the probabilistic method, and so it seems natural to continue with graph theory and probability.

10. The first two sections of Chapter 10 reproduce two fragments of the Erdős–Rényi theory of random graphs; the next section reports without proofs the fascinating results on the evolution of random graphs, with an emphasis on the double jump and its critical window; the concluding section puts the preceding material in its natural context of finite probability spaces.

11. Chapter 11 is more of an appendix than a genuine chapter: its theme, Hamiltonian graphs, was far from central among Erdős's interests in discrete mathematics. I have taken the liberty of recounting in its first section how a result of mine was directly inspired by Erdős's delightful algorithmic proof of Turán's theorem and presenting in the second section my first joint paper with Erdős. (Please note that I have displayed admirable restraint by not mentioning our third joint paper anywhere in this book. Except here.) A brief survey of results on Hamilton cycles in random graphs rounds up this chapter.

I regret the omission of two brilliant and important results, Lovász Local Lemma [249] and Szemerédi's Regular Partition Lemma [257]. I could not find a way of weaving them smoothly into the narrative.

Non-mathematical parts of the text are set in sans serif against a lightly shaded background like this.

Definitions that are used more than once are collected in Appendix B.

1   A Glorious Beginning: Bertrand's Postulate

In 1845, Joseph Bertrand (1822–1900) conjectured [29] that *for every integer n greater than 3 there is at least one prime p such that $n < p < 2n-2$*. The slightly weaker proposition,

> *for every positive integer n*
> *there is at least one prime p such that $n < p \leq 2n$,*

is known as *Bertrand's postulate* [211, Theorem 418]. (As all primes except 2 are odd, its constraint $n < p \leq 2n$ amounts to $n < p < 2n$ except when $n = 1$.) In 1852, it was proved [67] by Pafnuty Chebyshev (1821–94).

In March 1931, the 18-year-old Erdős found an elegant elementary proof of Bertrand's postulate; the following year, this proof appeared in his first publication [106].[a] Later, Erdős became fond of quoting Nathan Fine's couplet that celebrated this achievement:

> Chebyshev said it and I say it again:
> There is always a prime between *n* and *2n*.

The first draft of [106] was rewritten by László Kalmár (1905–76), a professor at the University of Szeged; as Erdős recalls in [131], he said in the introduction that Srinivasa Ramanujan (1887–1920) found [322] a somewhat similar proof. Erdős's proof and its background are described in the next five sections; Ramanujan's proof is sketched in section 1.7. Six years after Erdős's proof appeared, Godfrey Harold Hardy (1877–1947) and Edward Maitland Wright (1906–2005) included it in their textbook [211], a classic with its sixth edition appearing in 2008.

1.1   Binomial Coefficients

NOTATION: When *m* and *k* are nonnegative integers, the symbol $\binom{m}{k}$ – read "*m* choose *k*" – denotes the number of *k*-point subsets of a fixed *m*-point set. For

---

[a] Erdős must have considered his 1929 article [105] in a Hungarian mathematics and physics journal for high school students unimportant: In [131] he refers to [106] as "[my paper . . .] which was actually my very first."

example, $\{1, 2, \ldots, 5\}$ has precisely ten 3-point subsets, namely,

$$\{1, 2, 3\}, \{1, 2, 4\}, \{1, 2, 5\}, \{1, 3, 4\}, \{1, 3, 5\},$$
$$\{1, 4, 5\}, \{2, 3, 4\}, \{2, 3, 5\}, \{2, 4, 5\}, \{3, 4, 5\},$$

and so $\binom{5}{3} = 10$.

This combinatorial definition leads directly to a number of identities such as

$$\sum_{k=0}^{m} \binom{m}{k} = 2^m \tag{1.1}$$

(both sides count all subsets of a fixed m-point set, the left-hand side groups them by their size k),

$$\binom{m}{k} = \binom{m}{m-k} \tag{1.2}$$

(complementation $S \leftrightarrow T - S$ sets up a one-to-one correspondence between the set of all k-point subsets S of a fixed m-point set T and the set of all $(m-k)$-point subsets of T), and

$$\binom{m}{k} k = \binom{m}{k-1}(m - k + 1) \tag{1.3}$$

(for a fixed m-point set T, both sides count the number of pairs (S, x) such that $S \subseteq T$, $|S| = k$, and $x \in S$: the left-hand side chooses first S and then x, the right-hand side chooses first $S - \{x\}$ and then x).

Erdős's proof of Bertrand's postulate employs two standard inequalities which follow easily from these identities. First, (1.1) with $m = 2n + 1$ and (1.2) with $m = 2n + 1$, $k = n$ imply that

$$\binom{2n + 1}{n} \leq 4^n. \tag{1.4}$$

Second, (1.3) with $m = 2n$ guarantees that $\binom{2n}{n}$ is the largest of the $2n + 1$ numbers $\binom{2n}{k}$ with $k = 0, 1, \ldots, 2n$, and so it is the largest of the $2n$ terms in the sum $2 + \sum_{k=1}^{2n-1} \binom{2n}{k}$, which totals 4^n by (1.1) with $m = 2n$; we conclude that

$$\binom{2n}{n} \geq \frac{4^n}{2n} \text{ whenever } n \geq 1. \tag{1.5}$$

DEFINITION: The product $1 \cdot 2 \cdot \ldots \cdot m$ of the first m positive integers is called the *factorial of m* and denoted $m!$. The factorial of 0 is defined as $0! = 1$.

Induction on k using identity (1.3) shows that

$$\binom{m}{k} = \frac{m!}{k!(m-k)!}. \tag{1.6}$$

This formula is also used in Erdős's proof.

The quantities $\binom{m}{k}$ are referred to as the *binomial coefficients* since they are featured in the *binomial formula*

$$(a+b)^m = \sum_{k=0}^{m} \binom{m}{k} a^k b^{m-k}.$$

Validity of this formula can be perceived by contemplating how its left-hand side,

$$(a+b)(a+b)\cdots(a+b),$$

distributes into a sum of 2^m terms, each having the form $a^k b^{m-k}$. The binomial formula reduces to (1.1) by setting $a = b = 1$.

1.2 A Lemma

Bertrand's postulate asserts that, in a sense, primes appear in the sequence of positive integers relatively often. Paradoxically, Erdős's proof of the postulate relies on a lemma asserting that they do not appear too often: the product of all primes not exceeding a positive integer m is less than 4^m.

In number theory it is customary to reserve the letter p for primes; in particular, Erdős's lemma can be recorded as

$$\prod_{p \leq m} p < 4^m \quad \text{for every positive integer } m. \tag{1.7}$$

Some eight years after Erdős first proved (1.7), he and Kalmár found independently and almost simultaneously a simpler proof (see [131]). This proof goes by induction on m. The induction basis verifies (1.7) when $m \leq 2$. In the induction step, we consider an arbitrary integer m greater than 2 and assume that $\prod_{p \leq k} p < 4^k$ whenever $k < m$; then we distinguish between two cases. If m is even, then

$$\prod_{p \leq m} p = \prod_{p \leq m-1} p < 4^{m-1}.$$

If m is odd, then $m = 2n + 1$ with $n \geq 1$; since

$$\binom{2n+1}{n} = \frac{(2n+1) \cdot 2n \cdot (2n-1)\ldots \cdot (n+2)}{n!},$$

every prime in the range $n + 1 < p \leq 2n + 1$ divides $\binom{2n+1}{n}$, and so

$$\prod_{p < m} p = \left(\prod_{p \leq n+1} p \right) \cdot \left(\prod_{n+1 < p \leq 2n+1} p \right) \leq \left(\prod_{p \leq n+1} p \right) \cdot \binom{2n+1}{n}.$$

Using the induction hypothesis and (1.4), we conclude that

$$\prod_{p \leq m} p < 4^{n+1} \cdot 4^n = 4^m.$$

1.3   The Unique Factorization Theorem

Every child knows that a prime is a positive integer divisible by no positive integer other than itself and the integer 1. However, not all children may be aware that the integer 1 is decreed to be not a prime, even though it is divisible by no positive integer other than itself. Ruling this integer out of the set of all primes is not an arbitrary decision: ruling it in would ruin the following theorem, known as the *Fundamental Theorem of Arithmetic* or the *Unique Factorization Theorem*.

> *For every positive integer n and for all primes p,*
> *there are uniquely defined nonnegative integers e(p, n) such that*
>
> $$n = \prod_p p^{e(p,n)}.$$

(In the right-hand-side product, p runs through the infinite set of primes, but for every n only finitely many of the exponents $e(p, n)$ are nonzero: if $p > n$, then $e(p, n) = 0$.) Declaring 1 to be a prime would make the factorization no longer unique: $e(1, n)$ could assume any nonnegative integer value.

Some people attribute the Unique Factorization Theorem to Euclid [212, Proposition 14 of Book IX], whose *Elements* appeared around 300 BC, and others to Carl Friedrich Gauss (1777–1855), whose *Disquisitiones Arithmeticae* [184] appeared in the summer of 1801. The controversy is analyzed in [86].

1.4   Legendre's Formula

When n is the factorial $m!$, the exponents $e(p, n)$ in the unique factorization

$$n = \prod_p p^{e(p,n)}$$

can be calculated from a neat formula. To begin, for every choice of positive integers s and t we have

$$st = \left(\prod_p p^{e(p,s)}\right) \cdot \left(\prod_p p^{e(p,t)}\right) = \prod_p p^{e(p,s)+e(p,t)},$$

and so

$$e(p, \, st) = e(p, \, s) + e(p, \, t).$$

It follows that

$$e(p, m!) = e(p, 1) + e(p, 2) + \ldots + e(p, m).$$

We are going to express the right-hand-side sum in a more transparent way. Let us begin with the example of $p = 2$ and $m = 9$. Here,

$$e(2, 1) + e(2, 2) + \ldots + e(2, 9) = 0 + 1 + 0 + 2 + 0 + 1 + 0 + 3 + 0.$$

Of the nine terms,

- every second one contributes at least one unit to the total, and there are four such terms,
- every fourth one contributes at least two units to the total, and there are two such terms,
- every eighth one contributes at least three units to the total, and there is one such term,
- every 16th one contributes at least four units to the total, and there are no such terms.

These observations make it clear that

$$0 + 1 + 0 + 2 + 0 + 1 + 0 + 3 + 0 = 4 + 2 + 1 + 0.$$

This identity can be illustrated by the array

where column j holds a stack of $e(2, j)$ coins: the sum $0 + 1 + 0 + 2 + 0 + 1 + 0 + 3 + 0$ of the heights of the nine stacks counts the total number of coins, and the sum $4 + 2 + 1 + 0$ counts the same number row by row. In general, for any choice of p and m, there are stacks $1, 2, \ldots, m$, and stack j holds $e(p, j)$ coins. Counting the total number $e(p, 1) + e(p, 2) + \ldots + e(p, m)$ of coins row by row, we end up with the sum $\lfloor m/p \rfloor + \lfloor m/p^2 \rfloor + \lfloor m/p^3 \rfloor + \cdots$ (where, as usual, $\lfloor x \rfloor$ denotes x rounded down to the nearest integer): a coin appears in row i and column j if and only if $e(p, j) \geq i$, which is the case if and only if j is a multiple of p^i. It follows that

$$e(p, m!) = \sum_{i=1}^{\infty} \left\lfloor \frac{m}{p^i} \right\rfloor$$

(where only finitely many terms in the infinite sum are not zero). This formula was presented by Adrien-Marie Legendre (1752–1833) in the second edition of his book [273] published in 1808.

1.5 Erdős's Proof of Bertrand's Postulate

1.5.1 The Plan

Given a positive integer n, we shall choose a positive integer N and prove that

$$\prod_{p \leq n} p^{e(p,N)} < \prod_{p \leq 2n} p^{e(p,N)}, \tag{1.8}$$

which obviously implies Bertrand's postulate. Our choice is $N = \binom{2n}{n}$. Since formula (1.6) with $m = 2n$ and $k = n$ reads

$$N = \frac{2n \cdot (2n-1) \cdot (2n-2)\ldots \cdot (n+1)}{n!},$$

it is clear that all prime divisors of N are at most $2n$, and so

$$\prod_{p \le 2n} p^{e(p,N)} = N.$$

We propose to prove that

$$\prod_{p \le n} p^{e(p,N)} < \frac{4^n}{2n}. \tag{1.9}$$

Since (1.5) reads $4^n/2n \le N$, inequality (1.8) will then follow.

1.5.2   A Formula for $e(p, N)$

We will use the formula

$$e(p,N) = \sum_{i=1}^{\infty} \left(\left\lfloor \frac{2n}{p^i} \right\rfloor - 2 \left\lfloor \frac{n}{p^i} \right\rfloor \right), \tag{1.10}$$

which follows directly from (1.6) combined with Legendre's formula. Note that

$$\lfloor 2x \rfloor - 2\lfloor x \rfloor = \begin{cases} 0 & \text{if} \quad 0 \le x - \lfloor x \rfloor < 1/2, \\ 1 & \text{if} \quad 1/2 \le x - \lfloor x \rfloor < 1, \end{cases}$$

and so

$$\left\lfloor \frac{2n}{p^i} \right\rfloor - 2 \left\lfloor \frac{n}{p^i} \right\rfloor = \quad 0 \text{ or } 1 \text{ for all } i. \tag{1.11}$$

1.5.3   An Upper Bound on $p^{e(p, N)}$

Given p and n, consider the largest integer j such that $p^j \le 2n$. By (1.10) and (1.11), we have

$$e(p,N) = \sum_{i=1}^{j} \left(\left\lfloor \frac{2n}{p^i} \right\rfloor - 2 \left\lfloor \frac{n}{p^i} \right\rfloor \right) \le j,$$

and so

$$p^{e(p,N)} \le 2n. \tag{1.12}$$

1.5.4   Splitting the Left-Hand Side of (1.9)

We will partition the set of all primes not exceeding n into three classes:

- the set S of primes p such that $p \le \sqrt{2n}$,
- the set M of primes p such that $\sqrt{2n} < p \le 2n/3$,

- the set L of primes p such that $2n/3 < p \leq n$.

This classification reflects the size of $e(p, N)$: as we are about to prove,

$$p \in M \Rightarrow e(p, N) \leq 1, \tag{1.13}$$

$$p \in L \Rightarrow e(p, N) = 0. \tag{1.14}$$

Our proof of these implications relies on formula (1.10): since

$$p > \sqrt{2n} \text{ and } i \geq 2 \Rightarrow 2n/p^i < 1 \Rightarrow n/p^i < 1,$$

we have

$$p > \sqrt{2n} \Rightarrow e(p, N) = \left\lfloor \frac{2n}{p} \right\rfloor - 2 \left\lfloor \frac{n}{p} \right\rfloor. \tag{1.15}$$

Implication (1.13) follows directly from (1.15) and (1.11); implication (1.14) follows from (1.15) combined with the observation that $p \in L$ implies $\lfloor 2n/p \rfloor = 2$ and $\lfloor n/p \rfloor = 1$.

1.5.5 Putting the Pieces Together

By definition, we have

$$\prod_{p \leq n} p^{e(p, N)} = \prod_{p \in S} p^{e(p, N)} \cdot \prod_{p \in M} p^{e(p, N)} \cdot \prod_{p \in L} p^{e(p, N)};$$

by (1.12), we have

$$\prod_{p \in S} p^{e(p, N)} \leq (2n)^{\sqrt{2n} - 1};$$

by (1.13) and by (1.7) with $m = \lfloor 2n/3 \rfloor$, we have

$$\prod_{p \in M} p^{e(p, N)} \leq \prod_{p \in M} p \leq \prod_{p \leq 2n/3} p < 4^{2n/3};$$

by (1.14), we have

$$\prod_{p \in L} p^{e(p, N)} = 1;$$

altogether, we have

$$\prod_{p \leq n} p^{e(p, N)} < (2n)^{\sqrt{2n} - 1} \cdot 4^{2n/3}.$$

NOTATION: We let $\lg x$ stand for the binary logarithm $\log_2 x$.

To prove (1.9), we prove that

$$(2n)^{\sqrt{2n} - 1} \cdot 4^{2n/3} \leq \frac{4^n}{2n},$$

which can be written as

$$(2n)^{\sqrt{2n}} \leq 4^{n/3}$$

and then (taking binary logarithms of both sides) as $\sqrt{2n}\lg(2n) \leq 2n/3$, and finally as

$$3\lg(2n) \leq \sqrt{2n}.$$

A routine exercise in calculus shows that $3\lg x \leq \sqrt{x}$ whenever $x \geq 1024$, and so (1.9) holds whenever $n \geq 512$.

To complete the proof of Bertrand's postulate, we have to verify its validity for the remaining 511 values of n. To do this, just observe that each interval $(n, 2n]$ with $1 \leq n \leq 511$ includes at least one of the primes

$$5, 7, 11, 19, 31, 59, 113, 223, 443, 883. \tag{1.16}$$

Each prime in the sequence is less than twice its predecessor.

1.6   Proof of Bertrand's Original Conjecture

It is a routine matter to adjust Erdős's proof of Bertrand's postulate so as to prove Bertrand's stronger original conjecture. Let us spell out the details.

THEOREM 1.1   *For every integer n greater than 3, there is a prime p such that $n < p < 2n - 2$.*

Proof   As in Erdős's proof of Bertrand's postulate, write $N = \binom{2n}{n}$. Since $n < p < 2n$ implies $\lfloor 2n/p \rfloor = 1$ and $\lfloor n/p \rfloor = 0$, formula (1.15) shows that

$$n < p \leq 2n \;\Rightarrow\; e(p, N) = 1,$$

and so

$$\prod_{n<p<2n-2} p^{e(p,N)} = \frac{N}{\prod_{p\leq n} p^{e(p,N)} \cdot \prod_{2n-2\leq p\leq 2n} p^{e(p,N)}}$$

$$\geq \frac{N}{\prod_{p\leq n} p^{e(p,N)} \cdot (2n-1)};$$

as in Erdős's proof of Bertrand's postulate, we have

$$\frac{N}{\prod_{p\leq n} p^{e(p,N)}} > \frac{4^{n/3}}{(2n)^{\sqrt{2n}}}.$$

It follows that

$$\prod_{n<p<2n-2} p^{e(p,N)} > \frac{4^{n/3}}{(2n)^{1+\sqrt{2n}}}.$$

A routine exercise in calculus shows that

$$3\lg x < \sqrt{x} - 1 < \frac{x}{1+\sqrt{x}} \quad \text{whenever } x \geq 1024,$$

and so

$$\frac{4^{n/3}}{(2n)^{1+\sqrt{2n}}} > 1 \quad \text{whenever } n \geq 512.$$

If $3 < n < 512$, then the interval $(n, 2n - 2)$ includes at least one of the primes (1.16). □

1.7 Earlier Proofs of Bertrand's Postulate

NOTATION: We let $\ln x$ stand for the natural logarithm $\log_e x$.

1.7.1 Chebyshev

In his proof [67], published in 1852, Pafnuty Chebyshev introduced functions

$$\theta(x) = \sum_{p \leq x} \ln p,$$

$$\psi(x) = \sum_{i=1}^{\infty} \theta(x^{1/i})$$

(here, only finitely many terms in the infinite sum are nonzero) and proved the identity

$$\sum_{j=1}^{\infty} \psi\left(\frac{x}{j}\right) = \ln\left(\lfloor x \rfloor!\right) \tag{1.17}$$

(again, only finitely many terms in the infinite sum are nonzero). With the notation

$$a(i, j, p, x) = \begin{cases} 1 & \text{if } jp^i \leq x, \\ 0 & \text{otherwise,} \end{cases}$$

his argument can be stated as

$$\sum_{j=1}^{\infty} \psi\left(\frac{x}{j}\right) = \sum_{j=1}^{\infty}\sum_{i=1}^{\infty} \theta\left(\left(\frac{x}{j}\right)^{1/i}\right) = \sum_{j=1}^{\infty}\sum_{i=1}^{\infty}\sum_{p} a(i, j, p, x) \ln p$$

$$= \sum_{p}\sum_{i=1}^{\infty}\sum_{j=1}^{\infty} a(i, j, p, x) \ln p = \sum_{p}\sum_{i=1}^{\infty} \left\lfloor \frac{x}{p^i} \right\rfloor \ln p$$

$$= \sum_{p}\sum_{i=1}^{\infty} \left\lfloor \frac{\lfloor x \rfloor}{p^i} \right\rfloor \ln p = \sum_{p} e(p, \lfloor x \rfloor!) \ln p = \ln(\lfloor x \rfloor!).$$

From (1.17) and from Stirling's approximation (A.6)

$$0 < \ln(n!) - \left(n \ln n - n + \tfrac{1}{2} \ln(2\pi n)\right) < \frac{1}{12n},$$

he deduced by a lengthy arithmetical argument that $\theta(2n) - \theta(n) > 0$ (and so there is a prime p such that $n < p \leq 2n$) whenever $n > 160$.

1.7.2   Landau

Chebyshev's proof of Bertrand's postulate reappeared, with slight modifications, in §17–§20 of a monograph [271] written by Edmund Landau (1877–1938) and published in 1909. One of Landau's shortcuts involves the observation (inequalities (1), (2) in §18 of [271]) that (1.17) implies

$$\ln\left(\lfloor x \rfloor!\right) - 2\ln\left(\lfloor \tfrac{1}{2}x \rfloor!\right)$$
$$= \psi(x) - \psi\left(\tfrac{1}{2}x\right) + \psi\left(\tfrac{1}{3}x\right) - \psi\left(\tfrac{1}{4}x\right) + \psi\left(\tfrac{1}{5}x\right) - \psi\left(\tfrac{1}{6}x\right) + \cdots,$$

and so, as ψ is nondecreasing and nonnegative,

$$\psi(x) - \psi\left(\tfrac{1}{2}x\right) \;\le\; \ln\left(\lfloor x \rfloor!\right) - 2\ln\left(\lfloor \tfrac{1}{2}x \rfloor!\right) \;\le\; \psi(x). \tag{1.18}$$

This observation is reminiscent of Chebyshev's observation (inequalities (6) in §5 of [67]) that

$$\psi(x) - \psi\left(\sqrt{x}\right) = \theta(x) + \theta(x^{1/3}) + \theta(x^{1/5}) + \cdots$$
$$\psi(x) - 2\psi\left(\sqrt{x}\right) = \theta(x) - \theta(x^{1/2}) + \theta(x^{1/3}) - \theta(x^{1/4}) + \theta(x^{1/5}) - \theta(x^{1/6}) + \cdots,$$

and so, as θ is nondecreasing and nonnegative,

$$\psi(x) - 2\psi\left(\sqrt{x}\right) \;\le\; \theta(x) \;\le\; \psi(x) - \psi\left(\sqrt{x}\right). \tag{1.19}$$

1.7.3   Ramanujan

In [322], published in 1919, Srinivasa Ramanujan started out with a refinement of (1.18),

$$\psi(x) - \psi\left(\tfrac{1}{2}x\right) \;\le\; \ln\left(\lfloor x \rfloor!\right) - 2\ln\left(\lfloor \tfrac{1}{2}x \rfloor!\right) \;\le\; \psi(x) - \psi\left(\tfrac{1}{2}x\right) + \psi\left(\tfrac{1}{3}x\right), \tag{1.20}$$

and a cruder version of (1.19),

$$\psi(x) - 2\psi\left(\sqrt{x}\right) \;\le\; \theta(x) \;\le\; \psi(x). \tag{1.21}$$

He argued that Stirling's approximation (A.6) implies

$$\ln\left(\lfloor x \rfloor!\right) - 2\ln\left(\lfloor \tfrac{1}{2}x \rfloor!\right) < \tfrac{3}{4}x \quad \text{whenever } x > 0,$$
$$\ln\left(\lfloor x \rfloor!\right) - 2\ln\left(\lfloor \tfrac{1}{2}x \rfloor!\right) > \tfrac{2}{3}x \quad \text{whenever } x > 300;$$

these bounds, combined with (1.20), give

$$\psi(x) - \psi\left(\tfrac{1}{2}x\right) < \tfrac{3}{4}x \quad \text{whenever } x > 0, \tag{1.22}$$
$$\psi(x) - \psi\left(\tfrac{1}{2}x\right) + \psi\left(\tfrac{1}{3}x\right) > \tfrac{2}{3}x \quad \text{whenever } x > 300. \tag{1.23}$$

Then he noted that

$$\psi(x) < \tfrac{3}{2}x \quad \text{whenever } x > 0. \tag{1.24}$$

This inequality holds trivially when $0 < x < 2$ and can be verified by induction on $\lfloor \lg x \rfloor$ when $x > 2$, with (1.22) taking care of the induction step. If $n \geq 162$, then (1.21), (1.23), and (1.24) guarantee that

$$\theta(2n) - \theta(n) \geq \psi(2n) - 2\psi\left(\sqrt{2n}\right) - \psi(n) > \tfrac{4}{3}n - \psi\left(\tfrac{2}{3}n\right) - 2\psi\left(\sqrt{2n}\right)$$
$$> \tfrac{1}{3}n - 3\sqrt{2n} \geq 0.$$

1.8 Further Results and Problems Concerning Primes

1.8.1 Landau's Problems

In his invited address at the fifth International Congress of Mathematicians, held at Cambridge in 1912, Landau mentioned four conjectures, which he declared to be "unattackable at the present state of science":

1. The conjecture that there are infinitely many primes of the form $n^2 + 1$.
2. *The Goldbach conjecture:* Every even integer greater than 2 is the sum of two primes.
3. *The twin prime conjecture:* There are infinitely many primes p such that $p + 2$ is prime.
4. Legendre's conjecture that for every integer n there is a prime between n^2 and $(n + 1)^2$.

These four conjectures are now known as *Landau's problems* and they remain open.

1.8.2 Small Gaps between Consecutive Primes

In number theory, it is customary to let p_n denote the n-th prime:

$$p_1 = 2, \ p_2 = 3, \ p_3 = 5, \ p_4 = 7, \ p_5 = 11, \ p_6 = 13, \ p_7 = 17, \ p_8 = 19, \ \ldots .$$

In this notation, Bertrand's postulate asserts that

$$p_{n+1} - p_n \leq p_n \text{ for all } n.$$

In 1930, Guido Hoheisel (1894–1968) proved an asymptotically stronger result [216]:

$$p_{n+1} - p_n \leq p_n^{32999/33000} \text{ for all sufficiently large } n.$$

There followed successive improvements of the exponent. The latest of these comes from 2001: Roger Baker, Glyn Harman, and János Pintz proved that [15]

$$p_{n+1} - p_n \leq p_n^{21/40} \text{ for all sufficiently large } n.$$

If this bound could be strengthened further to

$$p_{n+1} - p_n \leq 2p_n^{1/2} \text{ for all } n,$$

then Legendre's conjecture would follow: With p_n the largest prime less than m^2, we would have $m^2 < p_{n+1} < m^2 + 2m$. Conversely, Legendre's conjecture implies that

$$p_{n+1} - p_n < 4p_n^{1/2} + 4 \text{ for all } n.$$

Given p_n, consider the unique m such that $(m-1)^2 < p_n < m^2$ and note that $p_{n+1} < (m+1)^2$ implies $p_{n+1} - p_n < 4m$.

The twin prime conjecture asserts that

$$p_{n+1} - p_n = 2 \text{ for infinitely many } n.$$

In progress toward proving this conjecture, an epoch-making breakthrough was made in April 2013 by Yitang Zhang [377]:

$$p_{n+1} - p_n \leq 70,000,000 \text{ for infinitely many } n.$$

The challenge of reducing this upper bound was answered in April 2014 by an online collaborative project, *Polymath 8* [315]:

$$p_{n+1} - p_n \leq 246 \text{ for infinitely many } n.$$

1.8.3   Large Gaps between Consecutive Primes

Since $m! + k$ is divisible by k whenever $2 \leq k \leq m$, it is evident that the interval $[m!+2, m!+m]$ includes no primes. It follows that gaps $p_{n+1} - p_n$ between consecutive primes p_n and p_{n+1} can be arbitrarily large. In 1938, Robert Rankin (1915–2001) proved [324] that for some positive c and infinitely many n

$$p_{n+1} - p_n > c \cdot \frac{\ln p_n \ln \ln p_n \ln \ln \ln \ln p_n}{(\ln \ln \ln p_n)^2}. \tag{1.25}$$

In 1990, Erdős wrote [133]

> I offered (perhaps somewhat rashly) \$ 10 000
> for a proof that (1.25) holds for every c.

In 2014, two distinct proofs were found simultaneously and independently by the team of Kevin Ford, Ben Green, Sergei Konyagin, and Terence Tao [167] and by James Maynard [291]. The prize of \$5,000 for the team of four and \$5,000 for Maynard alone was paid out in Erdős's stead by Ron Graham (1935–2020), Erdős's close friend and joint author of 32 papers written with him. During Erdős's life, Ron managed his finances, cashing Erdős's honoraria cheques and sending out cheques to solvers of Erdős's problems; after Erdős's passing, he controlled a small fund left by Erdős to reward future solvers of his problems.

1.8.4 Primes in Arithmetic Progressions

An *arithmetic progression* is a (finite or infinite) sequence of distinct numbers

$$a, \ a+d, \ a+2d, \ a+3d, \ a+4d, \ \ldots . \tag{1.26}$$

and its *length* is the number of its terms. For instance, $5, 11, 17, 23, 29$ is an arithmetic progression of length five (and it consists exclusively of primes). The set of all primes contains no infinite arithmetic progressions: if (1.26) has at least $a+1$ terms and if its a and d are positive integers, then its term $a+ad$ is composite. Nevertheless, Ben Green and Terence Tao proved [199] that

> *the set of all primes contains arbitrarily long arithmetic progressions.*

This was an old conjecture, implicit in investigations carried out by Joseph-Louis Lagrange (1736–1813) and Edward Waring (1736–98) around 1770 and subsumed in a special case of the "first Hardy–Littlewood conjecture" [210].

In 1974, Erdős [119, page 204] offered \$2,500 for a proof or disproof of his

CONJECTURE 1.2   *Every increasing sequence $a_1, a_2, a_3, \ldots$ of positive integers such that $\sum_{i=1}^{\infty} 1/a_i = \infty$ contains arbitrarily long arithmetic progressions.*

In 2020, Thomas Bloom and Olof Sisask [32, Corollary 1.2] proved that every increasing sequence $a_1, a_2, a_3, \ldots$ of positive integers such that $\sum_{i=1}^{\infty} 1/a_i = \infty$ contains arithmetic progressions of length three. Other than that, the conjecture remains open; since the sum of the reciprocals of prime numbers diverges (a classical result of Euler), its validity would imply the Green–Tao theorem. A weaker conjecture,

> *If a set of positive integers includes, for some positive ε and all sufficiently large n, at least εn of the first n positive integers, then it contains arbitrarily long arithmetic progressions,*

was made [121, page 296] by Erdős and his close friend Paul Turán (1910–76) in 1936; later, Erdős offered \$1,000 for its resolution. This conjecture was proved [356] by Endre Szemerédi in 1972, and we shall return to it in Section 6.4.

1.8.5 On revient toujours à ses premières amours

Number theory remained one of Erdős's most important interests throughout his life. The inventory [160] of his papers up to 1998 lists

- 229 on extremal problems and Ramsey theory,
- 191 on additive number theory,
- 176 on graph theory,
- 158 on multiplicative number theory,
- 149 on analysis,
- 77 on geometry,
- 69 on combinatorics,

- 52 on set theory,
- 41 on probability.[b]

Jointly with János Surányi (1918–2006), he wrote an introductory book on number theory [155], which had a great impact on generations of Hungarian mathematicians. Its English translation [156] appeared in 2003.

Erdős kept returning not only to the area of his first paper, but also to its proof technique. In this first paper, he established the existence of an object with specified properties (namely, a prime p such that $n < p \leq 2n$) without providing an efficient algorithm to find such an object. Such proofs of existence are called *non-constructive*. This particular non-constructive proof is a prototype of a scheme that Erdős used again and again in his subsequent papers unrelated to number theory. In the general setting, its condensed outline goes as follows:

A finite set Ω of objects is divided into disjoint subsets A (for 'acceptable') and B (for 'bad'). To prove the existence of an acceptable object, assign a nonnegative weight $w(p)$ to every object p in Ω and show that $\sum_{p \in B} w(p) < \sum_{p \in \Omega} w(p)$.

In the special case where all objects in Ω are assigned weight 1, this way of proving the existence of an acceptable object amounts to a computation showing that $|B| < |\Omega|$. Erdős used even this crudest variant with astounding success (see the proof of Theorem 3.4). Its enhancements (for examples, see the proofs of Lemma 9.12, Lemma 9.13, Lemma 9.21, Lemma 9.22, and Lemma 9.23) eventually developed into the *probabilistic method* [7, 153, 295].

In the special case of Bertrand's postulate, Ω consists of all prime divisors of $\binom{2n}{n}$; a p in Ω is acceptable if $n < p \leq 2n$ and bad otherwise; for every p in Ω, Erdős sets $w(p) = e(p, N) \log p$ with $N = \binom{2n}{n}$. We have $\sum_{p \in \Omega} w(p) = \log N$ by definition, and Erdős proves that $\sum_{p \in B} w(p) < \log N$. (This overview explains the paradox mentioned at the beginning of Section 1.2: in order to bound $\sum_{p \in A} w(p)$ from below, we bound $\sum_{p \in B} w(p)$ from above.)

Mathematicians don't have it easy. They spend their lives searching for Truth and Beauty and this search is not always entirely selfless. Vanity and competitiveness are frequent ingredients of their motives. *I have solved a problem which you could not solve. I am smarter than you. My daddy can beat up your daddy.* Can you blame them when a competitive nature is nurtured if not outright instilled in them from early on by problem-solving competitions?

Early aptitude for mathematics is paired up with an insidious trap. You are the best among your classmates in elementary school. You continue being the best at the next level. And so on, until being the best becomes a banal fact of life, something you routinely expect and feel entitled to. (Depending on whose side one takes, this could be called presumptuous arrogance or just a result of systematic conditioning.) But most people eventually climb to a level where a rude awakening lurks: suddenly the natural order of

[b] Papers co-authored with Erdős kept appearing even after 1998, and the list of his publications [202] compiled in January 2013 consists of 1525 items, the latest one dating from 2008.

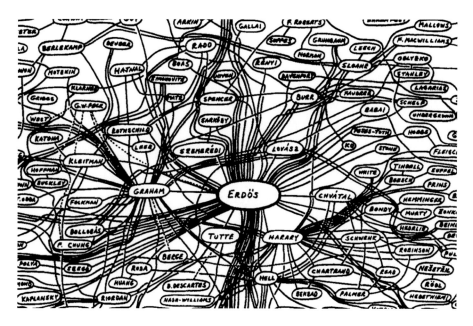

Figure 1.1 A portion of the *collaboration graph of mathematicians*. Cropped from Figure 1 of T. Odda, On properties of a well-known graph or what is your Ramsey number?, *Annals of the New York Academy of Sciences* **328** (1979), 166–172. ©1979, NYAS

the universe breaks down and you are no longer the smartest kid on the block. The later the disenchantment comes, the more it hurts.

Paul Erdős never suffered this shock. He grew up as a child prodigy in an environment where the concept of a *wunderkind* was ubiquitous. His photographs appeared in the tableaux of the winners of a correspondence mathematical competition for high school students in four consecutive school years, 1926–30. And he remained the smartest kid on the block for the rest of his life.

Despite this, he never looked down on any of us. He did not rank people by their mathematical achievements. He was not a snob. What is more, he taught us by example that cooperation instead of rivalry is what makes mathematics gratifying, that collaboration with friends is enjoyable, and that new friends can be made through collaboration. This attitude of his was infectious. It was a perfect antidote to the immature silliness of daddies beating up each other.

2 Discrete Geometry and Spinoffs

2.1 The Happy Ending Theorem

In the winter of 1932/33, 22-year-old Esther Klein (1910–2005) proved that

> *from any set of five points in the plane of which no three lie on the same line it is always possible to select four points that are vertices of a convex polygon.*

When she shared this discovery with a circle of her friends in Budapest, the following more general problem emerged:

> *Can we find for each integer n, greater than two, an integer N such that from any set of N points in the plane of which no three lie on the same line it is always possible to select n points that are vertices of a convex polygon?*

NOTATION: Let $N(n)$ denote the smallest integer N with this property.

To see that $N(4) \geq 5$, consider the three vertices of a triangle and a fourth point in the interior of this triangle; Esther Klein's proposition is $N(4) \leq 5$. Next, Endre Makai (1915–87) proved that $N(5) = 9$. (To see that $N(5) \geq 9$, consider the eight points in Figure 2.1; Makai did not publish his proof of the reverse inequality.[a]) A few weeks later, George Szekeres (1911–2005) proved the existence of $N(n)$ for all n and soon afterwards, Erdős came up with a different proof. In December 1934, Erdős and Szekeres submitted for publication a manuscript containing both proofs; the paper [157] appeared in 1935. Esther Klein and George Szekeres were married on 13 June 1937, and Paul Erdős took to referring to the theorem asserting the existence of $N(n)$ for all n as *The Happy Ending Theorem*.

Figure 2.1 No convex pentagon here.

---

[a] Proofs of Makai's $N(5) \leq 9$ appeared decades later in [233] and [279, Problem 14.31(c)].

Szekeres's proof produced very large upper bounds $N^*(n)$ on $N(n)$: for instance, $N^*(5) \approx 2^{10000}$. By contrast, Erdős's proof led to much smaller upper bounds on $N(n)$: for instance, $N(5) \le 21$. We will present Szekeres's proof in Section 3.4; Erdős's proof follows here.

THEOREM 2.1   *From any set of $\binom{2n-4}{n-2} + 1$ points in the plane of which no three lie on the same line it is always possible to select n points that are vertices of a convex polygon.*

Proof   Given a finite set S of points in the plane, choose the two perpendicular coordinate axes so that no two points in S have the same first coordinate; call a sequence

$$[x_1, y_1], \ [x_2, y_2], \ \ldots [x_k, y_k]$$

of points in S *convex* if $x_1 < x_2 < \ldots < x_k$ and

$$\frac{y_2 - y_1}{x_2 - x_1} < \frac{y_3 - y_2}{x_3 - x_2} < \cdots < \frac{y_k - y_{k-1}}{x_k - x_{k-1}};$$

call the sequence *concave* if $x_1 < x_2 < \ldots < x_k$ and

$$\frac{y_2 - y_1}{x_2 - x_1} > \frac{y_3 - y_2}{x_3 - x_2} > \cdots > \frac{y_k - y_{k-1}}{x_k - x_{k-1}}.$$

We will prove that, for all choices of integers i, j such that $i \ge 2$ and $j \ge 2$,

$$|S| > \binom{i+j-4}{i-2} \text{ implies that } S \text{ contains} \tag{2.1}$$
$$\text{a convex sequence of } i \text{ points or a concave sequence of } j \text{ points.}$$

If $i = 2$ or $j = 2$, then (2.1) is immediate: the lower bound on $|S|$ implies that $|S| \ge 2$ and every sequence of two points in S is both convex and concave. To prove (2.1) in its generality, we will use induction on the minimum of i and j: as we have just observed, the induction basis $i = j = 2$ is immediate, and in the induction step we may assume that $i \ge 3, j \ge 3$.

Let A denote the set of all points of S where a convex sequence of $i - 1$ points ends and let B denote the set of all points of S where a concave sequence of $j - 1$ points begins. We shall distinguish between three cases:

CASE 1: $|S - A| > \binom{i+j-5}{i-3}$. By definition of A, there is no convex sequence of $i - 1$ points in $S - A$, and so the assumption of this case and the induction hypothesis guarantee that $S - A$ contains a concave sequence of j points.

CASE 2: $|S - B| > \binom{i+j-5}{i-2}$. By definition of B, there is no concave sequence of $j - 1$ points in $S - B$, and so the assumption of this case and the induction hypothesis guarantee that $S - B$ contains a convex sequence of i points.

CASE 3: $|S - A| \le \binom{i+j-5}{i-3}$ and $|S - B| \le \binom{i+j-5}{i-2}$. In this case,

$$|S - A| + |S - B| < |S|,$$

and so $A \cap B \ne \emptyset$. This means that there is a sequence

$$[x_1, y_1], \ [x_2, y_2], \ \ldots [x_{i+j-3}, y_{i+j-3}]$$

of points in S such that $x_1 < x_2 < \ldots < x_{i+j-3}$ and

$$\frac{y_2 - y_1}{x_2 - x_1} < \frac{y_3 - y_2}{x_3 - x_2} < \cdots < \frac{y_{i-1} - y_{i-2}}{x_{i-1} - x_{i-2}},$$

$$\frac{y_i - y_{i-1}}{x_i - x_{i-1}} > \frac{y_{i+1} - y_i}{x_{i+1} - x_i} > \cdots > \frac{y_{i+j-3} - y_{i+j-4}}{x_{i+j-3} - x_{i+j-4}}.$$

If

$$\frac{y_{i-1} - y_{i-2}}{x_{i-1} - x_{i-2}} < \frac{y_i - y_{i-1}}{x_i - x_{i-1}},$$

then $[x_1, y_1]$, $[x_2, y_2]$, $\ldots [x_i, y_i]$ is a convex sequence of i points; else (since $[x_{i-2}, y_{i-2}]$, $[x_{i-1}, y_{i-1}]$, $[x_i, y_i]$ do not lie on the same line)

$$\frac{y_{i-1} - y_{i-2}}{x_{i-1} - x_{i-2}} > \frac{y_i - y_{i-1}}{x_i - x_{i-1}},$$

and so $[x_{i-2}, y_{i-2}], [x_{i-1}, y_{i-1}], \ldots, [x_{i+j-3}, y_{i+j-3}]$ is a concave sequence of j points. $\qquad\square$

NOTATION: When f and g are nonnegative real-valued functions defined on positive integers, we write $f(n) = O(g(n))$ to mean that $f(n) \le cg(n)$ for some constant c and all sufficiently large n.

The challenge of improving the upper bound

$$N(n) \le \binom{2n-4}{n-2} + 1$$

of Theorem 2.1 remained unanswered for 64 years, until 1998, when Fan Chung and Ron Graham [71] improved the bound to

$$N(n) \le \binom{2n-4}{n-2}. \tag{2.2}$$

Note that the right-hand side of (2.2) is $O(4^n)$. Having proved this inequality, Chung and Graham offered \$100 for the first proof that $N(n) = O(c^n)$ for some constant c smaller than 4. In 2016, Andrew Suk [350] proved a far stronger result:

$$N(n) \le 2^{n+6n^{2/3}\log n}$$

for all sufficiently large n. Later, the error term in the exponent was improved [217]:

$$N(n) \le 2^{n+O(\sqrt{n\log n})}.$$

As for lower bounds, Szekeres [352] wrote (with N standing for $N(n)$):

> Paul's method contained implicitly that $N > 2^{n-2}$, and this result appeared some thirty-five years later [158] in a joint paper, after Paul's first visit to Australia. …Of course, we firmly believe that $N = 2^{n-2} + 1$ is the correct value.

Erdős [134] stated that he would certainly pay $500 for a proof that

$$N(n) \overset{?}{=} 2^{n-2} + 1.$$

We have noted that Klein proved $N(4) = 5$ and Makai proved $N(5) = 9$. Szekeres and Peters [353] proved with the aid of a computer that $N(6) = 17$.

Not everybody may agree with Szekeres that Erdős's proof of Theorem 2.1 contains implicitly the lower bound $N(n) > 2^{n-2}$. We are now going to paraphrase the constructive proof of this bound that appears in [158]. First, let us note that the lower bound on $|S|$ in (2.1) is the best possible.

LEMMA 2.2   *For every choice of integers i and j greater than 1, there is a set $E(i,j)$ of points in the plane such that*

- *no three points of $E(i,j)$ lie on the same line and no two have the same first coordinate,*
- *$E(i,j)$ contains neither a convex sequence of i points nor a concave sequence of j points,*
- *$|E(i,j)| = \binom{i+j-4}{j-2}$,*
- *all lines passing through two points of $E(i,j)$ have positive slopes.*

Proof   By induction on the minimum of i and j. The induction basis, $i = 2$ or $j = 2$, is trivial; for the induction step, assume that both i and j are at least three. Now we take for $E(i,j)$ the union of $E(i-1,j)$ and the image $E'(i,j-1)$ of $E(i,j-1)$ under a translation $[x,y] \mapsto [x+\Delta_x, y+\Delta_y]$ with Δ_x, Δ_y chosen so that

- $x < x'$ and $y < y'$ whenever $[x,y] \in E(i-1,j)$ and $[x',y'] \in E'(i,j-1)$,
- each line passing through a point of $E(i-1,j)$ and a point of $E'(i,j-1)$
  has a greater slope than
     all lines passing through two points of $E(i-1,j)$ and
     all lines passing through two points of $E'(i,j-1)$.   □

THEOREM 2.3   *For every integer n greater than 1, there are 2^{n-2} points in the plane of which no three lie on the same line and no n are vertices of a convex polygon.*

Proof   We start with any sequence

$$[x_2, y_2], [x_3, y_3], \ldots, [x_n, y_n]$$

of points in the plane such that $x_2 < x_3 < \ldots < x_n$, $y_2 > y_3 > \ldots > y_n$ and such that, for all choices of subscripts r, s, t with $2 \le r < s < t \le n$, point $[x_s, y_s]$ lies below the line passing through points $[x_r, y_r]$ and $[x_t, y_t]$. (One way of meeting these conditions is setting $x_s = s$, $y_s = 1/s$ for all s.) Around each of these points $[x_s, y_s]$, we place a small square,

$$Q_s = [x_s - \varepsilon, x_s + \varepsilon] \times [y_s - \varepsilon, y_s + \varepsilon] \quad \text{with a positive } \varepsilon.$$

If ε is sufficiently small, then

- $x < x'$ and $y > y'$ whenever $[x,y] \in Q_s$, $[x',y'] \in Q_{s+1}$, and $2 \le s \le n-1$,

- for all choices of subscripts r, s, t with $2 \leq r < s < t \leq n$,
  the entire square Q_s lies below all lines intersecting both squares Q_r and Q_t.

Assume that this is the case and let sets $E(i,j)$ be as in Lemma 2.2. There is a number M such that all of the sets $E(n+2-s, s)$ with $s = 2, 3, \ldots, n$ are contained in the square $[-M, M] \times [-M, M]$; for each s in this range, the transformation $[x, y] \mapsto [x_s + (\varepsilon/M)x, y_s + (\varepsilon/M)y]$ maps $E(n+2-s, s)$ onto a subset $E^*(n+2-s, s)$ of Q_s. Writing

$$S = E^*(n, 2) \cup E^*(n-1, 3) \cup \ldots \cup E^*(2, n),$$

we note that $|S| = 2^{n-2}$ and that no three points of S lie on the same line; we claim that no n of them are vertices of a convex polygon. To justify this claim, consider an arbitrary nonempty subset C of S that consists of vertices of a convex polygon; let r be the smallest subscript such that $C \cap Q_r \neq \emptyset$ and let t be the largest subscript such that $C \cap Q_t \neq \emptyset$.

CASE 1: $r = t$. In this case, $C \subseteq E^*(n+2-r, r)$; since C is the union of a concave sequence and a convex sequence that share their first point as well as their last point, we have

$$|C| \leq (r-1) + (n+1-r) - 2 = n - 2.$$

CASE 2: $r < t$. No four vertices of a convex polygon have the property that one of them is inside the triangle formed by the other three. In particular (since $C \cap Q_r \neq \emptyset$ and $C \cap Q_t \neq \emptyset$ and since all lines passing through two points of the same $C \cap Q_s$ have positive slopes),

- $C \cap Q_r$ contains no convex sequence of three points;
- there is no subscript s such that $r < s < t$ and $|C \cap Q_s| \geq 2$;
- $C \cap Q_t$ contains no concave sequence of three points.

Since $C \cap Q_r$ contains no convex sequence of three points, its points form a concave sequence; since $C \cap Q_r \subseteq E^*(n+2-r, r)$, we conclude that $|C \cap Q_r| \leq r - 1$. Since $C \cap Q_t$ contains no concave sequence of three points, its points form a convex sequence; since $C \cap Q_t \subseteq E^*(n+2-t, t)$, we conclude that $|C \cap Q_t| \leq n + 1 - t$. To sum it up,

$$|C| = \sum_{s=r}^{t} |C \cap Q_s| \leq (r-1) + (t-r-1) + (n+1-t) = n - 1. \qquad \square$$

2.2   The Sylvester–Gallai Theorem

In 1933, while Erdős was reading the book [213], the following conjecture occurred to him:

> *if a finite number of points in the plane do not lie all on a line,*
> *then some line goes through precisely two of them.*

He writes [128] that he expected this problem to be easy, but to his great surprise and disappointment he could not find a proof. He told this to his friend Tibor Gallai (1912–92), who very soon found an ingenious proof. Seven years later and unaware of Gallai's result, Eberhard Melchior (1912–?) deduced from the assumption that not all the points lie on a line a stronger conclusion: at least three lines go through precisely two points [292]. In 1943, in turn unaware of Melchior's result, Erdős proposed the problem in the *American Mathematical Monthly* [107] and additional proofs were given by R. C. Buck (1920–98), N. E. Steenrod (1910–71), and Robert Steinberg (1922–2014). At that time, L. M. Kelly (1914–2002) noted that the same problem had been proposed also by James Joseph Sylvester (1814–97) in the March 1893 issue of *Educational Times* [351]. (The May 1893 issue of the same journal reported a four-line "solution" submitted by H. J. Woodall, A.R.C.S.,[b] followed by a comment pointing out two flaws in the argument and sketching another line of enquiry, which "is equally incomplete, but may be worth notice.") Even though it is not clear whether Sylvester had a proof or not, the statement is now commonly referred to as the *Sylvester–Gallai Theorem*. We now reproduce its proof by L. M. Kelly [90].

THEOREM 2.4 *If S is a set of finitely many points in the plane that do not lie all on a line, then some line goes through precisely two points of S.*

Proof Among all triples x, y, z of points in S such that z is not on the line $\overline{xy}$ passing through x and y (there are such triples: this is where we use the assumption that points of S do not lie all on a line), choose one which minimizes the distance from z to $\overline{xy}$. We claim that the line $\overline{xy}$ goes through precisely two points of S. To justify this claim, assume the contrary: at least three distinct points of S lie on the line $\overline{xy}$. On this line, locate the point p (not necessarily an element of S) that is closest to z. This point splits $\overline{xy}$ into two half-lines that overlap in p; at least one of these half-lines contains at least two points of S; label these points as p_1, p_2 in such a way that p_1 is closer to p (and possibly equal to p). Now p_1 is closer to $\overline{zp_2}$ than z is to $\overline{xy}$, a contradiction. □

Erdős [107] noted that Theorem 2.4 has the following corollary:

COROLLARY 2.5 *If V is a set of finitely many points in the plane that do not lie all on a line, then the number of lines that go through at least two points of V is at least $|V|$.*

Proof By induction on $|V|$. Theorem 2.4 guarantees the existence of points x, y of V such that the line $\overline{xy}$ passing through them contains no other point of V. If the points of $V - \{x\}$ do not lie all on a line, then the induction hypothesis guarantees that at least $|V| - 1$ distinct lines go through at least two points of $V - \{x\}$; by the choice of x, all of these lines are distinct from $\overline{xy}$. If the points of $V - \{x\}$ do lie all on a line, then this line and the $|V| - 1$ lines $\overline{xz}$ with $z \in V - \{x\}$ are all distinct. □

[b] the 'A.R.C.S' after H. J. Woodall's name stands for 'Associate of the Royal College of Science'.

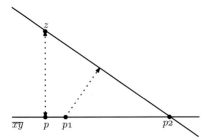

Figure 2.2 Kelly's proof of the Sylvester–Gallai theorem.

Willy Moser and Paul Erdős

© Geňa Hahn

2.3   A De Bruijn–Erdős Theorem

Nicolaas Govert de Bruijn (1918–2012) and Erdős wrote six papers together. Two of their joint results [93, 94], mutually unrelated, are referred to as "the De Bruijn–Erdős theorem"; one of these generalizes Corollary 2.5:

THEOREM 2.6 ([93])   *Let m and n be positive integers such that $m \geq 2$; let V be a set of n points; let E be a family of m subsets of V such that every two distinct points of V belong to precisely one member of E. Then $m \geq n$, with equality if and only if (i) E is of the type $\{p_1, \ldots, p_{n-1}\}$, $\{p_1, p_n\}$, $\{p_2, p_n\}$, $\ldots$, $\{p_{n-1}, p_n\}$ or (ii) $n = k(k-1)+1$ with each member of E containing precisely k points of V and each point of V contained in precisely k members of E.*

Lest the reader get confused, let us stress that 'points' in Theorem 2.6 mean not points in the plane, but abstract elements of V. For future reference, note that

---

[c] Author and co-author of [297], [244], [52], [56], and other works in geometry.

assumptions of this theorem imply

$$|L| \le n - 1 \text{ whenever } L \in E \tag{2.3}$$

and that we may also assume

$$|L| \ge 2 \text{ whenever } L \in E \tag{2.4}$$

since assumptions of the theorem remain satisfied if all sets of size less than 2 are removed from E.

LEMMA 2.7   *Under the assumptions of Theorem 2.6, let $d(p)$ denote the number of members of E that contain a point p in V. Then*

$$p \in V, \ L \in E, \ p \notin L \ \Rightarrow \ d(p) \ge |L|. \tag{2.5}$$

Furthermore, if $m \le n$, then members of E can be enumerated as $L_1, L_2, \ldots, L_m$ and points of V can be enumerated as $p_1, p_2, \ldots, p_n$ in such a way that

$$|L_i| \le d(p_i) \text{ for all } i = 1, 2, \ldots, m. \tag{2.6}$$

Proof   Consider a point p in V and a member L of E such that $p \notin L$; given any point x of L, let $F(x)$ denote the unique member of E that contains p and x and note that $F(x) \ne L$ as $p \in F(x)$, $p \notin L$. Next, consider an arbitrary point x in L and any point x' in $F(x) \cap L$. Since $x \in F(x) \cap L$ and $F(x) \ne L$ and since every two distinct points of V belong to precisely one member of E, we must have $x' = x$. It follows that $F(x) \cap L = \{x\}$, and so $x \ne y \Rightarrow F(x) \ne F(y)$, and so $|L| \le d(p)$. This proves (2.5).

Note that

$$d(p) \ge 2 \text{ for all } p:$$

to see this, take any point p' of V such that $p' \ne p$ and consider the unique member L of E that contains both p and p'. By (2.3), some point p'' of V lies outside L; the unique member of E that contains both p and p'' is distinct from L.

Next, let k denote the smallest $d(p)$ and let p^* be a point of V for which $d(p^*) = k$. Enumerate the members of E as $L_1, L_2, \ldots, L_m$ in such a way that all of $L_1, L_2, \ldots, L_k$ contain p^* and none of $L_{k+1}, L_{k+2}, \ldots, L_m$ do. By (2.4), the k sets

$$L_1 - \{p^*\}, L_2 - \{p^*\}, \ldots, L_k - \{p^*\}$$

are nonempty. Since $k \ge 2$, we can find points $p_1, p_2, \ldots, p_k$ of V such that

$$p_1 \in L_2 - \{p^*\}, \quad p_2 \in L_3 - \{p^*\}, \quad \ldots, \quad p_{k-1} \in L_k - \{p^*\}, \quad p_k \in L_1 - \{p^*\}.$$

Since the k sets $L_1 - \{p^*\}, L_2 - \{p^*\}, \ldots, L_k - \{p^*\}$ are pairwise disjoint, $p_1, p_2, \ldots, p_k$ are pairwise distinct and

$$p_1 \notin L_1, \quad p_2 \notin L_2, \quad \ldots, \quad p_{k-1} \notin L_{k-1}, \quad p_k \notin L_k.$$

Now (2.5) guarantees that

$$|L_i| \le d(p_i) \text{ for all } i = 1, 2, \ldots, k. \tag{2.7}$$

Enumerate the remaining $n - k$ points of V as $p_{k+1}, p_{k+2}, \ldots, p_n$. Fact (2.5) guarantees that $|L_i| \le d(p^*)$ for all $i = k + 1, k + 2, \ldots, m$, and our choice of p^*

guarantees that $d(p^*) \leq d(p_i)$ for all $i = 1, 2, \ldots, n$. It follows that, as long as $m \leq n$,

$$|L_i| \leq d(p^*) \leq d(p_i) \quad \text{for all } i = k+1, k+2, \ldots, m. \qquad (2.8)$$

Conjunction of (2.7) and (2.8) proves (2.6). $\qquad\qquad\square$

Proof of Theorem 2.6   We may assume that $m \leq n$: otherwise the conclusion is trivial. Enumerate members of E and points of V as in Lemma 2.7. Counting all pairs (L_i, p_j) such that $1 \leq i \leq m, 1 \leq j \leq n,\ p_j \in L_i$ in two different ways (first grouping them by L_i and then grouping them by p_j), we find that

$$\sum_{i=1}^{m} |L_i| = \sum_{j=1}^{n} d(p_j). \qquad (2.9)$$

Comparing (2.9) and (2.6), we conclude that

$$|L_i| = d(p_i) \quad \text{for all } i = 1, 2, \ldots, m$$

and that $m = n$.

Finally, choosing a subscript r which maximizes $|L_r|$, let us distinguish between two cases.

CASE 1: *$|L_s| < |L_r|$ whenever $s \neq r$.* In this case, $d(p_s) = |L_s| < |L_r|$ whenever $s \neq r$, and so (2.5) guarantees that $p_s \in L_r$ whenever $s \neq r$. It follows that $|L_r| = n - 1$; in turn, this implies that E is of the type $\{p_1, p_2, \ldots, p_{n-1}\}, \{p_1, p_n\}, \{p_2, p_n\}, \ldots, \{p_{n-1}, p_n\}$.

CASE 2: *There is a subscript s such that $s \neq r$ and $|L_s| = |L_r|$.* In this case, let k denote $|L_r|$. By our choice of r, we have

$$|L_i| = d(p_i) \leq k \quad \text{for all } i = 1, 2, \ldots, m \qquad (2.10)$$

with equality for $i = r$, for $i = s$, and possibly also for other values of i. If $p_i \notin L_r \cap L_s$, then $|L_i| = d(p_i) \geq k$ by (2.5) with $L = L_r$ or $L = L_s$, and so $|L_i| = d(p_i) = k$ by (2.10). Since $L_r \cap L_s$ contains at most one point, it follows that there is a subscript t such that

$$|L_i| = d(p_i) = k \quad \text{whenever } i \neq t.$$

If $|L_t| = d(p_t) \geq n - 1$, then $|L_t| \geq k$ by (2.3) with $L = L_r$; if $|L_t| = d(p_t) < n - 1$, then p_t lies outside some L_i with $i \neq t$, and so $|L_t| = d(p_t) \geq k$ by (2.5) with $L = L_i$. In either case, we have $|L_t| = d(p_t) \geq k$, and so $|L_t| = d(p_t) = k$ by (2.10). We conclude that

$$|L_i| = d(p_i) = k \quad \text{for all } i = 1, 2, \ldots, m.$$

This means that each member of E contains k points of V and each point of V is contained in k members of E. Since each of the $n(n-1)/2$ unordered pairs of distinct points of V belongs to precisely one of the n members of E and since each member of E contains precisely $k(k-1)/2$ of these pairs, we have $n(n-1)/2 = nk(k-1)/2$. $\quad\square$

Families of type (i) in Theorem 2.6 are called *near-pencils*. They can be represented by configurations of points and lines in the plane: take points $p_1, \ldots, p_{n-1}$ all

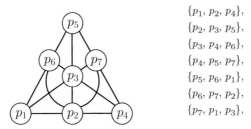

$$\{p_1, p_2, p_4\},$$
$$\{p_2, p_3, p_5\},$$
$$\{p_3, p_4, p_6\},$$
$$\{p_4, p_5, p_7\},$$
$$\{p_5, p_6, p_1\},$$
$$\{p_6, p_7, p_2\},$$
$$\{p_7, p_1, p_3\},$$

Figure 2.3 The seven lines of the Fano plane.

on a line and a point p_n off this line. The Sylvester–Gallai theorem shows the nonexistence of such geometric representations of families of type (ii) with $k \geq 3$. (The unique family of type (ii) with $k = 2$ is a near-pencil.)

Given a family of type (ii) in Theorem 2.6, let us refer to elements of V as *points* and to members of E as *lines*. In these terms,

(A) there are precisely $k^2 - k + 1$ points,
(B) every line has precisely k points,
(C) every two distinct points belong to precisely one line.

When $k \geq 3$, a collection of points and lines with properties (A),(B),(C) is called a *projective plane of order $k - 1$*. (As we have noted, the statement of the Sylvester–Gallai theorem becomes false when transferred to the context of these planes: here, the $k^2 - k + 1$ points do not lie all on a line and yet no line has precisely two points.) When their order is immaterial, projective planes of some order are called simply *finite projective planes*. The smallest finite projective plane is the unique projective plane of order 2; this plane is also known as the *Fano plane* [163]; it is presented in Figure 2.3 with its seven points labeled as p_1, p_2, p_3, p_4, p_5, p_6, p_7. More on projective planes of prescribed orders will be revealed in Section 5.3.2.

2.4   Other Proofs of the De Bruijn–Erdős Theorem

2.4.1   Hanani

Haim Hanani (1912–91) begins the English summary of [205] with the statement of the De Bruijn–Erdős theorem and says in a footnote

> The problem was raised by Th. Motzkin in 1938,
> and in that year the following solution was given.

Then he outlines his proof of the theorem. In a subsequent paper [206], he elaborates on the details of the proof and concludes with a footnote saying

> This theorem has been proved directly (without use of the Lemma) by N. G. de Bruijn and P. Erdős [here he cites [93]] and independently by the author [here he cites [205]].

Here is a paraphrase of his argument:

Let A denote the largest member of E and let B denote the second largest member of E; write $a = |A|$ and $b = |B|$. By (2.4), we have $b \geq 2$. As in Lemma 2.7, let $d(p)$ denote the number of members of E that contain a point p in V.

First, we propose to prove that

$$p \in A \implies d(p) - 1 \geq \frac{n-a}{b-1}. \tag{2.11}$$

For this purpose, let P denote the set of pairs (L, q) such that $L \in E$, $L \ni p$ and $q \in L - A$. For each point q in $V - A$, we have $(L, q) \in P$ if and only if $L \in E$ and $p, q \in L$, which is the case for precisely one L. It follows that

$$|P| = n - a. \tag{2.12}$$

If $(L, q) \in P$, then $q \in L - A$ forces $L \neq A$. In particular, $|L| \leq b$, and so $p \in L \cap A$ implies $|L - A| \leq b - 1$. Since for each of the $d(p) - 1$ choices of L such that $L \in E$, $L \ni p$, $L \neq A$ there are at most $b - 1$ choices of q such that $q \in L - A$, we have

$$|P| \leq (d(p) - 1)(b - 1). \tag{2.13}$$

Conjunction of (2.12) and (2.13) proves (2.11).

As $m \geq 1 + \sum_{p \in A}(d(p) - 1)$, we deduce from (2.11) that

$$m \geq 1 + \frac{a(n-a)}{b-1}. \tag{2.14}$$

If $A \cap B \neq \emptyset$, then let p denote the unique point of $A \cap B$; else let p denote a point of A. For each of the $(a - 1)$ points x in $A - \{p\}$ and for each of the b or $(b - 1)$ points y in $B - \{p\}$, a unique member L of E includes both x and y. These L are pairwise distinct, and none of them include p. From (2.11) again, we deduce that

$$m \geq 1 + \frac{n-a}{b-1} + (a-1)(b-1). \tag{2.15}$$

Now set $c = (b-2)/(a-1)$. Adding c times (2.14) to $1 - c$ times (2.15), we get the inequality

$$m \geq n + (b-2)(a-b), \tag{2.16}$$

which guarantees that $m \geq n$.

Finally, consider the case of $m = n$. Here, (2.16) guarantees that $b = 2$ or $a = b$. Furthermore, (2.16) is tight; since it is a nonnegative multiple of (2.14) plus a positive multiple of (2.15), inequality (2.15) must be tight, too:

$$n = 1 + \frac{n-a}{b-1} + (a-1)(b-1). \tag{2.17}$$

CASE 1: $b = 2$. As all members of E except (possibly) A have size two, we have $n = m = 1 + \binom{n}{2} - \binom{a}{2}$, and so $a = n - 1$. Now E is a near-pencil.

CASE 2: $a = b$. Here (2.17) reduces to $n = a(a - 1) + 1$. Given a point p of V, enumerate the members of E containing p as $L_1, L_2, \ldots, L_d$. Since $L_1 - \{p\}, L_2 - \{p\}, \ldots, L_d - \{p\}$ are pairwise disjoint and their union is $V - \{p\}$, we have $d = a$.

2.4.2   Motzkin

Theodore Motzkin (1908–70) writes in [298, 0.1, p. 451]

> H. Hanani gave in 1938 a combinatorial proof of [the De Bruijn-Erdős theorem]

and presents another paraphrase of Hanani's proof in [298, 4.4, p. 462]. Then he goes on to prove the following statement [298, 4.6, p. 463]:

LEMMA 2.8   *If an $m \times n$ matrix with zero-one entries a_{ij} satisfies*

$$\textstyle\sum_{j=1}^{n} a_{ij} < n \text{ for all } i$$

and $\sum_{i=1}^{m} a_{ij} > 0$ for all j and

$$a_{rs} = 0 \;\Rightarrow\; \textstyle\sum_{i=1}^{m} a_{is} \geq \sum_{j=1}^{n} a_{rj}, \tag{2.18}$$

then $m \geq n$.

In the context of Theorem 2.6, hypotheses of this lemma can be satisfied by setting

$$a_{ij} = \begin{cases} 1 & \text{if } p_j \in L_i, \\ 0 & \text{if } p_j \notin L_i \end{cases} \tag{2.19}$$

where $p_1, p_2, \ldots, p_n$ are the points of V, and $L_1, L_2, \ldots, L_m$ are the members of E. In particular, (2.18) is (2.5).

Motzkin's proof of Lemma 2.8 relies on a theorem of Frobenius (1849–1917) concerning determinants. Four decades later, Jeff Kahn and Paul Seymour published the following stronger assertion and referred to it as *Motzkin's Lemma* [232, (2.1)].

LEMMA 2.9   *If an $m \times n$ matrix with zero-one entries a_{ij} satisfies*

$$\textstyle\sum_{j=1}^{n} a_{ij} < n \text{ for all } i,$$

then there are subscripts r and s such that $a_{rs} = 0$ and

$$m\textstyle\sum_{j=1}^{n} a_{rj} \geq n\sum_{i=1}^{m} a_{is}.$$

Deducing Lemma 2.8 from Lemma 2.9 is a straightforward matter: Assumptions of Lemma 2.8 subsume assumptions of Lemma 2.9; the conclusion of Lemma 2.9 and the assumption (2.18) of Lemma 2.8 together imply

$$m\textstyle\sum_{j=1}^{n} a_{rj} \geq n\sum_{i=1}^{m} a_{is} \geq n\sum_{j=1}^{n} a_{rj};$$

under the assumptions of Lemma 2.8, we have $\sum_{i=1}^{m} a_{is} > 0$. It follows that $\sum_{j=1}^{n} a_{rj} > 0$, and so $m \geq n$.

Proof of Lemma 2.9   Order the columns of the matrix so that

$$\textstyle\sum_{i=1}^{m} a_{ij} < m \quad \text{for } j = 1, 2, \ldots, k,$$
$$\textstyle\sum_{i=1}^{m} a_{ij} = m \quad \text{for } j = k+1, k+2, \ldots, n$$

and note that

$$\sum_{j=1}^{k} a_{ij} = \sum_{j=1}^{n} a_{ij} - (n-k) < k \quad \text{for all } i.$$

Since

$$\sum_{r=1}^{m}\sum_{s=1}^{k}(1-a_{rs})\frac{m\sum_{j=1}^{k}a_{rj} - k\sum_{i=1}^{m}a_{is}}{(m-\sum_{i=1}^{m}a_{is})(k-\sum_{j=1}^{k}a_{rj})}$$

$$= \sum_{r=1}^{m}\sum_{s=1}^{k}(1-a_{rs})\left(\frac{k}{k-\sum_{j=1}^{k}a_{rj}} - \frac{m}{m-\sum_{i=1}^{m}a_{is}}\right)$$

$$= \sum_{r=1}^{m}\left(\frac{k}{k-\sum_{j=1}^{k}a_{rj}}\sum_{s=1}^{k}(1-a_{rs})\right) - \sum_{s=1}^{k}\left(\frac{m}{m-\sum_{i=1}^{m}a_{is}}\sum_{r=1}^{m}(1-a_{rs})\right)$$

$$= \sum_{r=1}^{m}k - \sum_{s=1}^{k}m = 0,$$

the terms

$$\frac{m\sum_{j=1}^{k}a_{rj} - k\sum_{i=1}^{m}a_{is}}{(m-\sum_{i=1}^{m}a_{is})(k-\sum_{j=1}^{k}a_{rj})}$$

ranging over r and s such that $a_{rs} = 0$ sum up to zero, and so the largest of these terms is nonnegative. The corresponding subscripts r and s satisfy the conclusion of the lemma as

$$m\sum_{j=1}^{n}a_{rj} = m\left(\sum_{j=1}^{k}a_{rj} + (n-k)\right) \ge k\sum_{i=1}^{m}a_{is} + (n-k)m \ge n\sum_{i=1}^{m}a_{is}.$$

(An essentially identical argument was used some 24 years earlier in [19]. We will review it on page 31.) □

2.4.3   Ryser

Herbert Ryser (1923–85) proved [336, Theorem 1.1]:

THEOREM 2.10   *Let m, n, and λ be positive integers such that $n \ge 2$; let V be a set of n points; let E be a family of m subsets of V such that every point of V belongs to more than λ members of E and every two distinct points of V belong to precisely λ members of E. Then $m \ge n$, with equality only if*

> *(i) there are distinct integers r_1, r_2 such that*
>> *every L in E has size r_1 or r_2,*
>> *at least one L in E has size r_1 and at least one L in E has size r_2,*
>
> *or*
>> *(ii) $n = k(k-1) + 1$ with*
>>> *each member of E containing k points of V and*
>>> *each point of V contained in k members of E.*   □

The special case $\lambda = 1$ of Theorem 2.10 nearly subsumes Theorem 2.6: all that is missing is the specification $\{r_1, r_2\} = \{2, n - 1\}$ in (i). Families of type (ii) in Theorem 2.10 are called *symmetric block designs*.

Proof of the inequality $m \geq n$ in Theorem 2.10   Enumerate the points of V as $p_1, p_2, \ldots, p_n$; enumerate the members of E as $L_1, L_2, \ldots, L_m$; define an $m \times n$ matrix A with entries a_{ij} by (2.19). We are going to show that

$$Ax = 0 \ \Rightarrow \ x = 0,$$

which means that the n columns of A are linearly independent, and so $n \leq m$.

If $Ax = 0$, then $x^T(A^T A)x = (Ax)^T(Ax) = 0$; let us prove that

$$x^T(A^T A)x = 0 \Rightarrow x = 0. \tag{2.20}$$

The entry in the i-th row and the j-th column of $A^T A$ equals the number of members of E that contain both p_i and p_j; by assumption, this number equals λ when $i \neq j$ and it is greater than λ when $i = j$. It follows that

$$x^T(A^T A)x \geq \lambda \left(\textstyle\sum_{i=1}^n x_i \right)^2 + \sum_{i=1}^n x_i^2 \geq \sum_{i=1}^n x_i^2,$$

and so $x^T(A^T A)x = 0 \Rightarrow x = 0$. $\qquad\qquad\qquad\qquad\qquad\qquad\qquad \square$

To prove (2.20), Ryser (and, as he notes, Kulendra Nath Majumdar in an earlier paper [289])[d] evaluates the determinant of $A^T A$. The shortcut taken here appears in [218].

2.4.4   Basterfield, Kelly, Conway

In [19], Basterfield and Kelly present yet another proof of the inequality $m \geq n$ in Theorem 2.6 and write

> We are indebted to J. Conway for the simplicity of the present formulation of the proof of Theorem 2.1.

Conway's argument relies on the fact that for all lines L (these satisfy $|L| < n$ by (2.3)) and for all points p such that $d(p) < m$, we have

$$\frac{m|L| - nd(p)}{(n - |L|)(m - d(p))} = \frac{n}{n - |L|} - \frac{m}{m - d(p)}.$$

It follows that

$$\sum_L \sum_{p \notin L} \frac{m|L| - nd(p)}{(n - |L|)(m - d(p))} = \sum_L \sum_{p \notin L} \frac{n}{n - |L|} - \sum_p \sum_{L \not\ni p} \frac{m}{m - d(p)} = \sum_L n - \sum_p m = 0,$$

---

[d] Beginning in 1962, Kulendra Nath Majumdar's last name appears as 'Majindar' in his publications.

and so the largest summand in the first double sum must be nonnegative: there are a line L and a point p not on this line such that $m|L| \geq nd(p)$. Now $m \geq n$ follows from (2.5).

Once upon a time, in the Hungarian city of Győr, there lived a high school teacher named Dániel Arany (1863–1944). He created a mathematics and physics journal for high school students, *Középiskolai Mathematikai Lapok* (Mathematical Journal for Secondary Schools). Since its launching on January 1, 1894, it has published nine monthly issues per academic year continually, except for two breaks (1915–24 and 1940–1945) caused by the two world wars. Over the years, its title went through several changes. Currently it is called *Középiskolai Matematikai és Fizikai Lapok* and known under the abbreviation *KöMaL*.

An important part of KöMaL is its annual correspondence problem solving contest for 14–18 year-old students. When Andor Faragó (1877–1944) took the journal over, he began publishing photographs of its the most successful problem solvers (28 of them in 1925–1926 and more in the following years). Archives of these portrait galleries are available at `www.komal.hu/tablok/` and here are a few of the photographs:

Paul Erdős

Eszter Klein

George Szekeres

György Hajós

Paul Turán

Tibor Grünwald

Endre Makai

János Surányi

András Sárközy

Gyula Katona

János Komlós

Béla Bollobás

Gerzson Kéry

Lajos Pósa

Miklós Simonovits

Imre Bárány   László Lovász   László Babai

János Pintz   József Beck   János Kollár

János Pach   Gábor Tardos   Gábor Simonyi

Tibor Szabó

Source: KöMaL online

3   Ramsey's Theorem

3.1   Ramsey's Theorem for Graphs

The Eötvös Mathematics Competition for high school students (including those who just graduated) was founded in 1894 and renamed the Kürschák Mathematics Competition after World War II. Problem 2 of its 48th season in 1947 [18] read:

> *Prove that in any group of six people, either there are three people who know one another or three people who do not know one another. Assume that "knowing" is a symmetric relation.*

Six years later, the same problem was stated in different terms as Question 2 of the morning session of the 13th William Lowell Putnam Mathematical Competition held on March 23, 1953 [64]:

> *Six points are in general position in space (no three in a line, no four in a plane). The fifteen line segments joining them are drawn and then painted, some segments red, some blue. Prove that some triangle has all its sides the same color.*

Five years after that, it reappeared in the American Mathematical Monthly [54] in its original 1947 formulation and it was solved by a number of people, including Bush and Cheney.[a]

DEFINITIONS: A *graph* is an ordered pair (V, E) such that V is a set and E is a set of two-point subsets of V. Elements of V are called *vertices* (plural of *vertex*) and elements of E are called *edges* of the graph. We refer to an edge $\{v, w\}$ by writing simply vw. Vertices v and w are said to be *adjacent* if vw is an edge; otherwise they are said to be *nonadjacent*. A *subgraph* of a graph (V, E) is a graph (V', E') such that $V' \subseteq V$ and $E' \subseteq E$. A *complete graph* is a graph where every two vertices are adjacent. The *order* of a graph is the number of its vertices.

In these graph-theoretic terms, the problem is to prove that

whenever the edges of the complete graph of order six
are coloured red and blue,

---

[a] The solvers Bush and Cheney of the Monthly problem were *not* George Bush and Dick Cheney.

there is a complete subgraph of order three with all edges red
or a complete subgraph of order three with all edges blue.

To prove this, take one of the six vertices and call it u. Three of the remaining five vertices must be joined to u by edges that have all the same colour; switching red and blue if necessary, we may assume that this colour is red. Let uv_1, uv_2, uv_3 denote the three red edges. If any of the three edges v_iv_j is red, then the complete subgraph on vertices u, v_i, v_j has all edges red; if none of the three edges v_iv_j is red, then the complete subgraph on vertices v_1, v_2, v_3 has all edges blue.

NOTATION: Let $a \rightarrow (b_1, b_2)$ denote the statement that

whenever the edges of the complete graph of order a
are coloured red and blue,
there is a complete subgraph of order b_1 with all edges red
or a complete subgraph of order b_2 with all edges blue.

In this notation, our starting problem was proving that $6 \rightarrow (3,3)$. In 1928, Frank Ramsey (1903–30; a mathematician, philosopher, economist, and also a militant atheist with a saintly tolerance of his brother Michael, who eventually became the one hundredth Archbishop of Canterbury) proved a theorem [323] whose special case states that *for every choice of positive integers b_1, b_2 there is an integer a such that $a \rightarrow (b_1, b_2)$*.

We are going to prove this special case of Ramsey's theorem by an argument coming from Erdős and Szekeres [157]. To begin, let us generalize our proof of $6 \rightarrow (3,3)$:

LEMMA 3.1   *If $a_1 \rightarrow (b_1 - 1, b_2)$ and $a_2 \rightarrow (b_1, b_2 - 1)$, then $a_1 + a_2 \rightarrow (b_1, b_2)$.*

Proof   Given a colouring of the edges of the complete graph of order $a_1 + a_2$ by colours red and blue, we aim to find a complete subgraph of order b_1 with all edges red or a complete subgraph of order b_2 with all edges blue. With A standing for the vertex set of the complete graph, take one of the vertices, call it u, and colour each vertex v of $A - \{u\}$ red or blue according to the colour of uv. Since $|A - \{u\}| > (a_1 - 1) + (a_2 - 1)$, at least one of the following two conditions is satisfied:

(R) there is an a_1-point subset R of $A - \{u\}$ such that
    all vertices in R are red, and so all edges uv with $v \in R$ are red,
(B) there is an a_2-point subset B of $A - \{u\}$ such that
    all vertices in B are blue, and so all edges uv with $v \in B$ are blue.

If (R) holds, then assumption $a_1 \rightarrow (b_1 - 1, b_2)$ guarantees that the complete graph with vertex set R has a complete subgraph on a set R_0 of $b_1 - 1$ vertices with all edges red or a complete subgraph of order b_2 with all edges blue. In the former case, we are done since all edges of the complete subgraph with vertex set $\{u\} \cup R_0$ are red; in the latter case, we are done trivially.

If (B) holds, then assumption $a_2 \rightarrow (b_1, b_2 - 1)$ guarantees that the complete graph with vertex set B has a complete subgraph of order b_1 with all edges red or a complete subgraph on a set B_0 of $b_2 - 1$ vertices with all edges blue. In the former case, we are done trivially; in the latter case, we are done since all edges of the complete subgraph with vertex set $\{u\} \cup B_0$ are blue.   □

THEOREM 3.2 (Erdős and Szekeres [157])   *If b_1, b_2 are positive integers, then*

$$\binom{b_1 + b_2 - 2}{b_1 - 1} \to (b_1, b_2).$$

Proof   By induction on $b_1 + b_2$.

INDUCTION BASIS: $b_1 = 1$ *or* $b_2 = 1$. Here the conclusion, $1 \to (1, b_2)$ or $1 \to (b_1, 1)$, is trivial.

INDUCTION STEP: $b_1 \geq 2$ *and* $b_2 \geq 2$. Here the conclusion follows from the induction hypothesis, Lemma 3.1, and the fact that

$$\binom{b_1 + b_2 - 3}{b_1 - 1} + \binom{b_1 + b_2 - 3}{b_1 - 2} = \binom{b_1 + b_2 - 2}{b_1 - 1}.$$ □

3.2   Ramsey Numbers

NOTATION: The smallest a such that $a \to (b_1, b_2)$ is denoted as $R(b_1, b_2)$.

In this notation, Theorem 3.2 asserts that

$$R(b_1, b_2) \leq \binom{b_1 + b_2 - 2}{b_1 - 1} \tag{3.1}$$

and Lemma 3.1 asserts that

$$R(b_1, b_2) \leq R(b_1 - 1, b_2) + R(b_1, b_2 - 1). \tag{3.2}$$

In [200], Robert E. Greenwood (1911–93) and Andrew M. Gleason (1921–2008) observed that in some cases, inequality (3.2) can be slightly improved:

LEMMA 3.3   *If both $R(b_1 - 1, b_2)$ and $R(b_1, b_2 - 1)$ are even, then*

$$R(b_1, b_2) \leq R(b_1 - 1, b_2) + R(b_1, b_2 - 1) - 1.$$

Proof   Write

$$a = R(b_1 - 1, b_2) + R(b_1, b_2 - 1) - 1.$$

Given a colouring of the edges of the complete graph of order a by colours red and blue, we aim to find a complete subgraph of order b_1 with all edges red or a complete subgraph of order b_2 with all edges blue. For each of the a vertices u, partition the remaining $a - 1$ vertices into two sets:

$$R(u) = \{v : \text{edge } uv \text{ is red}\}, \quad B(u) = \{v : \text{edge } uv \text{ is blue}\}.$$

Since the sum of the a integers $|R(u)|$ counts twice the total number of red edges in the colouring, this sum is even; since the number a of the summands $|R(u)|$ is odd, at least one of these summands must be even. Let u be any vertex with $|R(u)|$ even. Since

$$|R(u)| + |B(u)| = R(b_1 - 1, b_2) + R(b_1, b_2 - 1) - 2$$

and since both of $|R(u)|$ and $R(b_1 - 1, b_2)$ are even, we have

$$|R(u)| \geq R(b_1 - 1, b_2) \text{ or } |B(u)| \geq R(b_1, b_2 - 1).$$

If $|R(u)| \geq R(b_1 - 1, b_2)$, then there is a red complete subgraph on u and $b_1 - 1$ vertices in $R(u)$ or a blue complete subgraph on b_2 vertices in $R(u)$.

If $|B(u)| \geq R(b_1, b_2 - 1)$, then there is a red complete subgraph of order b_1 in $B(u)$ or a blue complete subgraph on u and $b_2 - 1$ vertices in $B(u)$.   □

We have noted that $R(3, 3) \leq 6$. Actually, we have

$$R(3, 3) = 6 :$$

to see that $R(3, 3) > 5$, consider the colouring of the edges of the complete graph on vertices $0, 1, 2, 3, 4$ by colours red and blue, where edge ij is coloured red if $(i - j)$ mod $5 \in \{1, 4\}$ and blue if $(i - j)$ mod $5 \in \{2, 3\}$.

Since $R(4, 2) = 4$ and $R(3, 3) = 6$, the Greenwood–Gleason Lemma 3.3 guarantees that $R(4, 3) \leq 9$; in turn, (3.2) guarantees that $R(4, 4) \leq 18$. Actually, Greenwood and Gleason proved that

$$R(4, 4) = 18 :$$

to show that $R(4, 4) > 17$, they set

$$R = \{1, 2, 4, 8, 9, 13, 15, 16\}, \ B = \{3, 5, 6, 7, 10, 11, 12, 14\}$$

and coloured each edge ij of the complete graph on vertices $0, 1, \ldots, 16$ red if $(i - j)$ mod $17 \in R$ and blue if $(i - j)$ mod $17 \in B$. The eight elements of R are the eight integers n^2 mod 17 with $n = 1, 2, \ldots, 8$; they are called *quadratic residues modulo* 17.

In verifying that this colouring creates neither a red complete graph of order four nor a blue complete graph of order four, it is helpful to note first that

$$(x \text{ mod } 17 \in B \quad \text{and} \quad y \text{ mod } 17 \in B) \quad \Rightarrow \quad xy \text{ mod } 17 \in R,$$
$$(x \text{ mod } 17 \in B \quad \text{and} \quad y \text{ mod } 17 \in R) \quad \Rightarrow \quad xy \text{ mod } 17 \in B,$$
$$(x \text{ mod } 17 \in R \quad \text{and} \quad y \text{ mod } 17 \in R) \quad \Rightarrow \quad xy \text{ mod } 17 \in R.$$

In particular, the permutation $k \mapsto 3k$ mod 17 of the 17 vertices changes the colour of all edges, and so there is a red complete graph of order four if and only if there is a blue complete graph of order four. Next, suppose that there is a red complete graph of order four; choose two of its four vertices; call them i and j. Since the edge ij is red, we have $(j - i)$ mod $17 \subset R$; since $1 \cdot 1$ mod $17 = 9 \cdot 2$ mod $17 = 13 \cdot 4$ mod $17 = 15 \cdot 8$ mod $17 = 16 \cdot 16$ mod $17 = 1$, there is an x in R such that $x(j - i)$ mod $17 = 1$. Since the permutation $k \mapsto x(k - i)$ mod 17 of vertices preserves the colour of all edges and maps i to 0 and maps j to 1, we conclude that there is a red complete graph of order four that includes vertices 0 and 1. Each of its two remaining vertices

k must have $k \in R$ and $(k - 1) \bmod 17 \in R$, which implies that k is one of the three vertices $2, 9, 16$. But each edge joining two of these three vertices is blue, and so the red graph is nonexistent.

All we know about the value of $R(5, 5)$ are the bounds from [161] and [10]:

$$43 \leq R(5, 5) \leq 48. \tag{3.3}$$

Brendan McKay, Stanisław Radziszowski, and Geoffrey Exoo conjecture that $R(5, 5) = 43$ and support this conjecture by strong experimental evidence [287, Section 4].

DEFINITIONS: A *clique* in a graph is a set of pairwise adjacent vertices; the *clique number* $\omega(G)$ of a graph G is the number of vertices in its largest clique. A *stable set* in a graph is a set of pairwise nonadjacent vertices; the *stability number* $\alpha(G)$ of a graph G is the number of vertices in its largest stable set. (Stable sets are often referred to as *independent sets,* in which case $\alpha(G)$ is called the *independence number* of G.)

Numbers $R(b_1, b_2)$ are called *Ramsey numbers.* Since colouring the edges of a complete graph of order n by colours red and blue amounts to specifying the graph that consists of the n vertices and all the red edges, Ramsey number $R(b_1, b_2)$ is the smallest positive integer n such that every graph G of order n has $\omega(G) \geq b_1$ or $\alpha(G) \geq b_2$.

Erdős believed that computing $R(5, 5)$, although difficult, could be within our reach and that computing $R(6, 6)$ was out of the question. He liked to illustrate this distinction on a parable of a powerful monster who threatens to destroy the Earth unless mankind delivers a certain Ramsey number. If the monster asked for $R(5, 5)$, then our most rational response would be marshalling all our resources to compute this number; if the monster asked for $R(6, 6)$, then our most rational response would be marshalling all our resources to destroy the monster.

One of Erdős's most important results takes up, along with its proof, less than a page of the three-page note [108]:

THEOREM 3.4

$$R(k, k) > 2^{k/2} \quad \text{whenever } k \geq 3.$$

Proof Consider arbitrary integers k and n such that $k \geq 3$ and $n \leq 2^{k/2}$. Let Ω denote the set of all colourings of the edges of the complete graph on vertices $1, 2, \ldots, n$ by colours red and blue. Call a colouring *bad* if it creates a red complete graph of order k or a blue complete graph of order k (or both) and let B denote the set of all bad colourings in Ω. In this notation, the theorem asserts that

$$|B| < |\Omega|.$$

Obviously,

$$|\Omega| = 2^{\binom{n}{2}};$$

we will prove the theorem by showing that

$$|B| \leq \binom{n}{k} 2^{1+\binom{n}{2}-\binom{k}{2}} \tag{3.4}$$

and

$$\binom{n}{k} 2^{1-\binom{k}{2}} < 1. \tag{3.5}$$

To prove (3.4), consider the following plan for constructing bad colourings.

Step 1: Choose a set of k of the vertices $1, 2, \ldots, n$ and let S denote this set.
Step 2: Choose one of the colours red and blue and colour all edges of the complete graph on vertex set S by this colour.
Step 3: Colour the remaining edges of the complete graph on vertices $1, 2, \ldots, n$ by colours red and blue.

There are $\binom{n}{k}$ ways of implementing Step 1, there are 2 ways of implementing Step 2, and there are $2^{\binom{n}{2}-\binom{k}{2}}$ ways of implementing Step 3; it follows that the right-hand side of (3.4) counts the number of different implementations of the entire plan. Pointing out that each bad colouring is constructed in at least one of these implementations concludes the proof of (3.4).

Inequality (3.5) can be written as

$$\binom{n}{k} \cdot 2 < 2^{\binom{k}{2}};$$

since

$$\binom{n}{k} < \frac{n^k}{k!} \quad \text{and} \quad n \leq 2^{k/2},$$

proving it reduces to verifying that

$$k! > 2 \cdot 2^{k/2},$$

which can be done by straightforward induction on k. ☐

A more meticulous calculation shows that Erdős proved a little more than Theorem 3.4: the inequality $k! > 2(k/e)^k$, easily proved by induction on k (the induction step relies on the fact that $1 + x \leq e^x$ for all x), implies that $\binom{n}{k} \cdot 2 < (en/k)^k$, and so

$$R(k, k) > \frac{1}{e\sqrt{2}} \cdot k 2^{k/2}.$$

The best known lower bound on $R(k, k)$ is about twice this much. More precisely, Joel Spencer proved [346] that

$$R(k, k) > c(k) \cdot k 2^{k/2} \text{ with } \lim_{k \to \infty} c(k) = \frac{\sqrt{2}}{e}. \tag{3.6}$$

Erdős offered \$100 for a *constructive* proof of

$$R(k, k) > (1 + \varepsilon)^k \text{ with a positive constant } \varepsilon. \tag{3.7}$$

The first construction of graphs showing that $R(k, k)$ grows faster than all polynomials in k came from Péter Frankl [169]. In each of his graphs, vertices are all the

$2s^2$-point subsets of a fixed t-point set; two vertices are adjacent if and only if their intersection consists of $2as + b$ points, where $0 \le a, b < s$.

Next, Péter Frankl and Richard M. Wilson [172, Theorem 8] constructed graphs showing that

$$R(k, k) > \exp(c \log^2 k / \log \log k), \text{ with a positive constant } c \qquad (3.8)$$

(actually, any c less than $1/4$ will work here). In each of their graphs, vertices are all the $(p^2 - 1)$-point subsets of a fixed p^3-point set, where p is a prime; two vertices are adjacent if and only if the size of their intersection is -1 modulo p.

DEFINITION: *Constructing a family of graphs* of varying orders n means providing a one-variable polynomial f and an algorithm that, given any two vertices of one of the graphs, takes time at most $f(n)$ in order to declare them either adjacent or nonadjacent. (Some authors limit the running time to $f(\log n)$.)

Boaz Barak, Anup Rao, Ronen Shaltiel, and Avi Wigderson [16] greatly improved the constructive bound (3.8) by a complex algorithm. This breakthrough was followed by a number of improvements. Currently, the best of these results comes from Gil Cohen [84]: There are a positive constant c and a construction of arbitrarily large graphs G with n vertices such that

$$\max\{\alpha(G), \omega(G)\} < (\log n)^{(\log \log \log n)^c}.$$

This is a constructive lower bound on $R(k, k)$, but expressing it by a formula in terms of k is not a straightforward matter. For comparison, note that (3.7) can be expressed as

$$\max\{\alpha(G), \omega(G)\} < c \log n \quad \text{with a positive constant } c.$$

Together, Theorem 3.4 and Theorem 3.2 guarantee that $2^{k/2} \le R(k, k) < 4^{k-1}$, and so

$$\sqrt{2} \le R(k, k)^{1/k} < 4$$

whenever $k \ge 3$. Erdős [127, p. 9] offered \$100 for a proof that $\lim R(k, k)^{1/k}$ exists and \$500 for the value of this limit.

The *off-diagonal Ramsey numbers*, $R(b_1, b_2)$ with $b_1 \ne b_2$ and, in particular, those with $b_1 = 3$, have also been the subject of intensive research. Now we know that there are positive constants c_1, c_2 such that

$$c_1 \frac{k^2}{\log k} \le R(3, k) \le c_2 \frac{k^2}{\log k}.$$

The upper bound was established by Miklós Ajtai, János Komlós, and Endre Szemerédi [1]; James B. Shearer [341] later proved that any number greater than 1 will work as c_2. The lower bound was established by Jeong Han Kim [247] (for this work, the Mathematical Programming Society and the American Mathematical Society awarded Kim in 1997 their joint Fulkerson Prize for outstanding papers in the area of discrete mathematics); later, Tom Bohman and Peter Keevash [33] and, independently, Gonzalo Fiz Pontiveros, Simon Griffiths, and Robert Morris [165] proved that any number less than $1/4$ will work as c_1.

Apart from the trivial cases where $b_1 \leq 2$ or $b_2 \leq 2$, only a few exact values of the off-diagonal Ramsey numbers are known. The proof of $R(4,4) = 18$ subsumes a proof of $R(3,4) = 9$. Furthermore, we have

$$R(3,5) = 14 \ [200],$$

$$R(3,6) = 18 \ [245],$$

$$R(3,7) = 23 \ \text{(lower bound in [235], upper bound in [197]),}$$

$$R(3,8) = 28 \ \text{(lower bound in [201], upper bound in [288]),}$$

$$R(3,9) = 36 \ \text{(lower bound in [235], upper bound in [201]),}$$

$$R(4,5) = 25 \ \text{(lower bound in [234], upper bound in [286]).}$$

3.3 A More General Version of Ramsey's Theorem

NOTATION: Let $a \to (b_1, b_2)^k$ denote the statement that

whenever the k-point subsets of an a-point set
are coloured red and blue,
there is a b_1-point set whose k-point subsets are all red
or a b_2-point set whose k-point subsets are all blue.

In this notation, a version of the pigeon-hole principle can be recorded as

$$b_1 + b_2 - 1 \to (b_1, b_2)^1,$$

Daniel Kleitman's "of three ordinary people, two must have the same sex" is $3 \to (2,2)^1$ and our old $a \to (b_1, b_2)$ is $a \to (b_1, b_2)^2$.

LEMMA 3.5 *If $a - 1 \to (a_1, a_2)^{k-1}$ and $a_1 \to (b_1 - 1, b_2)^k$, $a_2 \to (b_1, b_2 - 1)^k$, then $a \to (b_1, b_2)^k$.*

Proof Given a colouring of the k-point subsets of an a-point set A by colours red and blue, we aim to find a b_1-point set whose k-point subsets are all red or a b_2-point set whose k-point subsets are all blue. For this purpose, take one of the points in A, call it u, and colour each $(k-1)$-point subset V of $A - \{u\}$ red or blue according to the colour of $\{u\} \cup V$. Assumption $a - 1 \to (a_1, a_2)^{k-1}$ guarantees that at least one of the following two conditions is satisfied:

(R) there is an a_1-point subset R of $A - \{u\}$ such that
    all $(k-1)$-point subsets of R are red,
    and so all sets $\{u\} \cup V$ with $V \subseteq R$, $|V| = k - 1$ are red,
(B) there is an a_2-point subset B of $A - \{u\}$ such that
    all $(k-1)$-point subsets of B are blue,
    and so all sets $\{u\} \cup V$ with $V \subseteq B$, $|V| = k - 1$ are blue.

If (R) holds, then assumption $a_1 \rightarrow (b_1 - 1, b_2)^k$ guarantees that R has a $(b_1 - 1)$-point subset R_0 with all k-point subsets red or a b_2-point subset with all k-point subsets blue. In the former case, we are done since all k-point subsets of $R_0 \cup \{u\}$ are red; in the latter case, we are done trivially.

If (B) holds, then assumption $a_2 \rightarrow (b_1, b_2 - 1)^k$ guarantees that B has a b_1-point subset with all k-point subsets red or a $(b_2 - 1)$-point subset B_0 with all k-point subsets blue. In the former case, we are done trivially; in the latter case, we are done since all k-point subsets of $B_0 \cup \{u\}$ are blue. □

THEOREM 3.6 *For every choice of positive integers* b_1, b_2 *and* k, *there is an integer* a *such that*

$$a \rightarrow (b_1, b_2)^k.$$

Proof (Erdős and Szekeres [157]) Double induction using Lemma 3.5. □

NOTATION: Let $R(b_1, b_2; k)$ denote the smallest a such that $a \rightarrow (b_1, b_2)^k$.

Apart from the trivial cases where $b_1 \le k$ or $b_2 \le k$, only one exact value of the Ramsey numbers $R(b_1, b_2; k)$ with $k \ge 3$ is known,

$$R(4, 4; 3) = 13 :$$

Brendan McKay and Stanisław Radziszowski found by an exhaustive computer search [285] that $13 \rightarrow (4, 4)^3$ and a colouring constructed earlier [220] by John Isbell (1930–2005) shows that the 13 cannot be reduced to 12.

3.4   Applications to the Happy Ending Theorem

Szekeres derived the Happy Ending Theorem (see Section 2.1) from Theorem 3.6:

THEOREM 3.7 *If* $N \rightarrow (n, 5)^4$, *then from any set of* N *points in the plane of which no three lie on the same line it is always possible to select* n *points that are vertices of a convex polygon.*

Proof Consider an arbitrary set A of N points in the plane of which no three lie on the same line. Colour a four-point subset of A red if its four points are vertices of a convex quadrilateral; otherwise colour this four-point set blue. Since every five-point subset of A has at least one red four-point subset (see page 18), assumption $N \rightarrow (n, 5)^4$ guarantees the existence of an n-point subset of A with all its four-point subsets red. This set consists of vertices of a convex polygon, because

$$\begin{array}{r} n \text{ points are vertices of a convex polygon if and only if} \\ \text{every four of them are vertices of a convex quadrilateral.} \end{array} \tag{3.9}$$

□

In 1973, Michael Tarsi found another way of deriving the Happy Ending Theorem from Ramsey's theorem:[b]

THEOREM 3.8   If $N \to (n,n)^3$, then from any set of N points in the plane of which no three lie on the same line it is always possible to select n points that are vertices of a convex polygon.

Proof   Consider an arbitrary set of N points in the plane of which no three lie on the same line and label its N points by the N integers $1, 2, \ldots, N$. Given any triangle with vertices i, j, k such that $i < j < k$, orient its boundary by traveling from i to j to k to i. If this orientation is counterclockwise, then colour set $\{i, j, k\}$ red; if it is clockwise, then colour set $\{i, j, k\}$ blue. We claim that

> if four points in $\{1, 2, \ldots, N\}$ are not vertices of a convex quadrilateral, then two of the three-point sets formed from them have distinct colours. (3.10)

One way of justifying this claim (Tarsi used another) is to enumerate any four points in $\{1, 2, \ldots, N\}$ that are not vertices of a convex quadrilateral as i, j, k, x so that $i < j < k$ and point x lies inside triangle ijk. Claim (3.10) follows from observing that

- if $x < i$, then ijk and xik have distinct orientations,
- if $i < x < j$, then ijk and ixj have distinct orientations,
- if $j < x < k$, then ijk and jxk have distinct orientations,
- if $k < x$, then ijk and ikx have distinct orientations.

Assumption $N \to (n,n)^3$ guarantees the existence of an n-point subset C of $\{1, 2, \ldots, N\}$ with all its three-point subsets coloured by the same colour. By (3.10), every four points in C are vertices of a convex quadrilateral; in turn, (3.9) guarantees that the n points in C are vertices of a convex polygon.   □

Unaware of Tarsi's result, Scott Johnson [229] found later a completely different proof of Theorem 3.8:

Johnson's proof of Theorem 3.8   Consider an arbitrary set A of N points in the plane of which no three lie on the same line. Colour a three-point subset $\{a, b, c\}$ of A red if the triangle abc contains an even number of points of A in its interior; otherwise, colour this three-point subset of A blue. To justify claim (3.10) for this colouring, enumerate any four points in A that are not vertices of a convex quadrilateral as a, b, c, d so that point d lies inside triangle abc. With

- α the number of points of A in the interior of bcd,
- β the number of points of A in the interior of acd,
- γ the number of points of A in the interior of abd,

---

[b] At the time, Tarsi was an undergraduate at Technion. In a written final examination for a course in combinatorics, he was asked to deduce the Happy Ending Theorem from Ramsey's theorem. Unfortunately for him, he had been absent from the lecture that covered Theorem 3.7. Fortunately for him and for the rest of us, he created Theorem 3.8 on the spot: see [275].

the number of points of A in the interior of abc comes to $\alpha + \beta + \gamma + 1$. As the four integers $\alpha, \beta, \gamma, \alpha + \beta + \gamma + 1$ cannot have the same parity, (3.10) follows. Again, the proof is completed by invoking the assumption $N \to (n,n)^3$ along with (3.10) and (3.9). $\qquad\square$

3.5   Ramsey's Theorem in Its Full Generality

NOTATION: Let $a \to (b)_r^k$ denote the statement that

whenever the k-point subsets of an a-point set
are coloured by r colours,
there is a b-point set whose k-point subsets are all of the same colour.

In this notation, $a \to (b)_2^k$ is our old $a \to (b,b)^k$.

THEOREM 3.9 (Ramsey [323])   *For every choice of positive integers b, k, r, there is an integer a such that*

$$a \to (b)_r^k.$$

Proof   By induction on r. The induction basis ($r = 1$) is trivial. For the induction step, we rely on Theorem 3.6 combined with the following claim:

If $a \to (a_0, b)^k$ and $a_0 \to (b)_{r-1}^k$, then $a \to (b)_r^k$.

To justify this claim, consider an arbitrary colouring of k-point subsets of an a-point set by colours $1, 2, \ldots, r$; think of a set as red if it has colour r and think of it as blue if it has one of the colours $1, 2, \ldots, r - 1$. $\qquad\square$

NOTATION: Let $R_r(b)$ stand for the smallest a such that $a \to (b)_r^2$.

Apart from the trivial $R_r(2) = 2$ for all r, only one exact value of the Ramsey numbers $R_r(b)$ with $r \geq 3$ is known [200]:

$$R_3(3) = 17.$$

Fan Chung and Charles Grinstead [72, page 35] pointed out that

$$R_{s+t}(3) - 1 \geq (R_s(3) - 1)(R_t(3) - 1),$$

which implies that $R_r(3)^{1/r}$ tends to a limit as r tends to infinity. Erdős [70, page 23] offered \$100 for proving his conjecture that this limit is finite and \$250 for finding its value.

Stanisław Radziszowski periodically updates his excellent survey of information on small Ramsey numbers [320].

3.6   A Self-Centered Supplement: Self-Complementary Graphs

DEFINITIONS: Two graphs are said to be *isomorphic* if some bijection between their sets of vertices maps pairs of adjacent vertices onto pairs of adjacent vertices and it maps pairs of nonadjacent vertices onto pairs of nonadjacent vertices. The *complement* $\overline{G}$ of a graph G has the same vertices as G; two vertices are adjacent in $\overline{G}$ if and only if they are nonadjacent in G. A graph is *self-complementary* if it is isomorphic to its own complement.

By definition, $R(k, k)$ denotes the largest n such that some graph G of order $n - 1$ has $\alpha(G) < k$ and $\omega(G) < k$. The lower bound $R(3, 3) \geq 6$ is established by the graph with vertices $0, 1, 2, 3, 4$, where i and j are adjacent if and only if $(i - j)$ mod 5 is a quadratic residue modulo 5 (which means one of 1 and 4). This graph is referred to as the *cycle of length five* and denoted by C_5. It is self-complementary: the permutation $i \mapsto 2i$ mod 5 maps pairs of adjacent vertices onto pairs of nonadjacent vertices and vice versa.

The lower bound $R(4, 4) \geq 18$ is established by the Greenwood–Gleason graph with vertices $0, 1, \ldots, 16$, where i and j are adjacent if and only if $(i - j)$ mod 17 is a quadratic residue modulo 17 (which means one of $1, 2, 4, 8, 9, 13, 15, 16$). This graph is also self-complementary: the permutation $i \mapsto 3i$ mod 17 maps pairs of adjacent vertices onto pairs of nonadjacent vertices and vice versa.

With $R^*(k)$ standing for the largest n such that some self-complementary graph G of order $n - 1$ has $\alpha(G) = \omega(G) < k$, we have $R^*(k) \leq R(k, k)$ for all k and, as we have noted just now, equality holds here whenever $k \leq 4$.

The year 1971 was a magical one for me. In the fall, I started teaching at McGill and moved into an apartment on Lorne Street. My love affair with Montreal had begun.

A few weeks into the semester, I acquired a girlfriend Nancy with an apartment in the N.D.G. and the morning commute to Burnside Hall became more challenging. On the fifth day of this romance, I learned that Paul Erdős was in town. "You. Mean. To tell me," said Nancy in a glacial tone a little later, "that rather than spending the evening with me, you want to go and see some *mathematician*?"

Forty minutes later, I was sitting across a low table from the PGOM[c] in *Baraka II* on Crescent Street. This was the beginning of the end of my affair with Nancy. It was also the beginning of my second joint paper with Erdős.

Over pastilla and mint tea, I told him about my nagging fantasy that $R^*(k) = R(k, k)$ for all k and about my ambition to bound $R^*(k)$ from below by some modification of his proof that $R(k, k) > 2^{k/2}$. In response, he made up on the spot a proof that $R^*(k)$ cannot grow much more slowly than $R(k, k)$. I wrote it up and, after an addition of findings by my former teacher Zdeněk Hedrlín (1933–2018) relating Ramsey numbers to coding theory, the paper [82] was on its way.

---

[c] PGOM, short for Poor Great Old Man, was the title Erdős awarded himself before reaching the age of 60.

Here is a rudimentary version of the main result of [82]:

THEOREM 3.10

$$R^*(2t - 1) \geq 4R(t, t) - 3.$$

Proof   Given any graph G of order n, we shall construct a self-complementary graph H of order $4n$ such that $\alpha(H) = \omega(H) = \max\{\alpha(G) + \omega(G), 2\alpha(G)\}$. Vertices of H are ordered pairs (u, i) such that u is a vertex of G and $i \in \{1, 2, 3, 4\}$; two vertices (u, i) and (v, j) of H are adjacent if and only if

- $\{i, j\}$ is one of $\{1, 2\}$, $\{2, 3\}$, $\{3, 4\}$ or
- $i = j = 1$ or 4 and u, v are adjacent in G or
- $i = j = 2$ or 3 and u, v are nonadjacent in G.

Since the permutation

$$(w, 1) \mapsto (w, 3), \; (w, 2) \mapsto (w, 1), \; (w, 3) \mapsto (w, 4), \; (w, 4) \mapsto (w, 2)$$

of vertices of H maps pairs of adjacent vertices onto pairs of nonadjacent vertices and vice versa, H is self-complementary; verifying that $\alpha(H) = \omega(H) = \max\{\alpha(G) + \omega(G), 2\alpha(G)\}$ is a routine matter.

The theorem follows by applying this construction to a graph G of order $R(t, t) - 1$ with $\alpha(G) = \omega(G) \leq t - 1$: here, the self-complementary graph H has order $4R(t, t) - 4$ and $\alpha(H) = \omega(H) \leq 2t - 2$.   ☐

Together, Theorem 3.10 and Theorem 3.4 give

$$R^*(k) > 4 \cdot 2^{k/4}.$$

Vojtěch Rödl and Edita Šiňajová [331] improved this to

$$R^*(k) > c \cdot k2^{k/2}$$

with a constant c. This matches the best known lower bound (3.6) on $R(k, k)$ up to the constant factor. Later, Colin McDiarmid and Angelika Steger [284] improved the bound on $R^*(k)$ further to match even the constant:

$$R^*(k) > c(k) \cdot k2^{k/2} \text{ with } \lim_{k \to \infty} c(k) = \frac{\sqrt{2}}{e}.$$

We have $R^*(2) = R(2, 2) = 2$, $R^*(3) = R(3, 3) = 6$, and $R^*(4) = R(4, 4) = 18$. In [82], I wrote (with the approval of my co-authors)

It does not seem unlikely that $[R^*(k) = R(k, k)$ for all $k]$ ...

This impression turned out to be false:[d] nowadays we know that $R(5, 5) \geq 43$ and $R^*(5) \leq 42$. The lower bound $R(5, 5) \geq 43$ comes from [161]; in establishing the upper bound $R^*(5) \leq 42$, we shall use the following fact.

---

[d] *Beware of the law of small numbers:* $2^2 - 1 = 3$, $2^3 - 1 = 7$, $2^5 - 1 = 31$, $2^7 - 1 = 127$. All of these are primes. But $2^{11} - 1 = 2047 = 23 \cdot 89$. Similarly, $1^2 - 15 \cdot 1 + 97 = 83$, $2^2 - 15 \cdot 2 + 97 = 71$, ...$47^2 - 15 \cdot 47 + 97 = 1601$. All of these are primes. But $48^2 - 15 \cdot 48 + 97 = 1681 = 41^2$.

PROPOSITION 3.11   *The order of every self-complementary graph is $4k$ or $4k + 1$ with a non-negative integer k. Furthermore, every self-complementary graph of order $4k + 1$ has a self-complementary subgraph of order $4k$.*

Proof   Since a self-complementary graph of order n has the same number of edges as its complement, it has precisely $n(n-1)/4$ edges; this fraction has an integer value if and only if $n \bmod 4$ equals 0 or 1. Next, consider a self-complementary graph G of order $4k + 1$, let V denote its vertex set, and consider a permutation π of V that maps pairs of adjacent vertices onto pairs of nonadjacent vertices and vice versa. Like every finite permutation, π consists of pairwise disjoint *cycles,* which are sequences $v_1, v_2, \ldots v_t$ of pairwise distinct vertices such that

$$\pi(v_1) = v_2, \quad \pi(v_2) = v_3, \quad \ldots \quad \pi(v_{t-1}) = v_t, \quad \pi(v_t) = v_1.$$

If the *length t* of the cycle is at least 2, then in the sequence

$$\{v_1, v_2\}, \quad \{v_2, v_3\}, \quad \ldots \quad \{v_{t-1}, v_t\}, \quad \{v_t, v_1\}, \quad \{v_1, v_2\}$$

pairs of adjacent vertices alternate with pairs of nonadjacent vertices, and so t is even. Since the sum $4k + 1$ of the length of the cycles is odd, it follows that π must have a cycle of length 1. To put it differently, there must be a vertex v_1 such that $\pi(v_1) = v_1$. But then π permutes the vertices of $V - \{v_1\}$, and so G with v_1 removed is self-complementary.   □

By the first part of Proposition 3.11, every $R^*(k) \bmod 4$ is 1 or 2. Since $R^*(5) \leq R(5,5) \leq 48$, it follows that $R^*(5)$ is 45 or 46 or else at most 42. If $R^*(5)$ is 45 or 46 then, by the second part of Proposition 3.11, there is a self-complementary graph G of order 44 with $\alpha(G) = \omega(G) \leq 4$. Geoffrey Exoo found by repeated exhaustive computer searches that there is no such graph.

Ivars Zarins worked at the City Lights Bookstore. From there, he would bring me books that caught his interest and were likely to catch mine, too. Two of those he brought me in the spring of 1972 were *Post Office* and *Erections, Ejaculations, Exhibitions, and General Tales of Ordinary Madness* by Charles Bukowski. We marvelled at this discovery and smugly basked in the knowledge that we belonged to an exclusive club of those in the know. Years later Bukowski became a celebrity and I felt betrayed. Exclusivity all gone. Our cult hero now possessed by the vulgar masses.

With Paul Erdős I went through a similar experience, though it was on a different, more personal level. Here was our friend, someone we cherished, an elder of our tribe. And suddenly they make him a public property and parade his peculiarities in print for everyone's amusement. They sell Erdős numbers on eBay. Did I feel betrayed? Is the pope Catholic? In retrospect, all this publicity was only to be expected, but when it happened, it rankled all the same.

When confronted with a widely acclaimed personality, people may feel intimidated or jealous. A popular remedy is selecting a few of the legend's quirks and zapping them with a merciless spotlight. Saying that he was dishevelled and wore a ratty raincoat will make you look better by comparison. So will saying that he had a thick Hungarian accent. (What kind of a fool does not choose English as his mother tongue?) Every disparaging epithet helps. Don't say that he held his arm out to the side. Say instead that that he held his arm out to the side like a scarecrow. There. Feeling better already?

By focusing on Erdős's peripheral traits, such petty sneers distract attention from his fundamental nature, which was a profound kindness. He empathized with people, he was generous with his money, he was generous with his ideas. Michał Karoński said about him in [92]

> What strikes me is that he is very good, a very good man. If I could fix a standard for a good man, then Paul probably would be it, a definite standard.

All of us who were privileged to spend time with him (with the possible exception of his caricaturists) would agree.

In St. Gregory of Nyssa Episcopal church located at 500 De Haro Street in San Francisco, a series of murals show a procession of ninety dancing saints. A part of the statement that accompanies these murals reads

> Our broad idea of sainthood comes from both the Bible and Gregory of Nyssa's writings. The Hebrew concept of holiness originally had no moral content, but simply meant having God's stamp on you; being marked and set apart as God's own.

Along with Charles Darwin, Ella Fitzgerald, Gandhi, and 86 others, one of these saints is Paul Erdős. The commentary characterizes him as *an itinerant mathematical angel*. Is it not apt that he has been canonized in SF?[e]

Source: www.saintgregorys.org/the-dancing-saints.html
Permission: St. Gregory of Nyssa Episcopal Church,
    500 De Haro Street, San Francisco CA, 94107, USA
Artist: Mark Dukes
Photographer: David Sanger

[e] SF, short for Supreme Fascist, was Erdős's sobriquet for God.

4 Delta-Systems

4.1 Δ-Systems of Erdős and Rado

In [146], Paul Erdős and Richard Rado (1906–89) introduced the notion of a Δ-*system*. This is a family of sets such that the intersection of any two distinct sets in the family is always the same set: formally, $\mathcal{F}$ is a Δ-system if there is a set C such that

$$S, T \in \mathcal{F}, S \neq T \implies S \cap T = C.$$

(A Δ-system with a nonempty C is like a sunflower with centre C and petals $S - C$, one for each S in the system; for this reason, Δ-systems are often referred to as *sunflowers*. A Δ-system with an empty C is a family of pairwise disjoint sets.) Erdős and Rado were concerned with infinite as well as finite sets. A weaker but simpler version of their results restricted to the finite case goes as follows.

THEOREM 4.1 *For every choice of positive integers m and k, there is a positive integer M such that every family of more than M pairwise distinct sets of size k contains a Δ-system of more than m sets.[a] With $\phi(m, k)$ standing for the smallest such M, we have*

$$m^k \leq \phi(m, k) \leq k! \, m^k. \tag{4.1}$$

Proof We will prove the upper bound on $\phi(m, k)$ by induction on k. The induction basis ($k = 1$) is trivial. For the induction step, assume that $k > 1$ and let $\mathcal{F}$ be an arbitrary family of distinct sets of size k such that $|\mathcal{F}| > k! \, m^k$. Choose any maximal subfamily $\mathcal{F}_0$ of $\mathcal{F}$ that consists of pairwise disjoint sets. If $|\mathcal{F}_0| > m$, then we are done since $\mathcal{F}_0$ is a Δ-system. If $|\mathcal{F}_0| \leq m$, then let X_0 denote the union of all the members of $\mathcal{F}_0$. Since every member of $\mathcal{F}$ includes at least one point of X_0, some point of X_0 is included in at least $|\mathcal{F}|/|X_0|$ members of $\mathcal{F}$. Let x^* denote this point and let $\mathcal{F}^*$ denote the family of all sets $S - \{x^*\}$ such that $S \in \mathcal{F}$ and $x^* \in S$. Since

$$|\mathcal{F}^*| \geq \frac{|\mathcal{F}|}{|X_0|} \geq \frac{|\mathcal{F}|}{km} > (k-1)! \, m^{k-1},$$

---

[a] Sets in the families considered by Erdős and Rado are not necessarily distinct. Under this convention (which we do not adopt), a version of the pigeon-hole principle asserts that every family of more than m^2 sets of size 1 contains a Δ-system of more than m sets.

Paul Erdős and Richard Rado in the early 1950s
Reproduced with permission from the
Center for Excellence in Mathematical Education, Colorado Springs

the induction hypothesis guarantees that $\mathcal{F}^*$ contains a Δ-system $\mathcal{G}$ of more than m sets, and so $\{T \cup \{x^*\}: \ T \in \mathcal{G}\}$ is a Δ-system of more than m members of $\mathcal{F}$.

To prove the lower bound, take pairwise disjoint sets $A_1, A_2, \ldots, A_k$ of size m and consider the family $\mathcal{F}$ of all sets that intersect each A_i in precisely one point. Clearly, $|\mathcal{F}| = m^k$. Now consider an arbitrary Δ-system $\mathcal{G}$ contained in $\mathcal{F}$: we want to show that $|\mathcal{G}| \leq m$. For this purpose, we may assume that $\mathcal{G}$ includes two distinct sets (else $|\mathcal{G}| \leq 1$ and we are done); since these two sets of size k are distinct, their intersection C has size less than k, and so it is disjoint from some A_i. Each member of $\mathcal{G}$ intersects A_i in precisely one point; since distinct members of $\mathcal{G}$ intersect each other in C, they intersect A_i in distinct points, and so $|\mathcal{G}| \leq |A_i|$. $\qquad\square$

The lower bound in Theorem 4.1 is THEOREM II of [146]; the upper bound is a weakening of THEOREM III of [146], which asserts that

$$\phi(m, k) \ \leq \ k!\, m^k \left(1 - \sum_{i=1}^{k-1} \frac{i}{(i+1)!\, m^i}\right). \qquad (4.2)$$

The factor in parentheses may look awesome at first, but it does not reduce the bound all that much: it gets closer and closer to 1 as m gets larger and larger. More precisely, we have

$$\sum_{i=1}^{k-1} \frac{i}{(i+1)!\, m^i} \ < \ \sum_{i=1}^{k-1} \frac{1}{i!\, m^i} \ < \ \sum_{i=1}^{\infty} \frac{1}{i!\, m^i} \ = \ e^{1/m} - 1.$$

The proof of the stronger bound (4.2) is much like the proof of its counterpart in (4.1). Only the computation is more careful: we argue that some x^* in X_0 is included in at least

$$\frac{|\mathcal{F} - \mathcal{F}_0|}{|X_0|}$$

members of $\mathcal{F} - \mathcal{F}_0$. Now

$$|\mathcal{F}^*| \geq 1 + \frac{|\mathcal{F} - \mathcal{F}_0|}{|X_0|} = 1 + \frac{|\mathcal{F} - \mathcal{F}_0|}{k|\mathcal{F}_0|} = 1 + \frac{|\mathcal{F}|}{k|\mathcal{F}_0|} - \frac{|\mathcal{F}_0|}{k|\mathcal{F}_0|} = \frac{|\mathcal{F}|}{k|\mathcal{F}_0|} + \frac{k-1}{k}$$

and so, in case $|\mathcal{F}_0| \leq m$,

$$|\mathcal{F}| > k!\, m^k \left(1 - \sum_{i=1}^{k-1} \frac{i}{(i+1)!\, m^i} \right)$$

$$\Rightarrow \quad |\mathcal{F}^*| > (k-1)!\, m^{k-1} \left(1 - \sum_{i=1}^{k-2} \frac{i}{(i+1)!\, m^i} \right).$$

Erdős and Rado (([146], p. 86) wrote:

> It is not improbable that in the upper bound the factor $k!$ can be replaced by c^k, for some absolute positive constant c. Such a sharpened bound would have some applications in the theory of numbers, and in fact these applications originally gave rise to the present investigations.

Later, Erdős [122] offered 1000 Swiss francs (or 3 ounces of gold, whichever is worth more) for a proof or disproof that

$$\phi(2, k) \overset{?}{<} \alpha^k \text{ for some constant } \alpha.$$

Ryan Alweiss, Shachar Lovett, Kewen Wu, and Jiapeng Zhang [8] made an important breakthrough in progress toward resolving this conjecture when they proved that

$$\phi(m, k) < (\log k)^k (m \log \log k)^{\beta k} \text{ for some constant } \beta.$$

Anup Rao [325] then simplified their proof and improved the upper bound to

$$\phi(m, k) < (\gamma m \log(mk))^k \text{ for some constant } \gamma.$$

4.2   Ramsey's Theorem and Weak Δ-Systems

REMINDER: As in Section 3.5, $a \to (b)_r^k$ denotes the statement that

whenever the k-point subsets of an a-point set
are coloured by r colours,
there is a b-point set whose k-point subsets are all of the same colour.

The existence of $\phi(m, k)$ in Theorem 4.1 can be derived from Ramsey's theorem:

THEOREM 4.2 ([73], THEOREM 7)   *If m, k, M are positive integers such that*

$$M \to (m+1)_k^2, \tag{4.3}$$

and

$$m \geq 1 + \max_\lambda \binom{k}{\lambda}(k-\lambda), \tag{4.4}$$

then every family of M pairwise distinct sets of size k contains a Δ-system of more than m sets.

The proof relies on an observation of independent interest:

LEMMA 4.3 ([73], LEMMA 9)   *For every choice of integers k and λ such that $0 < \lambda < k$, there is a positive integer m with the following property:*

If $\mathcal{G}$ is a family of more than m sets of size k such that

$$S, T \in \mathcal{G}, \ S \neq T \ \Rightarrow \ |S \cap T| = \lambda, \tag{4.5}$$

then $\mathcal{G}$ is a Δ-system.

With $p_\lambda(k)$ standing for the smallest such m, we have

$$p_\lambda(k) \ \leq \ 1 + \binom{k}{\lambda}(k-\lambda).$$

Proof   Let $\mathcal{G}$ be a family of more than $1 + \binom{k}{\lambda}(k-\lambda)$ sets of size k, which satisfies (4.5). Choose a set S_0 in $\mathcal{G}$; for each λ-point subset C of S_0, let $\mathcal{G}_C$ denote the family of all the sets in $\mathcal{G}$ that intersect S_0 in C. Since each set in $\mathcal{G} - \{S_0\}$ belongs to precisely one $\mathcal{G}_C$, the lower bound on $|\mathcal{G}|$ guarantees that $|\mathcal{G}_C| > k - \lambda$ for some C; we will complete the proof by proving that $\mathcal{G} = \mathcal{G}_C \cup \{S_0\}$.

For this purpose, consider an arbitrary set S in $\mathcal{G}$ other than S_0: we want to show that $S \in \mathcal{G}_C$. Since $|S - S_0| = k - \lambda < |\mathcal{G}_C|$ and since the sets $T - S_0$ with $T \in \mathcal{G}_C$ are pairwise disjoint, $S - S_0$ must be disjoint from at least one $T - S_0$ with $T \in \mathcal{G}_C$. But then $|S \cap T| = \lambda$ dictates $S \cap T = C$, and so $S \in \mathcal{G}_C$.   □

Proof of Theorem 4.2   Let $\mathcal{F}$ be an arbitrary family of M pairwise distinct sets of size k. Consider the complete graph whose vertices are the members of $\mathcal{F}$; colour its edges by integers $0, 1, \ldots k - 1$ so that edge $\{S, T\}$ gets the colour $|S \cap T|$. Condition (4.3) guarantees that $\mathcal{F}$ has a subfamily $\mathcal{G}$ such that $|\mathcal{G}| = m + 1$ and such that, for some integer λ, every two distinct members of $\mathcal{G}$ intersect in precisely λ points. Condition (4.4) and Lemma 4.3 together guarantee that $\mathcal{G}$ is a Δ-system.   □

A *weak Δ-system* is a family $\mathcal{G}$ of sets that satisfies (4.5) with some λ. For contrast, a Δ-system as originally defined by Erdős and Rado is called a *strong Δ-system*. (In a strong Δ-system, the intersection of any two distinct sets is always the same set; in a weak Δ-system, the intersection of any two distinct sets has always the same size.) In these terms,

every weak Δ-system of more than $\max_\lambda p_\lambda(k)$ sets of size k is strong.

The family of lines in a projective plane of order $k - 1$ (see page 27) is a weak but not strong Δ-system; this family consists of $k^2 - k + 1$ sets of size k. Paul Erdős and László Lovász conjectured [122, page 406] that there is is no larger weak but not strong Δ-system of sets of size k:

$$\max_\lambda \; p_\lambda(k) \; \overset{?}{\leq} \; k^2 - k + 1. \tag{4.6}$$

Erdős offered \$100 for a proof or disproof and the conjecture was proved [98] by Michel Deza (1939–2016). Actually, Deza simply observed that validity of the Erdős-Lovász conjecture follows from an earlier theorem of his own in coding theory [97], which states that

$$p_k(2k) \; \leq \; k^2 + k + 1. \tag{4.7}$$

To deduce (4.6) from (4.7), he pointed out that

$$p_\lambda(k) \; \leq \; p_{k-1}(2k - 2) : \tag{4.8}$$

given sets $S_1, S_2, \ldots, S_m$ that satisfy

$$|S_i| = k \text{ for all } i \text{ and } |S_i \cap S_j| = \lambda \text{ whenever } i \neq j$$

(here $0 < \lambda < k$ as in the definition of $p_\lambda(k)$) without forming a strong Δ-system, we can construct sets $T_1, T_2, \ldots, T_m$ that satisfy

$$|T_i| = 2k - 2 \text{ for all } i \text{ and } |T_i \cap T_j| = k - 1 \text{ whenever } i \neq j$$

without forming a strong Δ-system. To do this, take sets $X, Y_1, Y_2, \ldots, Y_m$ with $|X| = k - 1 - \lambda$ and $|Y_i| = \lambda - 1$ $(1 \leq i \leq m)$ which are pairwise disjoint and disjoint from all $S_1, S_2, \ldots, S_m$; then set $T_i = S_i \cup X \cup Y_i$ for all i.

Bound (4.8) can be strengthened as

$$p_\lambda(k) \; \leq \; \begin{cases} p_{k-\lambda}(2k - 2\lambda) & \text{whenever } 2\lambda \leq k, \\ p_\lambda(2\lambda) & \text{whenever } 2\lambda \geq k \end{cases}$$

by variants of this construction: use $|X| = k - 2\lambda$, $Y_i = \emptyset$ when $2\lambda \leq k$ and use $X = \emptyset$, $|Y_i| = 2\lambda - k$ when $2\lambda \geq k$.

4.3   Deza's Theorem

THEOREM 4.4 ([97])   *If $\mathcal{G}$ is a family of $k^2 + k + 2$ sets of size $2k$ such that*

$$S, T \in \mathcal{G}, \; S \neq T \; \Rightarrow \; |S \cap T| = k, \tag{4.9}$$

then $\mathcal{G}$ is a strong Δ-system.

Let us paraphrase Deza's original proof along the lines of [79].

Proof of Theorem 4.4   Write $m = k^2 + k + 2$, enumerate the members of $\mathcal{G}$ as $S_1, S_2, \ldots, S_m$, and, for each nonempty subset A of $\{1, 2, \ldots, m\}$, let x_A denote the size of the "atom"

$$\bigcap_{i \in A} S_i - \bigcup_{i \notin A} S_i.$$

The assumption that $|S_i| = 2k$ for all i can be expressed as

$$\sum (x_A : i \in A) = 2k \quad \text{for all } i, \tag{4.10}$$

the assumption that $|S_i \cap S_j| = k$ whenever $i \neq j$ can be expressed as

$$\sum (x_A : i \in A, j \in A) = k \quad \text{whenever } i \neq j, \tag{4.11}$$

and the desired conclusion that $\mathcal{G}$ is a strong Δ-system can be expressed as

$$x_B = 0 \text{ for all } B \text{ such that } 2 \leq |B| \leq m - 1. \tag{4.12}$$

Given any subset B of $\{1, 2, \ldots, m\}$ such that $1 < |B| < m$, set

$$v = \frac{1}{|B|}, \quad w = \frac{1}{m + 1 - |B|}.$$

The linear combination of the equations (4.10), (4.11) with

multiplier v^2 at each of the equations (4.10) with $i \in B$,
multiplier w^2 at each of the equations (4.10) with $i \notin B$,
multiplier v^2 at each of the equations (4.11) with $i \in B, j \in B$,
multiplier w^2 at each of the equations (4.11) with $i \notin B, j \notin B$,
multiplier $-2vw$ at each of the equations (4.11) with $i \in B, j \notin B$

reads

$$\sum \left((v \cdot |A \cap B| - w \cdot |A - B|)^2 x_A : A \subseteq \{1, 2, \ldots, m\}, A \neq \emptyset \right) = \frac{k(m + 1)}{|B|(m + 1 - |B|)}.$$

Since the coefficient at each x_A is nonnegative, the coefficient at x_B equals 1, and the right-hand side is strictly less than 1 whenever $k + 2 \leq |B| \leq m - 1 - k$, we conclude that every nonnegative integer solution x of (4.10), (4.11) satisfies

$$x_B = 0 \text{ for all } B \text{ such that } k + 2 \leq |B| \leq m - 1 - k. \tag{4.13}$$

(This is nearly but not quite the desired conclusion (4.12).)

Now let $d(x)$ denote the number of members of $\mathcal{G}$ to which point x belongs and let C denote the set of points x such that $d(x) > k + 1$. By (4.13), we have

$$d(x) \geq m - k \quad \text{whenever } x \in C,$$
$$d(x) \leq k + 1 \quad \text{whenever } x \notin C.$$

We will complete the proof by showing that

$$|X \cap C| \geq k \text{ whenever } X \in \mathcal{G}, \tag{4.14}$$

$$|C| \leq k: \tag{4.15}$$

together, (4.14), (4.15), and (4.9) imply that $\mathcal{G}$ is a strong Δ-system.

In proving (4.14) and (4.15), we shall make use of the identity

$$\sum_{x \in X} d(x) = \sum_{S \in \mathcal{G}} |S \cap X|$$

satisfied by every set X: both the left-hand side and the right-hand side count the number of pairs (x, S) such that $x \in X$, $S \in \mathcal{G}$, and $x \in S$.

To prove (4.14), consider an arbitrary X in $\mathcal{G}$: we have

$$\sum_{S \in \mathcal{G}} |S \cap X| = 2k + (m-1)k,$$

$$\sum_{x \in X} d(x) \leq |X \cap C| \cdot m + (2k - |X \cap C|) \cdot (k+1),$$

and so

$$|X \cap C| \geq \frac{2k + (m-1)k - 2k(k+1)}{m - (k+1)} > k - 1.$$

To prove (4.15), consider an arbitrary X such that $|X| = k+1$: we have

$$\sum_{S \in \mathcal{G}} |S \cap X| \leq (k+1) + (m-1)k,$$

$$\sum_{x \in X} d(x) \geq |X \cap C| \cdot (m-k),$$

and so

$$|X \cap C| \leq \frac{(k+1) + (m-1)k}{m-k} < k+1.$$

It follows that $|C| \leq k$.   □

In the summer of 1965, my Prague mentor Zdeněk Hedrlín gave me the Erdős–Rado paper along with his notes containing minor improvements of the bounds on $\phi(m, k)$ and other observations on Δ-systems. To these, I added minor improvements and observations of my own. On the strength of the resulting manuscript, Hedrlín engineered an appointment for me to introduce myself to Erdős during his coming visit to the Slovak Academy of Sciences in Bratislava.

What happened next was an extraordinary stroke of luck. Through a heaven-sent scheduling blunder, Erdős's visit clashed with an administrative meeting of the Academy and as a stopgap I was asked to step in and play Erdős's host for the duration of his stay. A chauffeured limousine came with the job.

Each morning I made a rags-to-riches journey from my stark dormitory to Erdős's opulent hotel. A sumptuous breakfast on the hotel terrace, the silver sugar bowls darkly gleaming and the linen napkins whitely crisp. The car seats plush and the big engine softly purring on our jaunt to the town where Erdős's mother was born.

Against these backdrops, he offered me a dazzling feast of conjectures and theorems. He treated me with kindness, patience, tolerance. When the train was taking me back to Prague, I was one of the family.

68 IX 24

McGILL UNIVERSITY
MONTREAL

Dear Mr Chvátal,

My mother and I are in Canada since a few days. People here are trying to arrange a fellowship for you.

I enclosed three letters. Professor Hlawka and Prachar are at the University of Vienna Strudlhofgasse you probably know them by reputation. I write to both of them to give you 1000 shillings in case you need it and write them that I will return them the money immediately. The third letter (in Hungarian) is to a friend of our family (Stefi Tauszig), she will also give you 1000 shillings if you need it. You can return the money to me when you no longer need it.

I see from your letter that you got captured. Did the girl capture you who gave you the 21 ball point pens for your birthday?

How is your work getting on?

How many subsets of a set of n elements can you give so that no three of them should have pairwise the same intersection? (this problem is not due to me)

Kind regards

P. Erdős

A letter of mine will follow.
                              W G Brown.

Courtesy of Vašek Chvátal

5   Extremal Set Theory

5.1   Sperner's Theorem

The Erdős–Rado investigations of Δ-systems constitute a part of *extremal set theory*, which concerns extremal sizes of set families with prescribed properties (such as including only k-point sets and containing no Δ-system of more than m sets). A classic of this theory comes from Emanuel Sperner (1905–80):

THEOREM 5.1 (Sperner [347])   *Let n be a positive integer. If V is an n-point set and $\mathcal{E}$ is a family of subsets of V such that*

$$S, T \in \mathcal{E}, S \neq T \Rightarrow S \not\subseteq T, \tag{5.1}$$

then

$$|\mathcal{E}| \leq \binom{n}{\lfloor n/2 \rfloor}. \tag{5.2}$$

Furthermore, (5.2) holds as equality if and only if $\mathcal{E}$ consists of all subsets of V that have size $\lfloor n/2 \rfloor$ or all subsets of V that have size $\lceil n/2 \rceil$.

A family $\mathcal{E}$ of sets with property (5.1) is called an *antichain*.

The original proof of Theorem 5.1 involves an intermediate result of independent interest:

LEMMA 5.2   *Let n and k be integers such that $1 < k < n$; let V be an n-point set, let $\mathcal{F}$ be a nonempty family of k-point subsets of V, and let $\mathcal{F}^{(+)}$, $\mathcal{F}^{(-)}$ be defined by*

$$\mathcal{F}^{(+)} = \Big\{ T \subseteq V : |T| = k+1 \text{ and } S \subset T \text{ for some } S \text{ in } \mathcal{F} \Big\},$$

$$\mathcal{F}^{(-)} = \Big\{ R \subseteq V : |R| = k-1 \text{ and } R \subset S \text{ for some } S \text{ in } \mathcal{F} \Big\}.$$

Then

    (i)   *if $k < (n-1)/2$, then $|\mathcal{F}^{(+)}| > |\mathcal{F}|$,*
    (ii) *if $k = (n-1)/2$ and $|\mathcal{F}| < \binom{n}{k}$, then $|\mathcal{F}^{(+)}| > |\mathcal{F}|$,*
    (iii) *if $k > (n+1)/2$, then $|\mathcal{F}^{(-)}| > |\mathcal{F}|$,*
    (iv) *if $k = (n+1)/2$ and $|\mathcal{F}| < \binom{n}{k}$, then $|\mathcal{F}^{(-)}| > |\mathcal{F}|$.*

Proof   Let N denote the number of all pairs (S, T) such that $S \in \mathcal{F}$, $S \subset T \subseteq V$, and $|T| = k+1$. Each T in $\mathcal{F}^{(+)}$ appears in at most $k+1$ such pairs, and so

$N \leq |\mathcal{F}^{(+)}| \cdot (k+1)$; each S in $\mathcal{F}$ appears in precisely $n-k$ such pairs, and so $N = |\mathcal{F}| \cdot (n-k)$; we conclude that

$$\frac{|\mathcal{F}^{(+)}|}{|\mathcal{F}|} \geq \frac{N}{(k+1)|\mathcal{F}|} \geq \frac{n-k}{k+1}. \tag{5.3}$$

Proof of (i): Under the assumption $k < (n-1)/2$, we have $(n-k)/(k+1) > 1$ and the conclusion follows from (5.3).

Proof of (ii): Here, (5.3) is transformed into

$$\frac{|\mathcal{F}^{(+)}|}{|\mathcal{F}|} > \frac{N}{(k+1)|\mathcal{F}|} = 1 :$$

the assumption $k = (n-1)/2$ means that $(n-k)/(k+1) = 1$ and under the assumption $|\mathcal{F}| < \binom{n}{k}$, we will prove that $N < |\mathcal{F}^{(+)}| \cdot (k+1)$. To do this, we will find a T in $\mathcal{F}^{(+)}$ with at least one k-point subset outside $\mathcal{F}$. Since $|\mathcal{F}| > 0$, there is a k-point subset S of V such that $S \in \mathcal{F}$; since $|\mathcal{F}| < \binom{n}{k}$, there is a k-point subset S' of V such that $S' \notin \mathcal{F}$; replacing points in $S - S'$ by points in $S' - S$ one by one, we construct a sequence $S_0, S_1, \ldots, S_t$ of k-point subsets S_i of V such that $S_0 = S$, $S_t = S'$, and $|S_{i-1} \cap S_i| = k-1$ for all $i = 1, 2, \ldots, t$. If i is the smallest subscript such that $S_i \notin \mathcal{F}$, then $S_{i-1} \in \mathcal{F}$ and we can take $T = S_i \cup S_{i-1}$.

Proofs of (iii) and (iv): In (i) and (ii), replace $\mathcal{F}$ by $\{V - S : S \in \mathcal{F}\}$.   □

Proof of Theorem 5.1   Let n be a positive integer, let V be an n-point set, and let $\mathcal{E}$ be any largest antichain of subsets of V; write

$$k_{\min} = \min\{|S| : S \in \mathcal{E}\}, \quad k_{\max} = \max\{|S| : S \in \mathcal{E}\}$$

and

$$\mathcal{E}_{\min} = \{S \in \mathcal{E} : |S| = k_{\min}\}, \quad \mathcal{E}_{\max} = \{S \in \mathcal{E} : |S| = k_{\max}\}.$$

Since $\mathcal{E}$ is an antichain, $\mathcal{E}_{\min}^{(+)}$ is disjoint from $\mathcal{E}$ and $(\mathcal{E} - \mathcal{E}_{\min}) \cup \mathcal{E}_{\min}^{(+)}$ is an antichain; since $\mathcal{E}$ is a largest antichain of subsets of V, it follows that $|\mathcal{E}_{\min}^{(+)}| \leq |\mathcal{E}_{\min}|$; now part (i) of Lemma 5.2 guarantees that

(i) $k_{\min} \geq (n-1)/2$

and part (ii) of Lemma 5.2 guarantees that

(ii) if $k_{\min} = (n-1)/2$, then $|\mathcal{E}_{\min}| = \binom{n}{(n-1)/2}$.

Similar arguments show that

(iii) $k_{\max} \leq (n+1)/2$

and that

(iv) if $k_{\max} = (n+1)/2$, then $|\mathcal{E}_{\max}| = \binom{n}{(n+1)/2}$.

The conclusion of the theorem follows from (i), (ii), (iii), (iv).   □

5.1.1   A Simple Proof of Sperner's Theorem

Independently of each other, Koichi Yamamoto [376], Lev Dmitrievich Meshalkin (1934–2000) [293], and David Lubell [280] found the following inequality:

THEOREM 5.3 (The LYM inequality)   *If $\mathcal{E}$ is an antichain of subsets of an n-point set, then*

$$\sum_{S \in \mathcal{E}} \binom{n}{|S|}^{-1} \leq 1. \tag{5.4}$$

Proof   Let V be an arbitrary but fixed n-point set and let $\mathcal{E}$ be an arbitrary but fixed antichain of subsets of V. A *chain of length n* is any family $S_0, S_1, \ldots, S_n$ of subsets of V such that $|S_i| = i$ for all i and $S_0 \subset S_1 \subset \ldots \subset S_n$ (in particular, $S_0 = \emptyset$ and $S_n = V$). We will count in two different ways the number N of all pairs $(\mathcal{C}, S)$ such that $\mathcal{C}$ is a chain of length n and $S \in \mathcal{E} \cap \mathcal{C}$.

For each k-point subset S of V, there are precisely $k!$ choices of a sequence S_0, $S_1, \ldots, S_k$ of subsets of V such that $|S_i| = i$ for all i and $S_0 \subset S_1 \subset \ldots \subset S_k = S$; there are precisely $(n - k)!$ choices of a sequence $S_k, S_{k+1}, \ldots, S_n$ of subsets of V such that $|S_i| = i$ for all i and $S = S_k \subset S_{k+1} \subset \ldots \subset S_n$; it follows that each member S of our $\mathcal{E}$ participates in precisely $|S|!(n - |S|)!$ of our pairs $(\mathcal{C}, S)$, and so

$$N = \sum_{S \in \mathcal{E}} |S|!(n - |S|)! \,.$$

Since each chain contains at most one member of any antichain, each $\mathcal{C}$ participates in at most one of our pairs $(\mathcal{C}, S)$, and so N is at most the number of all chains of length n:

$$N \leq n! \,.$$

Comparing the exact formula for N with this upper bound, we get the inequality

$$\sum_{S \in \mathcal{E}} |S|!(n - |S|)! \leq n! \,,$$

which is just another way of writing (5.4). $\qquad\square$

Let us prove Theorem 5.1 along the lines of Theorem 5.3. The first part of Theorem 5.1 – inequality (5.2) – is a direct consequence of Theorem 5.3: every antichain $\mathcal{E}$ of subsets of an n-point set satisfies

$$|\mathcal{E}| = \sum_{S \in \mathcal{E}} 1 \leq \sum_{S \in \mathcal{E}} \binom{n}{\lfloor n/2 \rfloor} \binom{n}{|S|}^{-1} = \binom{n}{\lfloor n/2 \rfloor} \sum_{S \in \mathcal{E}} \binom{n}{|S|}^{-1} \leq \binom{n}{\lfloor n/2 \rfloor}. \tag{5.5}$$

To prove the second part of Theorem 5.1 – characterization of extremal antichains – along these lines, consider an arbitrary antichain $\mathcal{E}$ of subsets of an n-point set V that satisfies both inequalities in (5.5) with the sign of equality. The first of these equations implies that $\binom{n}{|S|} = \binom{n}{\lfloor n/2 \rfloor}$ for all S in $\mathcal{E}$, which means that $|S| = n/2$ when n is even and $|S| = (n \pm 1)/2$ when n is odd. If n is even, then we are done; if n is odd, then we will argue about the second inequality-turned-equation in (5.5). This equality means equality in (5.4); reviewing the proof of Theorem 5.3, we find that
• every chain of length n contains a member of $\mathcal{E}$,
and so (since all sets in $\mathcal{E}$ have size $(n \pm 1)/2$)

- for every two subsets S, T of V such that $|S| = (n-1)/2$, $|T| = (n+1)/2$, and $S \subset T$, precisely one of S and T belongs to $\mathcal{E}$.

It follows right away that

- $|S| = |S'| = (n-1)/2$, $S \in \mathcal{E}$, $S' \subset V$, $|S \cup S'| = (n+1)/2 \;\Rightarrow\; S' \in \mathcal{E}$

(consider $T = S \cup S'$) and then induction on $|S \cup S'|$ shows that

- $|S| = |S'| = (n-1)/2$, $S \in \mathcal{E}$, $S' \subset V \;\Rightarrow\; S' \in \mathcal{E}$.

5.1.2   The Bollobás Set Pairs Inequality

A year before Lubell's [280], Béla Bollobás discovered an inequality [34] that is more general than the LYM inequality:

THEOREM 5.4   *If $A_1, A_2, \ldots A_m$ and $B_1, B_2, \ldots B_m$ are finite sets such that $A_i \cap B_i = \emptyset$ for all i and $A_i \cap B_j \neq \emptyset$ whenever $i \neq j$, then*

$$\sum_{i=1}^{m} \binom{|A_i| + |B_i|}{|A_i|}^{-1} \leq 1.$$

To see that Theorem 5.3 is a special case of Theorem 5.4, enumerate the sets in the antichain $\mathcal{E}$ as $A_1, A_2, \ldots A_m$ and put $B_i = V - A_i$ with V the underlying n-point set. After Bollobás's paper appeared, Theorem 5.3 was rediscovered by François Jaeger and Charles Payan [223] and by Gyula Katona [237].

Katona's proof of Theorem 5.4. Let V denote the union of all $A_1, A_2, \ldots A_m$ and all $B_1, B_2, \ldots B_m$. Let n denote the size of V. We will count in two different ways the number N of all pairs $(\prec, i)$ such that $\prec$ is a linear order on V that places all points of A_i before all points of B_i

Specifying a linear order on V amounts to specifying the rank of each point of V, which means matching up the n points of V with the integers $1, 2, \ldots, n$. In particular, specifying a linear order on V that places all points of A_i before all points of B_i means first

1. choosing a set Img of $|A_i| + |B_i|$ integers in $\{1, 2, \ldots n\}$, then
2. matching up points of A_i with the $|A_i|$ smallest numbers in Img, then
3. matching up points of B_i with the remaining $|B_i|$ numbers in Img, and finally
4. matching up points of $V - (A_i \cup B_i)$ with the numbers in $\{1, 2, \ldots n\} - $ Img.

The total number of ways of doing this comes to

$$\binom{n}{|A_i| + |B_i|} \cdot |A_i|! \cdot |B_i|! \cdot (n - (|A_i| + |B_i|))!,$$

which equals

$$n! \binom{|A_i| + |B_i|}{|A_i|}^{-1}.$$

Each i participates in precisely this many of our pairs $(\prec, i)$, and so

$$N = n! \sum_{i=1}^{m} \binom{|A_i| + |B_i|}{|A_i|}^{-1}.$$

Next, consider a linear order $\prec$ on V and any two distinct pairs A_j, B_j and A_k, B_k. By assumption, there are a point x in $A_j \cap B_k$ and a point y in $A_k \cap B_j$. If $x \prec y$, then $\prec$ does not place all points of A_k before all points of B_k; if $y \prec x$, then $\prec$ does not place all points of A_j before all points of B_j. It follows that each $\prec$ participates in at most one of our pairs $(\prec, i)$, and so

$$N \leq n! \, .$$

Comparing the exact formula for N with this upper bound, we arrive at the theorem.

$\square$

Zsolt Tuza has written a two-part survey on the fundamentals of the set-pair method and on its many applications [365, 366].

5.2 The Erdős–Ko–Rado theorem

In 1938, Paul Erdős, Chao Ko, whose name is also transliterated as Ke Zhao (1910–2002), and Richard Rado proved a theorem that they published 23 years later [142, Theorem 1]. Its simplified version presented below is known as the *Erdős–Ko–Rado theorem*.

THEOREM 5.5   *Let n and k be positive integers such that $2k \leq n$. If V is an n-point set and $\mathcal{E}$ is a family of k-point subsets of V such that*

$$S, T \in \mathcal{E} \Rightarrow S \cap T \neq \emptyset, \tag{5.6}$$

then

$$|\mathcal{E}| \leq \binom{n-1}{k-1}. \tag{5.7}$$

A family $\mathcal{E}$ of sets with property (5.6) is called an *intersecting family*.

The original proof of Theorem 5.5 involves an intermediate result of independent interest:

LEMMA 5.6   *Let V be a set and let $\mathcal{E}$ be an intersecting family of subsets of V. Given two elements x, y of V, write*

$$\mathcal{E}^* = \{S \in \mathcal{E} : x \in S, \, y \notin S, \, (S - \{x\}) \cup \{y\} \notin \mathcal{E}\}$$

and define $f : \mathcal{E} \to 2^V$ by

$$f(S) = \begin{cases} (S - \{x\}) \cup \{y\} & \text{if } S \in \mathcal{E}^*, \\ S & \text{otherwise.} \end{cases}$$

Then $\{f(S) : S \in \mathcal{E}\}$ is an intersecting family of size $|\mathcal{E}|$.

Proof   We will verify that $S, T \in \mathcal{E} \Rightarrow f(S) \cap f(T) \neq \emptyset$ and that $S \neq T \Rightarrow f(S) \neq f(T)$.

CASE 1: $S, T \in \mathcal{E} - \mathcal{E}^*$. In this case, $f(S) = S$, $f(T) = T$, and so both implications hold trivially.

Erdős with the University of Manchester number theory group in 1937 or 1938.
Ke Zhao at far right.

Reproduced with permission from the
Center for Excellence in Mathematical Education, Colorado Springs

CASE 2: $S, T \in \mathcal{E}^*$. In this case, we have $f(S) \cap f(T) \neq \emptyset$ since $y \in f(S) \cap f(T)$. In addition, $S = (f(S) - \{y\}) \cup \{x\}$, $T = (f(T) - \{y\}) \cup \{x\}$, and so $f(S) = f(T) \Rightarrow S = T$.

CASE 3: $S \in \mathcal{E}^*$, $T \in \mathcal{E} - \mathcal{E}^*$. In this case, $f(S) = (S - \{x\}) \cup \{y\} \notin \mathcal{E}$ and $f(T) = T$; in particular, $f(S) \neq f(T)$ since $f(S) \notin \mathcal{E}$ and $f(T) \in \mathcal{E}$. It remains to prove that $f(S) \cap f(T) \neq \emptyset$. If $y \in T$, then $y \in f(S) \cap f(T)$ and we are done; now we may assume that

$$y \notin T,$$

and so $f(S) \cap f(T) = (S - \{x\}) \cap T$. If $x \notin T$, then $f(S) \cap f(T) = S \cap T$ and we are done again; now we may assume that

$$x \in T.$$

Since $S \in \mathcal{E}^*$, we have

$$x \in S \text{ and } y \notin S.$$

Now

$$f(S) \cap f(T) = (S - \{x\}) \cap T = S \cap (T - \{x\}) = S \cap ((T - \{x\}) \cup \{y\})$$

Since $x \in T$, $y \notin T$, and $T \in \mathcal{E} - \mathcal{E}^*$, we must have $(T - \{x\}) \cup \{y\} \in \mathcal{E}$; since $\mathcal{E}$ is an intersecting family, we conclude that

$$f(S) \cap f(T) = S \cap ((T - \{x\}) \cup \{y\}) \neq \emptyset. \qquad \square$$

Proof of Theorem 5.5   Let n and k be positive integers such that $2k \leq n$ and let $\mathcal{E}$ be any intersecting family of k-point subsets of $\{1, 2, \ldots, n\}$. We will use induction on n to show that $|\mathcal{E}| \leq \binom{n-1}{k-1}$. The induction basis, $n = 2$, is trivial; in the induction step, we assume that $n \geq 3$.

If $k = 1$, then we are done right away: $\mathcal{E}$, being an intersecting family of one-point sets, cannot contain two sets. If $2k = n$, then we are done again: in this case, $\mathcal{E}$ includes at most one set from each pair $(S, \{1, 2, \ldots, 2k\} - S)$, and so $|\mathcal{E}| \leq \frac{1}{2}\binom{2k}{k} = \binom{2k-1}{k-1}$. Now we may assume that

$$k \geq 2 \quad \text{and} \quad 2k \leq n - 1.$$

Define the *weight* $w(\mathcal{F})$ of a family $\mathcal{F}$ of subsets of $\{1, 2, \ldots, n\}$ as $\sum_{S \in \mathcal{F}} \sum_{x \in S} x$. We may assume that among all intersecting families $\mathcal{F}$ of k-point subsets of $\{1, 2, \ldots, n\}$ such that $|\mathcal{F}| = |\mathcal{E}|$, family $\mathcal{E}$ has the smallest weight. This assumption guarantees that

> for every two points x, y in $\{1, 2, \ldots, n\}$ such that $x > y$, we have
> $$S \in \mathcal{E}, \ x \in S, \ y \notin S \Rightarrow (S - \{x\}) \cup \{y\} \in \mathcal{E}:$$

(5.8)

since Lemma 5.6 transforms $\mathcal{E}$ into an intersecting family of k-point subsets of $\{1, 2, \ldots, n\}$ that has size $|\mathcal{E}|$ and weight $w(\mathcal{E}) - |\mathcal{E}^*| \cdot (x - y)$, minimality of $w(\mathcal{E})$ implies $\mathcal{E}^* = \emptyset$.

Finally, let us set

$$\mathcal{E}_k = \{S \in \mathcal{E} : n \notin S\} \quad \text{and} \quad \mathcal{E}_{k-1} = \{S - \{n\} : S \in \mathcal{E}, n \in S\}.$$

We will complete the proof by showing that $\mathcal{E}_{k-1}$ is an intersecting family: this assertion and the induction hypothesis imply that

$$|\mathcal{E}| = |\mathcal{E}_k| + |\mathcal{E}_{k-1}| \leq \binom{n-2}{k-1} + \binom{n-2}{k-2} = \binom{n-1}{k-1}.$$

To show that $\mathcal{E}_{k-1}$ is an intersecting family, consider arbitrary sets A, B in $\mathcal{E}_{k-1}$ and take any point y in $\{1, 2, \ldots, n-1\} - (A \cup B)$. By definition, $A \cup \{n\}$ and $B \cup \{n\}$ belong to $\mathcal{E}$; in turn, (5.8) with $x = n$ and $S = B \cup \{n\}$ guarantees that $B \cup \{y\}$ belongs to $\mathcal{E}$; since $\mathcal{E}$ is an intersecting family, we conclude that

$$A \cap B = (A \cup \{n\}) \cap (B \cup \{y\}) \neq \emptyset. \qquad \square$$

5.2.1   A Simple Proof of the Erdős–Ko–Rado Theorem

Gyula Katona [236] found a proof of Theorem 5.5 that imitates the proof of Theorem 5.3. We are going to paraphrase a variation on Katona's theme that comes from Chris Godsil and Gordon Royle [186]. Here, the key notion is a certain family of n sets of size k.

DEFINITION: An (n, k)-*ring* is defined by a cyclic order on its underlying set V that has size n: each member of the ring consists of k points of V that are consecutive in the cyclic order. To put it formally, when the elements of V are enumerated as $v_1, v_2, \ldots, v_n$, members $S_1, S_2, \ldots, S_n$ of the ring are defined by

$$S_i = \{v_i, v_{i+1}, \ldots, v_{i+k-1}\} \tag{5.9}$$

with subscript arithmetic modulo n (so that $v_{n+1} = v_1$, $v_{n+2} = v_2$, and so on).

LEMMA 5.7 *If n and k are positive integers such that $2k \le n$, then every intersecting subfamily of an (n, k)-ring consists of at most k sets.*

Proof Let n and k be positive integers such that $2k \le n$ and let $\mathcal{F}$ be an intersecting subfamily of an (n, k)-ring; let $S_1, S_2, \ldots, S_n$ be the members of the ring as in (5.9). If $\mathcal{F} = \emptyset$, then there is nothing to prove; else $\mathcal{F}$ includes at least one member of the ring and symmetry allows us to assume that $\mathcal{F}$ includes S_k. Since every S_i in $\mathcal{F}$ has $S_i \cap S_k \ne \emptyset$, its subscript i must be one of $1, 2, \ldots, 2k - 1$; since the two sets in each of the $k - 1$ pairs

$$(S_1, S_{k+1}), \ (S_2, S_{k+2}), \ \ldots, (S_{k-1}, S_{2k-1})$$

are disjoint, $\mathcal{F}$ includes at most one of these two sets, and so it includes at most k of the n sets S_i. □

Alternative proof of Theorem 5.5 Given any positive integers n and k such that $2k \le n$ and given an arbitrary but fixed n-point set V, let M denote the number of (n, k)-rings $\mathcal{R}$ on V and, for each k-point subset S of V, let $d(S)$ denote the number of (n, k)-rings $\mathcal{R}$ on V such that $S \in \mathcal{R}$. By symmetry, $d(S)$ is a constant d dependent on n and k but independent of the choice of S; counting in two different ways the pairs $(\mathcal{R}, S)$ such that $\mathcal{R}$ is an (n, k)-ring on V and $S \in \mathcal{R}$, we find that

$$Mn = \binom{n}{k}d. \tag{5.10}$$

Next, given an arbitrary but fixed intersecting family $\mathcal{E}$ of k-point subsets of V, we will count in two different ways the number N of all pairs $(\mathcal{R}, S)$ such that $\mathcal{R}$ on V is an (n, k)-ring and $S \in \mathcal{E} \cap \mathcal{R}$: since each S in $\mathcal{E}$ is featured in d such pairs and, by Lemma 5.7, each $\mathcal{R}$ is featured in at most k such pairs, we have

$$|\mathcal{E}|d = N \le Mk. \tag{5.11}$$

Together, (5.11) and (5.10) imply that $|\mathcal{E}| \le \frac{M}{d}k = \frac{k}{n}\binom{n}{k} = \binom{n-1}{k-1}$. □

5.2.2　Extremal Families in the Erdős–Ko–Rado Theorem

A family $\mathcal{E}$ of sets such that some point belongs to all members of $\mathcal{E}$ is called a *star*.

THEOREM 5.8 (Theorem 7.8.1 in [186]) *Let n and k be positive integers such that $2k < n$. If V is an n-point set and $\mathcal{E}$ is an intersecting family of k-point subsets of V such that*

$$|\mathcal{E}| = \binom{n-1}{k-1},$$

then $\mathcal{E}$ is a star.

The assumption $2k < n$ of Theorem 5.8 cannot be relaxed to the $2k \leq n$ of Theorem 5.5: for example, the family of all k-point subsets of $\{1, 2, \ldots, 2k - 1\}$ is intersecting, consists of $\binom{2k-1}{k-1}$ sets, and is not a star.

The remainder of the present section is devoted to a proof of Theorem 5.8.

LEMMA 5.9 (Lemma 7.7.1 in [186]) *If n and k are positive integers such that $2k \leq n$, then every intersecting subfamily of an (n, k)-ring that consists of k sets is a star.*

Proof Let $\mathcal{F}$ and $S_1, S_2, \ldots, S_n$ be as in the proof of Lemma 5.7; in particular, $\mathcal{F} \subset \{S_1, S_2, \ldots, S_{2k-1}\}$ and $S_k \in \mathcal{F}$. Assumption $|\mathcal{F}| = k$ implies that

(i) $\mathcal{F}$ includes precisely one of the two sets S_i, S_{k+i}
    in each of the $k - 1$ pairs (S_1, S_{k+1}), (S_2, S_{k+2}), $\ldots$, (S_{k-1}, S_{2k-1})

since the two sets in each of these pairs are disjoint, and so at most one of them can be included in $\mathcal{F}$. In addition,

(ii) $\mathcal{F}$ includes at most one of the two sets S_i, S_{k+i+1}
    in each of the $k - 2$ pairs (S_1, S_{k+2}), (S_2, S_{k+3}), $\ldots$, (S_{k-2}, S_{2k-1})

since the two sets in each of these pairs are disjoint.

If all k sets $S_1, S_2, \ldots, S_k$ belong to $\mathcal{F}$, then $\mathcal{F}$ is a star (point v_k belongs to all its members) and we are done; now we may assume that there is at least one subscript j such that $1 \leq j \leq k-1$ and $S_j \notin \mathcal{F}$; let j be the largest subscript with these properties. Now $S_{j+1}, S_{j+2}, \ldots, S_k \in \mathcal{F}$; by (i) and (ii), we have

$$S_i \notin \mathcal{F} \;\Rightarrow\; S_{k+i} \in \mathcal{F} \;\Rightarrow\; S_{i-1} \notin \mathcal{F};$$

referring to these implications with $i = j, j-1, \ldots, 1$, we find that $S_{k+j}, S_{k+j-1}, \ldots, S_{k+1} \in \mathcal{F}$. So $\mathcal{F}$ consists of the k sets $S_{j+1}, S_{j+2}, \ldots, S_{j+k}$, and that makes it a star (point v_{j+k} belongs to all its members). $\qquad\square$

Another ingredient of the proof of Theorem 5.8 is this:

LEMMA 5.10 *Let n and k be positive integers such that $2k < n$, let V be an n-point set, and let A, B, X be k-point subsets of V such that A, B intersect in precisely one point and X does not include this point. Then there is an (n, k)-ring $\mathcal{R}$ with the following properties:*

(i) $X \in \mathcal{R}$,
(ii) *if $\mathcal{F} \subset \mathcal{R}$ and $\mathcal{F}$ is a star of k sets
     and every member of $\mathcal{F}$ intersects both A and B, then $X \notin \mathcal{F}$.*

Proof Let w denote the single point of $A \cap B$ and let us write $A_0 = A - \{w\}$, $B_0 = B - \{w\}$; let us enumerate

the elements of $A_0 - X$	followed by
the elements of $X \cap A_0$	followed by
the elements of $X - (A_0 \cup B_0)$	followed by
the elements of $X \cap B_0$	followed by
the elements of $B_0 - X$	

as $v_1, v_2, \ldots, v_t$. (Since w is missing from this sequence, we have $t \le n - 1$; if $X \subseteq A_0 \cup B_0$, then $t = 2(k-1) \le n - 3$.) Then let us enumerate the remaining elements of V as $v_{t+1}, v_{t+2}, \ldots, v_n$ in such a way that

$$ w = \begin{cases} v_n & \text{if } X \nsubseteq A_0 \cup B_0, \\ v_{n-1} & \text{if } X \subseteq A_0 \cup B_0. \end{cases} $$

Finally, let $\mathcal{R}$ consist of the sets $S_1, S_2, \ldots, S_n$ defined by (5.9). By this definition, $X \in \mathcal{R}$; to prove that $\mathcal{R}$ has property (ii), note first that a subfamily of $\mathcal{R}$ is a star of k sets if and only if it is one of $\mathcal{F}_1, \mathcal{F}_2, \ldots, \mathcal{F}_n$ defined by

$$ \mathcal{F}_i = \{S_i, S_{i+1}, \ldots, S_{i+k-1}\} $$

with subscript arithmetic modulo n. Proving (ii) means proving that

> if an $\mathcal{F}_i$ includes X, then it includes a set disjoint from A or B (or both).

To prove this, we distinguish between two cases.

CASE 1: $X \nsubseteq A \cup B$. In this case, $w = v_n$. To begin, note that
- if i is one of $1, \ldots, k$, then S_k, disjoint from A, is included in $\mathcal{F}_i$;
- if i is one of $k, \ldots, n - k$, then S_i, included in $\mathcal{F}_i$, is disjoint from A;
- if i is one of $n - k + 2, \ldots, n$, then S_1, disjoint from B, is included in $\mathcal{F}_i$.

To summarize, if $i \ne n - k + 1$, then $\mathcal{F}_i$ includes a set disjoint from A or B (or both); we will complete the analysis of this case by showing that $X \cap A = \emptyset$ or else $X \notin \mathcal{F}_{n-k+1}$. To do this, consider the subscript j such that $X = S_j$. If $X \cap A \ne \emptyset$, then (as $v_n \notin X$) we have $1 \le j \le k - 1$, and so $S_j \notin \mathcal{F}_{n-k+1}$.

CASE 2: $X \subseteq A \cup B$. In this case, $w = v_{n-1}$. To begin, note that
- if i is one of $1, \ldots, k$, then S_k, disjoint from A, is included in $\mathcal{F}_i$;
- if i is one of $k, \ldots, n - k - 1$, then S_i, included in $\mathcal{F}_i$, is disjoint from A;
- if i is one of $n - k + 1, \ldots, n$, then S_n, disjoint from B, is included in $\mathcal{F}_i$.

To summarize, if $i \ne n - k$, then $\mathcal{F}_i$ includes a set disjoint from A or B (or both); we will complete the analysis of this case by showing that $X \cap A = \emptyset$ or else $X \notin \mathcal{F}_{n-k}$. To do this, consider the subscript j such that $X = S_j$. If $X \cap A \ne \emptyset$, then (as $v_{n-1} \notin X$) we have $j = n$ or $1 \le j \le k - 1$, and so $S_j \notin \mathcal{F}_{n-k}$. $\qquad \square$

Proof of Theorem 5.8   Let $n, k, V, \mathcal{E}$ satisfy the assumptions of the theorem. The argument used in the alternative proof of Theorem 5.5 shows that each (n, k)-ring on V includes precisely k members of $\mathcal{E}$ (else we would have $N < Mk$ in (5.11) and so $|\mathcal{E}| < \binom{n-1}{k-1}$); this and Lemma 5.9 together imply that

($\star$) if $\mathcal{R}$ is an (n, k)-ring on V, then $\mathcal{R} \cap \mathcal{E}$ is a star of k sets.

In particular, $\mathcal{E}$ includes distinct sets that intersect in precisely one point; let A, B denote them and let w denote the single point of their intersection.

We will complete the proof of the theorem by showing that w belongs to all members of $\mathcal{E}$: given any k-point subset X of V such that $w \notin X$, we will prove that $X \notin \mathcal{E}$. To do this, consider an (n, k)-ring $\mathcal{R}$ on V with properties (i) and (ii) of Lemma 5.10. Fact $(\star)$ guarantees that $\mathcal{R} \cap \mathcal{E}$ is a star of k sets; since every member of $\mathcal{R} \cap \mathcal{E}$ intersects both A and B, it follows from Lemma 5.10 that $X \in \mathcal{R}$ and $X \notin \mathcal{R} \cap \mathcal{E}$. $\square$

5.3   Turán Numbers

DEFINITIONS: A *hypergraph* is a set V along with a set E of subsets of V. Elements of V are the *vertices* of the hypergraph and members of E are its *hyperedges*. If, for some integer k, every hyperedge consists of k vertices, then the hypergraph is said to be k-*uniform*. Paul Turán asked [364] for the smallest number of hyperedges in a k-uniform hypergraph on n vertices in which every set of ℓ vertices contains at least one hyperedge. Today, these numbers are called *Turán numbers* and denoted $T(n, \ell, k)$.

All Turán numbers $T(n, \ell, 2)$ have been computed by Turán [362]; we shall consider them in Section 7.1. When $k \geq 3$, Turán numbers $T(n, \ell, k)$ are hard to compute. Turán [364] proposed

CONJECTURE 5.11

$$T(n, 4, 3) = \begin{cases} (2s-1)(s-1)s & \text{if } n = 3s, \\ (2s-1)s^2 & \text{if } n = 3s+1, \\ (2s+1)s^2 & \text{if } n = 3s+2 \end{cases}$$

and constructed hypergraphs showing that the left-hand side of this conjectured equation is at most its right-hand side. In this construction, the vertex set is split into parts V_1, V_2, V_3 that are as equally large as possible; three vertices form a hyperedge if and only if either they belong to the same V_i or else two of them belong to the same V_i and the third one belongs to V_{i+1}, where $V_4 = V_1$. As time progressed, larger and larger families of such hypergraphs have been constructed by Alexandr Kostochka in [261], by William Brown in [60], and by Dmitrii Germanovich Fon-Der-Flaass (1962–2010) in [166]. Abundance of these examples seems to suggest that the conjecture is difficult. Gyula Katona, Tibor Nemetz (1941–2006), and Miklós Simonovits [238] verified it for $n \leq 10$.

In the same paper [238], these three authors proved that

$$\frac{T(n, \ell, k)}{\binom{n}{k}} \geq \frac{T(n-1, \ell, k)}{\binom{n-1}{k}} \quad \text{whenever } n > \ell \geq k. \tag{5.12}$$

To verify (5.12), consider a k-uniform hypergraph with n vertices and $T(n, \ell, k)$ hyperedges in which every set of ℓ vertices contains at least one hyperedge; let N

denote the number of pairs (H, v) such that H is a hyperedge and v is a vertex outside H. Since every hyperedge appears in precisely $n - k$ of these pairs, we have

$$N = T(n, \ell, k)(n - k);$$

since every vertex appears in at least $T(n - 1, \ell, k)$ of our pairs, we have

$$N \geq nT(n - 1, \ell, k).$$

We conclude that $T(n, \ell, k)(n - k) \geq nT(n - 1, \ell, k)$, which is just another way of writing the inequality in (5.12).

For every choice of positive integers ℓ and k such that $\ell \geq k$, the sequence $T(n, \ell, k)/\binom{n}{k}$ with $n = \ell, \ell + 1, \ell + 2, \ldots$ is nondecreasing by (5.12) and bounded from above by 1, and so it tends to a limit; let $t(\ell, k)$ denote the value of this limit. In [326], Gerhard Ringel (1919–2008) constructed 3-uniform hypergraphs showing that

$$t(\ell, 3) \leq 4/(\ell - 1)^2 \tag{5.13}$$

for all ℓ. In his construction, the vertex set is split into $\ell - 1$ parts that are as equally large as possible and then the set of these parts is cyclically ordered; now three vertices form a hyperedge if and only if either they belong to the same part or else two of them belong to the same part and the third one belongs to the part that is next in the cyclic order. (When ℓ is odd, (5.13) also follows from another construction: split the vertex set into $\lfloor (\ell - 1)/(k - 1) \rfloor$ parts that are as equally large as possible and let k vertices form a hyperedge if and only if they belong to the same part.) Turán's conjecture about $T(n, 4, 3)$ implies $t(4, 3) = 4/9$; in addition, he conjectured that $t(5, 3) = 1/4$ (see [132, p. 13]); Erdős [127, p. 30] offered \$500 for the determination of even one $t(\ell, k)$ with $\ell > k \geq 3$.

5.3.1   A Lower Bound on $T(n, \ell, k)$

NOTATION: When f and g are real-valued functions defined on positive integers, we write $f(n) \sim g(n)$ to mean that $\lim_{n \to \infty} f(n)/g(n) = 1$.

For every choice of positive integers q, r, n such that $r \leq q \leq n$, Erdős and Hanani [141] defined

$\overline{\overline{m}}(q, r, n)$ as the largest number of hyperedges in a q-uniform hypergraph on n vertices in which every set of r vertices is contained in at most one hyperedge

and

$\overline{\overline{M}}(q, r, n)$ as the smallest number of hyperedges in a q-uniform hypergraph on n vertices in which every set of r vertices is contained in at least one hyperedge.

They noted that [141, inequality (1)]

$$\overline{\overline{m}}(q,r,n) \leq \frac{\binom{n}{r}}{\binom{q}{r}} \leq \overline{\overline{M}}(q,r,n) \quad \text{whenever } r \leq q \leq n$$

and they stated that it may be conjectured that

$$\overline{\overline{m}}(q,r,n) \sim \overline{\overline{M}}(q,r,n) \sim \frac{\binom{n}{r}}{\binom{q}{r}} \quad \text{whenever } r \leq q.$$

This conjecture remained open for over two decades until Vojtěch Rödl [330] proved it by an ingenious semi-random construction. His method became known as *Rödl nibble* and had a great impact on combinatorics (see [240] and the references in its section 1.2).

DEFINITION: The *complement* of a hypergraph with vertex set V and hyperedge set E is the hypergraph with vertex set V and hyperedge set $\{V - B : B \in E\}$.

Since

in a k-uniform hypergraph on n vertices,
every set of ℓ vertices contains at least one hyperedge
if and only if
in the complement of this hypergraph,
every set of $n - \ell$ vertices is contained in at least one hyperedge,

we have

$$T(n,\ell,k) = \overline{\overline{M}}(n-k,n-\ell,n).$$

Since

$$\binom{n}{n-\ell}\binom{\ell}{k} = \binom{n}{k}\binom{n-k}{n-\ell}$$

(both sides count the number of pairs (A,B) such that $A \subseteq B \subseteq \{1,2,\dots,n\}$ and $|A| = k$, $|B| = \ell$), the lower bound on $\overline{\overline{M}}(q,r,n)$ shows that

$$T(n,\ell,k) \geq \frac{\binom{n}{k}}{\binom{\ell}{k}} \quad \text{whenever } n \geq \ell \geq k \tag{5.14}$$

(which, besides following from (5.12) by induction on n, is also easy to prove directly). Rödl's theorem shows that this bound is asymptotically tight in the sense that

$$T(n,n-r,n-q) \sim \frac{\binom{n}{n-q}}{\binom{n-r}{n-q}} \quad \text{whenever } q \geq r. \tag{5.15}$$

5.3.2   Turán Numbers and Steiner Systems

A *Steiner system* with parameters (n, q, r) such that $n \geq q \geq r$ is a q-uniform hypergraph on n vertices in which every set of r vertices is contained in precisely one hyperedge. Deciding whether or not there exists a Steiner system with a prescribed triple of parameters may be a difficult problem, and this problem amounts to computing a Turán number:

THEOREM 5.12   *We have*

$$T(n, \ell, k) = \frac{\binom{n}{k}}{\binom{\ell}{k}}$$
(5.16)

if and only if there is a Steiner system with parameters $(n, n - k, n - \ell)$.

Proof   Given n, ℓ, k, let M denote the right-hand side of (5.16). On the one hand, lower bound (5.14) makes (5.16) equivalent to the claim that there exists a k-uniform hypergraph with n vertices and M hyperedges in which every set of ℓ vertices contains at least one hyperedge. On the other hand, the complement of a Steiner system with parameters $(n, n - k, n - \ell)$ is a k-uniform hypergraph with n vertices in which every set of ℓ vertices contains precisely one hyperedge.

Now consider an arbitrary k-uniform hypergraph $\mathcal{H}$ with n vertices in which every set of ℓ vertices contains at least one hyperedge. To complete the proof, we will show that $\mathcal{H}$ has precisely M hyperedges if and only if every set of ℓ vertices contains precisely one hyperedge. For this purpose, let m denote the number of hyperedges of $\mathcal{H}$ and let N denote the number of pairs (A, B) such that A is a hyperedge, B is a set of ℓ vertices, and $A \subseteq B$; in addition, given a set B of ℓ vertices, let $w(B)$ denote the number of hyperedges contained in B. In this notation,

$$N = m \binom{n - k}{\ell - k}$$

and

$$N = \sum_{B} w(B)$$

with B running through all sets of ℓ vertices. It follows that

$$m = \frac{\sum_{B} w(B)}{\binom{n-k}{\ell-k}} = \frac{\binom{n}{\ell} + \sum_{B}(w(B) - 1)}{\binom{n-k}{\ell-k}} = M + \frac{\sum_{B}(w(B) - 1)}{\binom{n-k}{\ell-k}}.$$

Since $w(B) - 1 \geq 0$ for all B, we conclude that $m = M$ if and only if $w(B) - 1 = 0$ for all B.   □

A Steiner system with parameters $(k^2 + k + 1, k + 1, 2)$ is a projective plane of order k (see page 27). For which values of k is there a projective plane of order k? On the one hand, Oswald Veblen (1880–1960) and William H. Bussey (1879–1962) found a way of constructing a projective plane of order k whenever k is a prime or a power of a prime [370] (see also [335, Theorem 4.2 on

p. 93]); in particular, there are projective planes of orders $2, 3, 4, 5, 7, 8, 9$. On the other hand, Richard H. Bruck (1914–91) and Herbert J. Ryser (referred to on page 30) proved [62] that when $k \bmod 4$ is 1 or 2, a projective plane of order k exists only if there are integers x, y such that $k = x^2 + y^2$ (which is the case if and only if every prime congruent to 3 modulo 4 occurs with an even exponent in the prime factorization of k); in particular, there is no projective plane of order 6.

It had been thought for a long time that the condition of the Bruck–Ryser Theorem, necessary for the existence of a projective plane of a prescribed order, might be also sufficient; in particular, this would imply that there is a projective plane of order 10. As John Conway (1937–2020) and Vera Pless (1931–2020) put it in [88],

> a question which has long tantalized mathematics is
> whether or not a projective plane of order 10 can exist.

Theorem 5.12 shows that this question amounts to asking whether $T(111, 109, 100) = 111$. Between 1957 and 1989, some 100 papers dealt with it. This era came to an end with the announcement [270] by Clement Lam, Larry Thiel, and Stan Swiercz (see also [269]): their computer search revealed that there is no such plane.

Apart from these results, we know nothing about the values of k for which there is a projective plane of order k. And this is just the tip of an iceberg: What are the values of n, q, r for which there is a Steiner system with parameters (n, q, r)?

THEOREM 5.13   *There is a Steiner system with parameters (n, q, r) only if*

$$\binom{q - i}{r - i} \text{ divides } \binom{n - i}{r - i} \text{ for all } i = 0, 1, \dots, r - 1. \tag{5.17}$$

Proof   We shall prove more: in every Steiner system with parameters (n, q, r), every set of i vertices such that $0 \le i < r$ is contained in precisely

$$\frac{\binom{n-i}{r-i}}{\binom{q-i}{r-i}}$$

hyperedges. For this purpose, given a set S of i vertices such that $0 \le i < r$, let R denote the family of all sets of r vertices that contain S, let Q denote the family of all hyperedges that contain S, and let N denote the number of pairs (A, B) such that $A \in R, B \in Q, A \subseteq B$. On the one hand, for each A in R there is a unique hyperedge B such that $A \subseteq B$; since $S \subseteq A \subseteq B$, we have $B \in Q$. It follows that

$$N = |R| = \binom{n - i}{r - i}.$$

On the other hand, for each B in Q there are precisely $\binom{q-i}{r-i}$ sets A in R such that $A \subseteq B$; it follows that

$$N = |Q| \binom{q-i}{r-i}.$$

Comparing the two expressions for N, we conclude that

$$|Q| = \frac{\binom{n-i}{r-i}}{\binom{q-i}{r-i}}. \qquad \square$$

Steiner systems with parameters $(n, 3, 2)$ are called *Steiner triple systems*. By Theorem 5.13, they exist only if $n \bmod 6$ is 1 or 3. Thomas Penyngton Kirkman (1806–95) proved [248] that this necessary condition for their existence is also sufficient. Haim Hanani proved in [207] that the necessary condition of Theorem 5.13 is also sufficient when $(q, r) = (4, 3)$ and added in [208] the cases $(q, r) = (4, 2)$ and $(q, r) = (5, 2)$. However, the necessary condition of Theorem 5.13 is not always sufficient: we have already noted that the Bruck–Ryser theorem implies nonexistence of $S(43, 7, 2)$. A theorem of Peter Keevash [240] subsumes the following special case:

> For every pair of positive integers q, r such that $q \geq r$
> there is a positive integer $n_0(q, r)$ such that
> Steiner systems with parameters (n, q, r)
> exist for all n satisfying (5.17) and $n \geq n_0(q, r)$.

The appearance of this result was a great breakthrough: until then, only finitely many Steiner systems with $r \geq 4$ were known and none of them had $r \geq 6$. For more on Steiner systems, see [85].

5.3.3   An Upper Bound on $T(n, \ell, k)$

REMINDER: We let $\ln x$ stand for the natural logarithm $\log_e x$.

THEOREM 5.14

$$T(n, \ell, k) \;\leq\; 1 + \frac{\binom{n}{k}}{\binom{\ell}{k}} \ln \binom{n}{\ell} \quad \text{whenever } n \geq \ell \geq k.$$

Proof   We will follow the argument presented in [75, pp. 435–436]. To begin, let us prove the inequality

$$T\left(\binom{n}{k}, \binom{n}{k} - T(n, \ell, k) + 1, \binom{\ell}{k}\right) \;\leq\; \binom{n}{\ell} \tag{5.18}$$

by exhibiting an $\binom{\ell}{k}$-uniform hypergraph $\mathcal{H}$ with $\binom{n}{k}$ vertices and $\binom{n}{\ell}$ hyperedges, in which every set of $\binom{n}{k} - T(n, \ell, k) + 1$ vertices contains at least one hyperedge. To describe $\mathcal{H}$, let $\binom{S}{i}$ denote the set of all i-point subsets of a set S. In this notation, the vertex set of $\mathcal{H}$ is $\binom{Y}{k}$ for some n-point set Y and the hyperedge set of $\mathcal{H}$ is in a one-to-one correspondence with $\binom{Y}{\ell}$: the hyperedge corresponding to a set X in $\binom{Y}{\ell}$ is $\binom{X}{k}$. Given any set A of $\binom{n}{k} - T(n, \ell, k) + 1$ vertices of $\mathcal{H}$, consider the k-uniform hypergraph $\mathcal{H}_0$ with vertex set Y and hyperedge set $\binom{Y}{k} - A$. Since $\left| \binom{Y}{k} - A \right| < T(n, \ell, k)$, some set X of ℓ vertices of $\mathcal{H}_0$ contains no hyperedge of $\mathcal{H}_0$, which means that $\binom{X}{k}$ is disjoint from $\binom{Y}{k} - A$, and so $\binom{X}{k} \subseteq A$, and so A contains a hyperedge of $\mathcal{H}$. This observation completes the proof of (5.18).

Inequality (5.18) is a device for transforming lower bounds on Turán numbers into upper bounds on Turán numbers. In particular, lower bound (5.14) guarantees that

$$T\left(\binom{n}{k}, \binom{n}{k} - T(n, \ell, k) + 1, \binom{\ell}{k} \right) \geq \frac{\binom{\binom{n}{k}}{\binom{\ell}{k}}}{\binom{\binom{n}{k} - T(n, \ell, k) + 1}{\binom{\ell}{k}}};$$

comparing this inequality with (5.18), we find that

$$\binom{n}{\ell} \geq \frac{\binom{\binom{n}{k}}{\binom{\ell}{k}}}{\binom{\binom{n}{k} - T(n, \ell, k) + 1}{\binom{\ell}{k}}}, \qquad \text{and so}$$

$$\binom{n}{\ell} \geq \left(\frac{\binom{n}{k}}{\binom{n}{k} - T(n, \ell, k) + 1} \right)^{\binom{\ell}{k}}, \qquad \text{and so}$$

$$\frac{\binom{n}{k} - T(n, \ell, k) + 1}{\binom{n}{k}} \geq \binom{n}{\ell}^{-1/\binom{\ell}{k}}, \qquad \text{and so}$$

$$T(n, \ell, k) \leq 1 + \binom{n}{k} \left(1 - \binom{n}{\ell}^{-1/\binom{\ell}{k}} \right).$$

Since $\ln x \leq x - 1$ for all positive x, we have

$$1 - \binom{n}{\ell}^{-1/\binom{\ell}{k}} \leq - \ln \left(\binom{n}{\ell}^{-1/\binom{\ell}{k}} \right) = \frac{1}{\binom{\ell}{k}} \ln \binom{n}{\ell};$$

this observation completes the proof of the theorem. $\qquad \square$

Improvements of Theorem 5.14 and more on Turán numbers can be found in [342] and [239].

5.4   Turán Functions

NOTATION: Given a k-uniform hypergraph F, let $\mathrm{ex}(F, n)$ denote the largest number of hyperedges in a k-uniform hypergraph on n vertices that contains no F.

The task of evaluating *Turán functions* $ex(F, n)$ subsumes the task of evaluating Turán numbers: we have $T(n, \ell, k) = \binom{n}{k} - ex(F, n)$, where F is the k-uniform hypergraph with ℓ vertices and $\binom{\ell}{k}$ hyperedges.

With F the k-uniform hypergraph on $2k$ vertices that has two disjoint hyperedges and only these two hyperedges, the Erdős–Ko–Rado theorem asserts that

$$ex(F, n) = \binom{n-1}{k-1} \quad \text{whenever } n \geq 2k.$$

Erdős [115] generalized this: if F is the k-uniform hypergraph on tk vertices that has t pairwise disjoint hyperedges and only these t hyperedges, then

$$ex(F, n) = \binom{n}{k} - \binom{n-t+1}{k} \quad \text{whenever } n \text{ is sufficiently large relative to } t \text{ and } k.$$

(For $k = 2$, this was proved earlier by Erdős and Gallai [137].) Here, the extremal hypergraphs (meaning hypergraphs with n vertices and $ex(F, n)$ hyperedges that contain no F) are constructed by first specifying a set of $t - 1$ vertices and then letting a set of k vertices be a hyperedge if and only if it includes at least one of these $t - 1$ vertices.

Problems of determining or at least estimating $ex(F, n)$ for prescribed uniform hypergraphs F constitute a rapidly developing area, which is rich in results. Here is a small sample:

- When F is the Fano plane, we have

$$ex(F, n) = \binom{n}{3} - \binom{\lfloor n/2 \rfloor}{3} - \binom{\lceil n/2 \rceil}{3}.$$

  The extremal hypergraphs are constructed by splitting the vertex set into two parts as equally large as possible (which means that their sizes are $\lfloor n/2 \rfloor$ and $\lceil n/2 \rceil$) and then letting a set of three vertices be a hyperedge if and only if it has at least one vertex in each of the two parts. This was conjectured in 1973 by Vera Sós [345] and proved for all sufficiently large n after more than three decades independently and simultaneously by two pairs of researchers: Peter Keevash and Benny Sudakov [242] and Zoltán Füredi and Miklós Simonovits [179]. Some thirteen years later, Louis Bellmann and Christian Reiher [24] removed the lower bound on n: they proved the conjecture for all n such that $n \geq 7$.
- Peter Keevash and Benny Sudakov [241] determined $ex(F, n)$ when F is the 'expanded triangle,' which means the $2k$-uniform hypergraph with vertex set $V_1 \cup V_2 \cup V_3$ and precisely three hyperedges, $V_1 \cup V_2$, $V_2 \cup V_3$, and $V_3 \cup V_1$, such that V_1, V_2, V_3 are pairwise disjoint sets of size k. For all n that are sufficiently large with respect to k, the extremal hypergraphs are constructed by splitting the vertex set into two parts and then letting a set of $2k$ vertices be a hyperedge if and only if it has an odd number of vertices in each of the two parts. (Maximizing the number of hyperedges requires the right choice of the sizes of the two parts, approximately $(n + \sqrt{3n - 4})/2$ and $(n - \sqrt{3n - 4})/2$.) This proves a conjecture of Frankl [170].

- When F has vertices $1, 2, 3, 4, 5$ and hyperedges $\{1, 2, 3\}, \{1, 2, 4\}, \{3, 4, 5\}$, we have

$$\mathrm{ex}(F, n) = \left\lfloor \frac{n}{3} \right\rfloor \cdot \left\lfloor \frac{n+1}{3} \right\rfloor \cdot \left\lfloor \frac{n+2}{3} \right\rfloor \quad \text{whenever } n \geq 3000:$$

  this is a theorem of Peter Frankl and Zoltán Füredi [171]. Extremal hypergraphs can be constructed by splitting the vertex set into three parts as equally large as possible (which means that their sizes are $\lfloor n/3 \rfloor$, $\lfloor (n + 1)/3 \rfloor$, and $\lfloor (n + 2)/3 \rfloor$) and then letting a set of three vertices be a hyperedge if and only if it has a vertex in each of the three parts.

More on $\mathrm{ex}(F, n)$ can be found in [239] and elsewhere.

5.5   Chromatic Number of Hypergraphs

In 1937, E. W. Miller [294] proposed to say that a family of sets has *property B* if there is a set that has a nonempty intersection with each of its members, but does not contain any of them; he used the letter B in honour of Felix Bernstein (1878–1956), whose work in the early years of the twentieth century involved this notion. Later on, Paul Erdős and András Hajnal (1931–2016) asked in [138, p. 119] for the smallest $m(k)$ such that there exists a k-uniform hypergraph with $m(k)$ hyperedges and without property B; they noted that the family of all k-point subsets of a $(2k-1)$-point set provides the upper bound $m(k) \leq \binom{2k-1}{k}$ and that $m(3) = 7$, with the upper bound provided by the Fano plane. (About a decade later, Paul Seymour [339] and Bjarne Toft [361] proved that $m(4) \leq 23$; nearly four decades after that, Patric Östergård [309] found by an exhaustive computer search that $m(4) \geq 23$.)

Erdős [112, 114] proved that

$$2^{k-1} < m(k) < k^2 2^{k+1} \tag{5.19}$$

and then Erdős and László Lovász [143, p. 610] stated that

it seems likely that $m(k)/2^k \to \infty$.

This hunch was confirmed by József Beck [22]: he proved that $m(k) \geq \frac{1}{5} 2^k \lg k$ whenever $k \geq 2^{100}$ and then improved this lower bound in [23]. More recently, Jaikumar Radhakrishnan and Aravind Srinivasan [319, Theorem 2.1] improved Beck's bounds even further, to $m(k) \geq \frac{7}{10} 2^k \sqrt{k / \ln k}$ for all sufficiently large k. A simpler proof of their result was found later by Danila D. Cherkashin and Jakub Kozik [68].

In [114], Erdős wrote

A reasonable guess seems to be that $m(k)$ is of the order $k\, 2^k$;

this conjecture remains open. Jaikumar Radhakrishnan and Saswata Shannigrahi proved that every k-uniform hypergraph without property B and with fewer than $k2^k$ hyperedges must have more than $k^2/4k \ln 2k$ vertices [318, Lemma 2].

DEFINITION: The *chromatic number* $\chi(H)$ of a hypergraph H where every hyperedge consists of at least two vertices is the smallest t such that the vertex set of H can be split into t sets, none of which contain a hyperedge.

This definition comes from Erdős and Hajnal [140], who used the term 'set-system' instead of 'hypergraph' and pointed out [140, p. 61] that $\chi(H) \leq 2$ if and only if the hyperedge set of H has property B.

NOTATION: Let $m(k, s)$ denote the smallest number of hyperedges in a k-uniform hypergraph of chromatic number greater than s.

In this notation, $m(k) = m(k, 2)$. We shall present Erdős's bounds (5.19) in the more general context of $m(k, s)$.

THEOREM 5.15 $s^{k-1} < m(k, s) < 1 + k^2 s^{k+1} \ln s$.

Proof Proving the lower bound on $m(k, s)$ means proving that the vertices of an arbitrary k-uniform hypergraph H with at most s^{k-1} hyperedges can be coloured by s colours in such a way that no hyperedge is monochromatic. To do this, let Ω denote the set of all colourings of the vertices of H by colours $1, 2, \ldots s$, call a colouring in Ω *bad* if it makes at least one edge monochromatic, and let B denote the set of all bad colourings. In this notation, our task is to prove that $|B| < |\Omega|$. For this purpose, let n denote the number of vertices of H and let m denote the number of hyperedges H. Since $|B| < |\Omega|$ is obvious when $m = 1$, we may assume that $m > 1$. Under this assumption, we propose to prove that

$$|B| < ms^{n-k+1}, \tag{5.20}$$

which implies the desired conclusion as $m \leq s^{k-1}$ and $s^n = |\Omega|$. To prove (5.20), consider the following plan for constructing bad colourings.

Step 1: Choose a hyperedge E.
Step 2: Choose one of the s colours and colour all vertices of E by this colour.
Step 3: Colour the remaining $n - k$ vertices of H by the s colours.

There are m ways of implementing Step 1, there are s ways of implementing Step 2, and there are s^{n-k} ways of implementing Step 3; it follows that the right-hand side of (5.20) counts the number of different implementations of the entire plan. Pointing out that each bad colouring is constructed in at least one of these implementations and that (as $m > 1$) colouring of all vertices by colour 1 is constructed in more than one implementation concludes the proof of (5.20).

To prove the upper bound on $m(k, s)$, let $\mu(n, k, s)$ stand for the smallest possible number of monochromatic k-point subsets of an n-point set whose points are coloured by s colours. The combinatorial content of the argument used by Erdős in [114] can be extracted in the claim that, as long as $\mu(n, k, s) > 0$, we have

$$T\left(\binom{n}{k}, \binom{n}{k} - m(k, s) + 1, \mu(n, k, s)\right) \leq s^n. \tag{5.21}$$

We will prove (5.21) by exhibiting a $\mu(n, k, s)$-uniform hypergraph $\mathcal{H}$ with $\binom{n}{k}$ vertices and s^n hyperedges, in which every set of $\binom{n}{k} - m(k, s) + 1$ vertices contains at least one hyperedge. To describe $\mathcal{H}$, let $\binom{S}{i}$ denote the set of all i-point subsets of a set S. In this notation, the vertex set of $\mathcal{H}$ is $\binom{Y}{k}$ for some n-point set Y and the hyperedge set of $\mathcal{H}$ is in a one-to-one correspondence with the set of colourings of Y by s colours: for each of these colourings, the corresponding hyperedge consists of some $\mu(n, k, s)$ of the (possibly many more) sets in $\binom{Y}{k}$ that are made monochromatic by the colouring. Given any set A of $\binom{n}{k} - m(k, s) + 1$ vertices of $\mathcal{H}$, consider the k-uniform hypergraph $\mathcal{H}_0$ with vertex set Y and hyperedge set $\binom{Y}{k} - A$. Since $\left|\binom{Y}{k} - A\right| < m(k, s)$, the chromatic number of $\mathcal{H}_0$ is at most s, which means that some hyperedge of $\mathcal{H}$ is disjoint from $\binom{Y}{k} - A$, and so A contains this hyperedge of $\mathcal{H}$. This observation completes the proof of (5.21).

The rest is counting. Its ingredients are

(i) $m(k, s) \leq 1 + \dfrac{\binom{n}{k} n \ln s}{\mu(n, k, s)}$ for all n such that $n > s(k - 1)$,

(ii) if s divides n and $n \geq sk$, then $\mu(n, k, s) = s\binom{n/s}{k}$,

(iii) $\dbinom{sk^2}{k} \leq s^{k+1} \dbinom{k^2}{k}$;

with $n = sk^2$, these three ingredients imply that

$$
m(k, s) \leq 1 + \frac{\dbinom{sk^2}{k} sk^2 \ln s}{\mu(sk^2, k, s)} = 1 + \frac{\dbinom{sk^2}{k} k^2 \ln s}{\dbinom{k^2}{k}} \leq 1 + k^2 s^{k+1} \ln s.
$$

To prove (i), note that $n > s(k - 1)$ implies $\mu(n, k, s) > 0$, and so (5.21) combined with the Katona-Nemetz-Simonovits lower bound (5.14) on Turán numbers implies that

$$
s^n \geq \frac{\binom{n}{k}}{\left(\dfrac{\binom{n}{k} - m(k, s) + 1}{\mu(n, k, s)}\right)}, \qquad \text{and so}
$$

$$
s^n \geq \left(\frac{\binom{n}{k}}{\binom{n}{k} - m(k, s) + 1}\right)^{\mu(n,k,s)}, \qquad \text{and so}
$$

$$
\frac{\binom{n}{k} - m(k, s) + 1}{\binom{n}{k}} \geq s^{-n/\mu(n,k,s)}, \qquad \text{and so}
$$

$$
m(k, s) \leq 1 + \binom{n}{k}\left(1 - s^{-n/\mu(n,k,s)}\right).
$$

Since $\ln x \leq x - 1$ for all positive x, we have

$$
1 - s^{-n/\mu(n,k,s)} \leq -\ln\left(s^{-n/\mu(n,k,s)}\right) = \frac{n \ln s}{\mu(n, k, s)};
$$

this observation completes the proof of (i).

To prove (ii), note that $\mu(n,k,s)$ is the minimum of $\sum_{r=1}^{s}\binom{d_r}{k}$ over all choices of nonnegative integers $d_1,\dots d_s$ such that $\sum_{r=1}^{s}d_r = n$. Therefore (ii) is a special case of Lemma A.1, but this special case can be also proved directly as follows. Consider nonnegative integers $d_1,\dots d_s$ whose average is an integer and at least k. Assuming that not all of these integers are equal, we shall find nonnegative integers $c_1,\dots c_s$ such that $\sum_{r=1}^{s}c_r = \sum_{r=1}^{s}d_r$ and $\sum_{r=1}^{s}\binom{c_r}{k} < \sum_{r=1}^{s}\binom{d_r}{k}$; of course, this will imply (ii). By assumption, some d_j is larger than average and some d_i is smaller than average; since the average is an integer, it follows that $d_j \geq d_i + 2$. Setting

$$
c_r = \begin{cases} d_r - 1 & \text{if } r = j, \\ d_r + 1 & \text{if } r = i, \\ d_r & \text{for all other } r \end{cases}
$$

and invoking the bound $d_j \geq k$, we find that

$$
\sum_{r=1}^{s}\binom{d_r}{k} - \sum_{r=1}^{s}\binom{c_r}{k} = \binom{d_i}{k} + \binom{d_j}{k} - \binom{d_i+1}{k} - \binom{d_j-1}{k} = \binom{d_j-1}{k-1} - \binom{d_i}{k-1} > 0.
$$

To prove (iii), note that

$$
\frac{s^{k+1}\binom{k^2}{k}}{\binom{sk^2}{k}} = s^{k+1}\prod_{i=0}^{k-1}\frac{k^2-i}{sk^2-i} = s^k\prod_{i=1}^{k-1}\frac{k^2-i}{sk^2-i}
$$

$$
= s\prod_{i=1}^{k-1}\frac{sk^2-si}{sk^2-i} = s\prod_{i=1}^{k-1}\left(1 - \frac{(s-1)i}{sk^2-i}\right) \geq s\left(1 - \frac{(s-1)(k-1)}{sk^2-k+1}\right)^{k-1}
$$

and that, since $(1-x)^{k-1} \geq 1 - (k-1)x$ for all nonnegative x,

$$
s\left(1 - \frac{(s-1)(k-1)}{sk^2-k+1}\right)^{k-1} \geq s\left(1 - \frac{(s-1)(k-1)^2}{sk^2-k+1}\right).
$$

The proof is concluded by observing that inequality

$$
s\left(1 - \frac{(s-1)(k-1)^2}{sk^2-k+1}\right) \geq 1
$$

is just a restatement of the self-evident

$$
(s-1)(ks + (k-1)(s-1)) \geq 0. \qquad \square
$$

Noga Alon [3] conjectured that $\lim_{s\to\infty} m(k,s)/s^k$ exists for every k.

Erdős's lower bound on the diagonal Ramsey numbers follows from the lower bound on $m(k,2)$ in Theorem 5.15. To see this, consider the hypergraph H with $\binom{n}{2}$ vertices that are the edges of an n-vertex complete graph G and with $\binom{n}{k}$ hyperedges, each of which consists of the $\binom{k}{2}$ edges of a k-vertex complete subgraph of G. If

$$
\binom{n}{k} \leq 2^{\binom{k}{2}-1},
$$

then $\chi(H) \leq 2$, which means that $r(k,k) > n$.

More on $m(k,s)$ and related problems can be found in [319].

'979 XI. 26

IMPERIAL COLLEGE OF SCIENCE AND TECHNOLOGY

Department of Mathematics
Exhibition Road, London SW7 2RH
Telephone: 01-589 5111 Telex: 261503

Dear Vaclav,

I hope you + your son are both well and your work is progressing satisfactorily. I leave for France the day after tomorrow and from Dec 10 until Jan 15 my address Technion Math Dept Haifa Israel. I hope to see you at the San Antonio meeting and (or) at the meeting at Baton Rouge February 9-12. What about your post in Montreal? What about our problems on games?

I thought about your paper London Math. Soc. J. Vol 9 dedicated to my memory. Perhaps the following problem is interesting: Denote by $f(n, k, m, t)$ the largest cardinality of an (n, k, m) set where $|A_i \cap A_j| \leq t$. For $t = k-1$ of course $f(n, k, m, t) = f(n, k, m)$. For example $f(n, 3, 2, 1)$ is the largest family of triples no two of which have two elements in common and every set of 6 elements contain at most two triples. $\{123\}, \{345\}, \{1,2,6\}$ is forbidden and the only forbidden thing). Szemerédi and Ruzsa proved $f(n, 3, 2, 1) = o(n^2)$ but $> n^{2-\varepsilon}$ this was an old problem of mine. Nothing is known about other values of $f(n, k, m, t)$.

I saw Szemerédi in Oberwolfach early in November, he swore eternal obedience and soon they will have a new ε.

I may be going to Prague late in Aug for a topology meeting.

Kind regards to all, au revoir

E. P.

Don't you think $f(n, k, m, t) = o(n^{k-1})$ if $t < k-1$. In fact perhaps it should be $o(n^{t+1})$.

Extremal set theory and Szemerédis' epsilon (who turned out to be twins).
Courtesy of Vašek Chvátal

6 Van Der Waerden's Theorem

6.1 The Theorem

One day in 1926, Bartel van der Waerden (1903–96) had lunch with Emil Artin (1898–1962) and Otto Schreier (1901–29). The day before, he told them of a conjecture he heard from the Dutch mathematician Han Baudet (1891–1921):

(i) *For every positive integer k,*
> *whenever all the positive integers are coloured red and white,*
> *there is an arithmetic progression of k distinct terms*
> *which are all of the same colour.*

After lunch, the three went into Artin's office in the Mathematics Department of the University of Hamburg and tried to prove the conjecture. Reminiscing about that fine Hamburg afternoon, van der Waerden [372] recounts the steps that gradually revealed a proof.

First, Schreier proposed working on a finite version of the conjecture:

(ii) *For every positive integer k*
> *there is a positive integer n such that*
> *whenever the first n positive integers are coloured red and white,*
> *there is an arithmetic progression of k distinct terms*
> *which are all of the same colour.*

Of course, (ii) implies (i); Schreier proved that the converse, (i) implies (ii), is also true. From then on, the trio tried to prove (ii) by induction on k.

Next, Artin observed that (ii) implies a multicolour version of itself

(iii) *For every choice of positive integers k and r*
> *there is a positive integer n such that*
> *whenever the first n positive integers are coloured by r colours,*
> *there is an arithmetic progression of k distinct terms*
> *which are all of the same colour.*

Conjecture (iii) provides a strong induction hypothesis: in proving it for a particular combination of k and r, one could assume that it has been proved for all smaller values of k combined with all values of r. In addition, Artin had the idea of applying the induction hypothesis to blocks of consecutive integers rather than to single integers: the first MN positive integers may be thought of as N consecutive blocks, each block consisting of M consecutive integers. When the MN integers are coloured by r colours, each of the N blocks is coloured in one of r^M possible ways; when N is large enough (with respect to k and r^M), the induction hypothesis with r^M colours guarantees the presence of an arithmetic progression of $k - 1$ equally coloured blocks; when M is large enough (with respect to k and r), the induction hypothesis with r colours guarantees the presence of an arithmetic progression of $k-1$ equally coloured integers in each block.

In turn, van der Waerden took the lead. First, he proved the special case $k = 3$, $r = 2$; next, he proved the special case $k = 3$, $r = 3$; then he deduced the special case $k = 4$, $r = 2$ from the induction hypothesis of $k = 3$ and all r.

DEFINITION: The *van der Waerden number* $W(k, r)$ is the smallest n such that for every colouring of the first n positive integers by r colours there is an arithmetic progression of k distinct terms which are all of the same colour.

Trivially, $W(k, 1) = k$ for all k and $W(1, r) = 1$, $W(2, r) = r + 1$ for all r.

6.1.1   Van der Waerden's proof of $W(3, 2) \leq 325$

(Direct case analysis shows that $W(3, 2) = 9$, but it provides no inspiration for dealing with arbitrary k and r. Van der Waerden's proof of the far weaker upper bound does provide such inspiration.) Let each of the first 325 positive integers be coloured red or white. Arrange these integers into an array with 5 rows and 65 columns:

$$
\begin{array}{cccccc}
1 & 6 & \ldots & \ldots & 321 \\
2 & 7 & \ldots & \ldots & 322 \\
3 & 8 & \ldots & \ldots & 323 \\
4 & 9 & \ldots & \ldots & 324 \\
5 & 10 & \ldots & \ldots & 325 \\
\end{array}
$$

Since there are only 2^5 distinct ways of colouring the five entries in a column, some two of the first 33 columns must be coloured in the same way. Let these columns be c and $c + \Delta$ with $\Delta > 0$. Some two of the first three entries in column c must have the same colour. Let these entries be a and $a + b_1$ with $b_1 > 0$. Switching the two colours if necessary, we may assume that

a and $a + b_1$ are red.

Since a and $a + b_1$ are among the first three of the five entries in column c, this column also includes entry $a + 2b_1$. If this entry is red, then $a, a + b_1, a + 2b_1$ is a red arithmetic progression and we are done; now we may assume that

$a + 2b_1$ is white.

Next, write $b_2 = 5\Delta$, so that each entry i in column c and the corresponding entry $i + b_2$ in column $c + \Delta$ have the same colour. In particular,

$$a, \qquad a + b_2,$$
$$a + b_1, \qquad a + b_1 + b_2 \qquad \text{are red and}$$

$$a + 2b_1, \quad a + 2b_1 + b_2 \qquad \text{are white.}$$

Since columns c and $c + \Delta$ are among the first 33 of the 65 columns in our array, this array also includes column $c + 2\Delta$. In particular, $a + 2b_1 + 2b_2 \le 325$. No matter how $a + 2b_1 + 2b_2$ is coloured, it completes a monochromatic arithmetic progression of three terms:

red $a, \qquad a + b_1 + b_2, \quad a + 2b_1 + 2b_2$ or
white $a + 2b_1, \; a + 2b_1 + b_2, \; a + 2b_1 + 2b_2.$

Now van der Waerden [372] says:

> After having found this proof in the special case $r = 2$ and $k = 3$, I explained it to Artin and Schreier. I felt sure that the same proof would work in the general case. They did not believe it, and so I proceeded to present the proof for the next higher case $r = 3, k = 3$.

6.1.2   Van der Waerden's proof of $W(3, 3) \le MN$ with $M = 7(2 \cdot 3^7 + 1)$ and $N = 2 \cdot 3^M + 1$

(The true value of $W(3, 3)$, found in 1969 by computer search [74], is 27.) Let each of the first MN positive integers be coloured red, white, or green. Arrange these integers in a three-dimensional array of N horizontal layers, where layer c is the two-dimensional array with 7 rows and $2 \cdot 3^7 + 1$ columns,

$$
\begin{array}{llll}
(c-1)M + 1 & (c-1)M + 8 & \ldots & \ldots \quad cM - 6 \\
(c-1)M + 2 & (c-1)M + 9 & \ldots & \ldots \quad cM - 5 \\
(c-1)M + 3 & (c-1)M + 10 & \ldots & \ldots \quad cM - 4 \\
(c-1)M + 4 & (c-1)M + 11 & \ldots & \ldots \quad cM - 3 \\
(c-1)M + 5 & (c-1)M + 12 & \ldots & \ldots \quad cM - 2 \\
(c-1)M + 6 & (c-1)M + 13 & \ldots & \ldots \quad cM - 1 \\
(c-1)M + 7 & (c-1)M + 14 & \ldots & \ldots \quad\quad cM \\
\end{array}
$$

Since there are only 3^M ways to colour the M entries in a layer, some two of the first $3^M + 1$ layers must be coloured in the same way. Let these layers be c_3 and $c_3 + \Delta_3$ with $\Delta_3 > 0$. Since there are only 3^7 ways to colour the 7 entries in a column of a layer, some two of the first $3^7 + 1$ columns in layer c_3 must be coloured in the same way. Let these columns be c_2 and $c_2 + \Delta_2$ with $\Delta_2 > 0$. Since there are only 3 ways to colour an entry in a column of a layer, some two of the first four entries in column c_2 of layer c_3 must have the same colour. Let these entries be a and $a + b_1$ with $b_1 > 0$.

Switching colours if necessary, we may assume that

a and $a + b_1$ are red.

Since a and $a + b_1$ are among the first four of the seven entries in column c_2 of layer c_3, this column also includes entry $a + 2b_1$. If this entry is red, then $a, a+b_1, a+2b_1$ is a red arithmetic progression and we are done; now we may assume, switching colours if necessary, that

$a + 2b_1$ is white.

Next, write $b_2 = 7\Delta_2$, so that each entry i in column c_2 of layer c_3 and the corresponding entry $i + b_2$ in column $c_2 + \Delta_2$ of layer c_3 have the same colour. In particular,

$$a, \qquad a + b_2,$$
$$a + b_1, \qquad a + b_1 + b_2 \qquad \text{are red and}$$

$$a + 2b_1, \qquad a + 2b_1 + b_2 \qquad \text{are white.}$$

If $a+2b_1+2b_2$ is red, then $a, a+b_1+b_2, a+2b_1+2b_2$ is a red arithmetic progression and we are done; if $a + 2b_1 + 2b_2$ is white, then $a + 2b_1, a + 2b_1 + b_2, a + 2b_1 + 2b_2$ is a white arithmetic progression and we are done; now we may assume that

$a + 2b_1 + 2b_2$ is green.

Finally, write $b_3 = M\Delta_3$, so that each entry i in layer c_3 and the corresponding entry $i + b_3$ in layer $c_3 + \Delta_3$ have the same colour. In particular,

$$a, \qquad a + b_3,$$
$$a + b_1, \qquad a + b_1 + b_3,$$
$$a + b_2, \qquad a + b_2 + b_3,$$
$$a + b_1 + b_2, \qquad a + b_1 + b_2 + b_3 \qquad \text{are red and}$$

$$a + 2b_1, \qquad a + 2b_1 + b_3,$$
$$a + 2b_1 + b_2, \qquad a + 2b_1 + b_2 + b_3 \qquad \text{are white and}$$

$$a + 2b_1 + 2b_2, \qquad a + 2b_1 + 2b_2 + b_3 \qquad \text{are green.}$$

Since layers c_3 and $c_3 + \Delta_3$ are among the first $3^M + 1$ layers of the $2 \cdot 3^M + 1$ layers in our three-dimensional array, this array also includes layer $c_3 + 2\Delta_3$. In particular, $a + 2b_1 + 2b_2 + 2b_3 \le MN$. No matter how $a + 2b_1 + 2b_2 + 2b_3$ is coloured, it completes a monochromatic arithmetic progression of three terms:

red	$a,$	$a + b_1 + b_2 + b_3,$	$a + 2b_1 + 2b_2 + 2b_3$ or
white	$a + 2b_1,$	$a + 2b_1 + b_2 + b_3,$	$a + 2b_1 + 2b_2 + 2b_3$ or
green	$a + 2b_1 + 2b_2,$	$a + 2b_1 + 2b_2 + b_3,$	$a + 2b_1 + 2b_2 + 2b_3.$

Van der Waerden [372] again:

After this, all of us agreed that the same kind of proof could be given for arbitrary r. However, Artin and Schreier still wanted to see the case $k = 4$.

6.1.3   Van der Waerden's proof of $W(4, 2) \leq MN$ with $M = \lfloor \frac{3}{2} W(3, 2) \rfloor$ and $N = \lfloor \frac{3}{2} W(3, 2^M) \rfloor$

(The true value of $W(4, 2)$, found in 1969 by computer search [74], is 35.) Let each of the first MN positive integers be coloured red or white. Arrange these integers into an array with M rows and N columns:

$$
\begin{array}{ccccc}
1 & M + 1 & \ldots & \ldots & (N - 1)M + 1 \\
2 & M + 2 & \ldots & \ldots & (N - 1)M + 2 \\
\ldots & \ldots & \ldots & \ldots & \ldots \\
M & 2M & \ldots & \ldots & NM
\end{array}
$$

Among the first $W(3, 2^M)$ columns of this array, some three equally spaced columns are coloured in the same way. Let these columns be

$$c, \ c + \Delta, \ c + 2\Delta$$

with $\Delta > 0$. Among the first $W(3, 2)$ entries of column c, some three equally spaced entries have the same colour. Let these entries be

$$a, \ a + b_1, \ a + 2b_1$$

with $b_1 > 0$. Switching the two colours if necessary, we may assume that

$a, \ a + b_1, \ a + 2b_1$ are red.

Since entries a and $a + 2b_1$ are among the first $W(3, 2)$ entries of the $\lfloor \frac{3}{2} W(3, 2) \rfloor$ entries in column c, this column also includes entry $a + 3b_1$. If this entry is red, then $a, a + b_1, a + 2b_1, a + 3b_1$ is a red arithmetic progression of four terms and we are done; now we may assume that

$a + 3b_1$ is white.

Next, write $b_2 = M\Delta$, so that each entry i in column c shares its colour with the corresponding entries $i + b_2$ in column $c + \Delta$ and $i + 2b_2$ in column $c + 2\Delta$. In particular,

$$
\begin{array}{lll}
a, & a + b_2, & a + 2b_2, \\
a + b_1, & a + b_1 + b_2, & a + b_1 + 2b_2, \\
a + 2b_1, & a + 2b_1 + b_2, & a + 2b_1 + 2b_2 \quad \text{are red and}
\end{array}
$$

$$a + 3b_1, \quad a + 3b_1 + b_2, \quad a + 3b_1 + 2b_2 \quad \text{are white.}$$

Since columns c and $c+2\Delta$ are among the first $W(3, 2^M)$ columns of the $\lfloor \frac{3}{2} W(3, 2^M) \rfloor$ columns in our array, this array also includes column $c + 3\Delta$. In particular, $a + 3b_1 +$

$3b_2 \leq MN$. No matter how $a + 3b_1 + 3b_2$ is coloured, it completes a monochromatic arithmetic progression of four terms:

red $a,$ $a + b_1 + b_2,$ $a + 2b_1 + 2b_2,$ $a + 3b_1 + 3b_2$ or

white $a + 3b_1,$ $a + 3b_1 + b_2,$ $a + 3b_1 + 2b_2,$ $a + 3b_1 + 3b_2.$

And van der Waerden [372] says

> Now it was clear to every one of us that the induction proof from $k - 1$ to k works for arbitrary k and for any fixed value of r

and, a little later,

> After the discussion with Artin and Schreier I worked out the details of proof and published it in [371].

A particularly pleasing presentation of the details has been designed by Ron Graham and Bruce Rothschild [195]. Its slightly streamlined version is our next subject.

6.2   A Proof

6.2.1   A Warm-up Example

Let us begin by comparing van der Waerden's reminiscences with his finished proof [371] on the example of $W(3, 3) \leq MN$, where $M = 7(2 \cdot 3^7 + 1)$ and $N = M(2 \cdot 3^M + 1)$.

Given an arbitrary colouring of the first MN positive integers by three colours, van der Waerden found positive integers a, b_1, b_2, b_3 such that $a + 2b_1 + 2b_2 + 2b_3 \leq MN$ and

$a,$	$a + b_3,$	
$a + b_1,$	$a + b_1 + b_3,$	
$a + b_2,$	$a + b_2 + b_3,$	
$a + b_1 + b_2,$	$a + b_1 + b_2 + b_3$	have the same colour and

$a + 2b_1,$	$a + 2b_1 + b_3,$	
$a + 2b_1 + b_2,$	$a + 2b_1 + b_2 + b_3$	have the same colour and

$a + 2b_1 + 2b_2,$ $a + 2b_1 + 2b_2 + b_3$ have the same colour.

He argued that one of the six arithmetic progressions

$(a,$ $a + b_1,$ $a + 2b_1),$
$(a,$ $a + b_1 + b_2,$ $a + 2b_1 + 2b_2),$
$(a,$ $a + b_1 + b_2 + b_3,$ $a + 2b_1 + 2b_2 + 2b_3),$
$(a + 2b_1,$ $a + 2b_1 + b_2,$ $a + 2b_1 + 2b_2),$
$(a + 2b_1,$ $a + 2b_1 + b_2 + b_3,$ $a + 2b_1 + 2b_2 + 2b_3),$
$(a + 2b_1 + 2b_2,$ $a + 2b_1 + 2b_2 + b_3,$ $a + 2b_1 + 2b_2 + 2b_3)$

must be monochromatic. (Colours of $a + b_3$, $a + b_1 + b_3$, $a + b_2$, $a + b_2 + b_3$, $a + 2b_1 + b_3$ are irrelevant to his argument.) In [371], he arrived at this conclusion in a way that is both more direct and more elegant than the way sketched in his reminiscences.

Reminiscences published in [372]: Call the colour of a *red*. If $a + 2b_1$ red, then a, $a + b_1$, $a + 2b_1$ is a red arithmetic progression and we are done; now we may assume that $a + 2b_1$ is not red. Call its colour *white*. If $a + 2b_1 + 2b_2$ is red, then a, $a + b_1 + b_2$, $a + 2b_1 + 2b_2$ is a red arithmetic progression and we are done; if $a+2b_1+2b_2$ is white, then $a+2b_1$, $a+2b_1+b_2$, $a+2b_1+2b_2$ is a white arithmetic progression and we are done; now we may assume that $a + 2b_1 + 2b_2$ is neither red nor white. Call its colour *green*. If $a+2b_1+2b_2+2b_3$ is red, then a, $a+b_1+b_2+b_3$, $a + 2b_1 + 2b_2 + 2b_3$ is a red arithmetic progression; if $a + 2b_1 + 2b_2 + 2b_3$ is white, then $a + 2b_1$, $a + 2b_1 + b_2 + b_3$, $a + 2b_1 + 2b_2 + 2b_3$ is a white arithmetic progression; if $a + 2b_1 + 2b_2 + 2b_3$ is green, then $a + 2b_1 + 2b_2$, $a + 2b_1 + 2b_2 + b_3$, $a + 2b_1 + 2b_2 + 2b_3$ is a green arithmetic progression.

Proof published in [371]: By the pigeon-hole principle, two of the four integers

$$a, a + 2b_1, a + 2b_1 + 2b_2, a + 2b_1 + 2b_2 + 2b_3$$

must have the same colour, which means that there are s and t with $0 \le s < t \le 3$ such that

$$a + 2\sum_{i=1}^{s} b_i \text{ and } a + 2\sum_{i=1}^{t} b_i \text{ have the same colour.} \tag{6.1}$$

The three-term arithmetic progression

$$a + 2\sum_{i=1}^{s} b_i, \ a + 2\sum_{i=1}^{s} b_i + \sum_{i=s+1}^{t} b_i, \ a + 2\sum_{i=1}^{s} b_i + 2\sum_{i=s+1}^{t} b_i$$

is monochromatic: its two-term prefix is one of the monochromatic pairs

$$(a, \qquad a + b_1),$$
$$(a, \qquad a + b_1 + b_2),$$
$$(a, \qquad a + b_1 + b_2 + b_3),$$
$$(a + 2b_1, \qquad a + 2b_1 + b_2),$$
$$(a + 2b_1, \qquad a + 2b_1 + b_2 + b_3),$$
$$(a + 2b_1 + 2b_2, \quad a + 2b_1 + 2b_2 + b_3)$$

and property (6.1) guarantees that the last term, too, has the colour of the first term.

6.2.2　An Overview of the Proof

NOTATION:　For every choice of positive integers k and d, let $C(k, d)$ denote the following claim:

> For every positive integer r, there is a positive integer n such that for every colouring of the first n positive integers by r colours, there are positive integers $a, b_1, b_2, \ldots, b_d$ with the properties that
>
> (i) $a + k \sum_{i=1}^{d} b_i \leq n$,
> (ii) for all s and t such that $0 \leq s < t \leq d$, the arithmetic progression
> $$(a + k \sum_{i=1}^{s} b_i) + x \sum_{i=s+1}^{t} b_i \quad (x = 0, 1, \ldots k - 1)$$
> is monochromatic.

Let $GR(k, d, r)$ denote the smallest n with the property featured in $C(k, d)$.

Since $C(k, 1)$ asserts that

> For every positive integer r, there is a positive integer n such that for every colouring of the first n positive integers by r colours, there are positive integers a and b with the properties that
>
> (i) $a + kb \leq n$,
> (ii) the arithmetic progression $a, \ a + b, \ \ldots, \ a + (k - 1)b$
> is monochromatic,

van der Waerden's theorem amounts to the claim that $C(k, 1)$ holds for all k. We have $W(1, r) = 1$, $GR(1, 1, r) = 2$, and

$$W(k, r) \ \leq \ GR(k, 1, r) \ < \ \frac{k}{k - 1} \cdot W(k, r) \ \text{whenever } k \geq 2:$$

the slack $GR(k, 1, r) - W(k, r)$ only makes room for the placeholder $a + kb$, whose colour is irrelevant.

We are going to prove validity of all $C(k, d)$ by double induction: we will

- observe that $C(1, d)$ holds for all d,
- deduce each $C(k + 1, 1)$ with $k \geq 1$ from $C(k, d)$ with all d, and
- deduce each $C(k, d + 1)$ with $d \geq 1$ from $C(k, d)$.

6.2.3 $C(1, d)$ Holds for All d

Claim $C(1, d)$ is the claim that

> For every positive integer r, there is a positive integer n such that for every colouring of the first n positive integers by r colours, there are positive integers $a, b_1, b_2, \ldots, b_d$ with the property that
> $a + \sum_{i=1}^{d} b_i \leq n$

and so $GR(1, d, r) = d + 1$ for all r.

6.2.4 $C(k, d)$ with All d Implies $C(k + 1, 1)$

Given a positive integer r, note that $GR(k, r, r)$ exists by claim $C(k, r)$. We will prove that

$$GR(k + 1, 1, r) \leq 2GR(k, r, r) :$$

given an arbitrary colouring of the first $GR(k, r, r)$ positive integers by r colours, we will find in $\{1, 2, \ldots, GR(k, r, r)\}$ a monochromatic arithmetic progression $\alpha, \alpha + \beta$, $\ldots, \alpha + k\beta$ (of course, $\alpha + (k + 1)\beta \leq 2GR(k, r, r)$). (When $k = 2$ and $r = 3$, our argument reduces to the argument given in Section 6.2.1.)

There are positive integers $a, b_1, b_2, \ldots, b_r$ such that $a + k\sum_{i=1}^{r} b_i \leq GR(k, r, r)$ and, for all s and t such that $0 \leq s < t \leq r$, the arithmetic progression

$$a + k\sum_{i=1}^{s} b_i + x\sum_{i=s+1}^{t} b_i \quad (x = 0, 1, \ldots k - 1) \tag{6.2}$$

is monochromatic. By the pigeon-hole principle, two of the $r + 1$ integers

$$a + k\sum_{i=1}^{j} b_i \quad (j = 0, 1, \ldots, r)$$

must have the same colour, which means that there are s and t with $0 \leq s < t \leq r$ such that

$$a + k\sum_{i=1}^{s} b_i \text{ and } a + k\sum_{i=1}^{t} b_i \text{ have the same colour.} \tag{6.3}$$

The arithmetic progression of the $k + 1$ terms

$$\left(a + k\sum_{i=1}^{s} b_i\right) + x\sum_{i=s+1}^{t} b_i \quad (x = 0, 1, \ldots k)$$

is monochromatic: its k-term prefix is one of the arithmetic progressions (6.2) and property (6.3) guarantees that the last term, too, has the colour of the first term.

6.2.5 $C(k, d)$ Implies $C(k, d + 1)$

Given a positive integer r, note that $GR(k, d, r)$ exists by claim $C(k, d)$ and set $M = GR(k, d, r)$. Then note that $GR(k, 1, r^M)$ exists by claim $C(k, 1)$, which is subsumed in $C(k, d)$, and set $N = GR(k, 1, r^M)$. We will prove that

$$GR(k, d + 1, r) \leq MN \text{ with } M = GR(k, d, r), N = GR(k, 1, r^M) :$$

given an arbitrary colouring of the first MN positive integers by r colours, we will find positive integers $a, b_1, b_2, \ldots, b_{d+1}$ such that $a + k\sum_{i=1}^{d+1} b_i \leq MN$ and, for all s and t such that $0 \leq s < t \leq d + 1$, the arithmetic progression

$$a + k\sum_{i=1}^{s} b_i + x\sum_{i=s+1}^{t} b_i \quad (x = 0, 1, \ldots k - 1)$$

is monochromatic.

For this purpose, let us arrange the MN integers into an array with M rows and N columns:

$$
\begin{array}{ccccc}
1 & M+1 & \cdots & \cdots & (N-1)M+1 \\
2 & M+2 & \cdots & \cdots & (N-1)M+2 \\
\cdots & \cdots & \cdots & \cdots & \cdots \\
\\
M & 2M & \cdots & \cdots & NM
\end{array}
$$

When entries in this array are coloured by r colours, each column is coloured in one of r^M distinct ways. Since $N = GR(k, 1, r^M)$, there are positive integers α, β such that $\alpha + k\beta \le N$ and the k columns in the arithmetic progression $\alpha, \alpha + \beta, \ldots, \alpha + (k-1)\beta$ are coloured in the same way. With b_{d+1} standing for βM, this means that

(i) for each entry c in column α and for each x in $\{0, 1, \ldots k - 1\}$,
the two integers c and $c + xb_{d+1}$ have the same colour.

Since $M = GR(k, d, r)$, there are an integer a in column α and positive integers $b_1, b_2, \ldots, b_d$ such that the integer $a + k\sum_{i=1}^{d} b_i$ belongs to column α and,

(ii) for all s and t such that $0 \le s < t \le d$, the arithmetic progression
$a + k\sum_{i=1}^{s} b_i + x\sum_{i=s+1}^{t} b_i$ $(x = 0, 1, \ldots k - 1)$ is monochromatic.

Since the integer $a + k\sum_{i=1}^{d} b_i$ belongs to column α, the integer $a + k\sum_{i=1}^{d+1} b_i$ belongs to column $\alpha + k\beta$, and so it is at most NM. It remains to be shown that

(iii) for all s and t such that $0 \le s < t \le d + 1$, the arithmetic progression
$a + k\sum_{i=1}^{s} b_i + x\sum_{i=s+1}^{t} b_i$ $(x = 0, 1, \ldots k - 1)$ is monochromatic.

If $t \le d$, then (iii) reduces to (ii); if $t = d + 1$, then (iii) follows from (ii) by substituting the integers

$$
a + k\sum_{i=1}^{s} b_i + x\sum_{i=s+1}^{d} b_i
$$

with $x = 0, 1, \ldots k - 1$ for c in (i).

6.3 Van der Waerden Numbers

6.3.1 Exact Values

Apart from the trivial $W(k, 1) = k$, $W(1, r) = 1$, and $W(2, r) = r + 1$, only seven values of $W(k, r)$ are known. These are collected in Table 6.1.

Table 6.1 Known values of $W(k, r)$.

	$r = 2$		$r = 3$		$r = 4$	
$k = 3$	9	(see [74])	27	(see [74])	76	(see [21])
$k = 4$	35	(see [74])	293	(see [264])		
$k = 5$	178	(see [349])				
$k = 6$	1132	(see [265])				

6.3.2   Upper Bounds

The following device for coping with large numbers has been introduced by Don Knuth [252].

NOTATION: $a \uparrow b$ stands for a^b and $a \uparrow b \uparrow c \uparrow \ldots \uparrow f$ is bracketed as

$$a \uparrow (b \uparrow (c \uparrow (\ldots \uparrow f))).$$

Timothy Gowers [190] proved that

$$W(k, r) < 2 \uparrow 2 \uparrow r \uparrow 2 \uparrow 2 \uparrow (k + 9). \tag{6.4}$$

Ron Graham [191] offered $1,000 for a proof that

$$W(k, 2) < 2 \uparrow k \uparrow 2.$$

A theorem of Thomas Bloom and Olof Sisask [32, Theorem 1.1] implies that

$$W(3, r) \leq \exp\left(r^{1-\varepsilon}\right) \quad \text{for some positive constant } \varepsilon. \tag{6.5}$$

6.3.3   Lower Bounds

NOTATION: When f and g are nonnegative real-valued functions defined on positive integers, we write $f(n) = \Omega(g(n))$ to mean that $f(n) \geq cg(n)$ for some positive constant c and all sufficiently large n.

THEOREM 6.1 (Erdős and Rado [145])

$$W(k + 1, r) > (2k)^{1/2} r^{k/2}. \tag{6.6}$$

Proof   Given positive integers k and n such that $k < n$, consider the hypergraph H with vertices $1, 2, \ldots, n$ whose hyperedges are arithmetic progressions with $k + 1$ terms. Each of these arithmetic progressions

$$a, \; a + d, \; \ldots, \; a + kd$$

is determined by its first term a and by the difference d between its successive terms; constraint $1 \leq a < a + kd \leq n$ limits the choices of d to $1, 2, \ldots D$ with D the largest integer such that $1 + kD \leq n$ and, once d has been chosen, the same constraint limits the choices of a to $1, 2, \ldots n - kd$. It follows that H has m hyperedges, where

$$m = \sum_{d=1}^{D} (n - kd) = nD - k\frac{D(D + 1)}{2}.$$

Since D is about n/k, the value of m is about $n^2/2k$. Actually, this rough estimate of m turns out to be a strict upper bound on m: we have

$$nD - k\frac{D(D+1)}{2} < nD - \frac{kD^2}{2} = \frac{n^2}{2k} - \frac{k}{2}\left(D - \frac{n}{k}\right)^2 \leq \frac{n^2}{2k}.$$

If $n \leq (2k)^{1/2}r^{k/2}$, then $m \leq r^k$, in which case the lower bound in Theorem 5.15 guarantees that $\chi(H) \leq r$, which means that $W(k+1, r) > n$. $\square$

This non-constructive proof of a lower bound on van der Waerden numbers goes just like Erdős's earlier non-constructive proof of a lower bound on Ramsey numbers. However, unlike the bound of Theorem 3.4, the bound of Theorem 6.1 is not uniformly stronger than known constructive lower bounds. For instance, Leo Moser (1921–70) proved in [296] by an explicit construction that

$$W(k+1, r) > kr^{\Omega(\log r)};$$

when r is sufficiently large compared to k, this bound is stronger than (6.6). Similarly, Thomas Blankenship, Jay Cummings, and Vladislav Taranchuk proved in [31] by an explicit construction that

$$W(k+1, r) > k^{r-1}2^k \quad \text{whenever } k \text{ is a prime such that } k \geq r. \qquad (6.7)$$

(Their proof relies on a previous result [26] of Elwyn Berlekamp (1940–2019), which is (6.7) restricted to $r = 2$.) When k is a prime and $r \leq 4$, bound (6.7) is stronger than (6.6).

Jakub Kozik and Dmitry Shabanov [267] proved that

$$W(k+1, r) = \Omega(r^k).$$

6.4   Szemerédi's Theorem

NOTATION: When f and g are real-valued functions defined on positive integers, we write $f(n) = o(g(n))$ to mean that $\lim_{n\to\infty} f(n)/g(n) = 0$.

In 1936, Erdős and Turán [159] defined $r_k(n)$ as the largest size of a subset of $\{1, 2, \ldots, n\}$ which does not contain any k-term arithmetic progression. They established, for every positive ε and for all n large enough with respect to this ε, the upper bound $r_3(n) \leq (\frac{3}{8} + \varepsilon)n$ and commented that it was probable that $r_3(n) = o(n)$. In addition, they pointed out the relationship between $r_k(n)$ and the van der Waerden numbers. Twenty-one years later, Erdős [109] reminisced:

> We had been motivated by the fact that the inequality $r_k(n) < n/2$ would imply van der Waerden's theorem [...]; but the problem itself seems to be much older (it seems likely that Schur gave it to Hildegard Ille, in the 1920's).

Eventually, Erdős and Turán conjectured that

$$r_k(n) = o(n) \quad \text{for all } k \qquad (6.8)$$

and Erdős [124, p. 287] offered \$1,000 for the proof or disproof of this conjecture. Klaus Roth (1925–2015) proved [334] that $r_3(n) = o(n)$, then Szemerédi proved [355] that $r_4(n) = o(n)$ and finally, in an extraordinary *tour de force,* Szemerédi [356] proved (6.8) in its entirety. The king of Norway awarded the 2012 Abel Prize to Endre Szemerédi for "his fundamental contributions to discrete mathematics and theoretical computer science, and in recognition of the profound and lasting impact of these contributions on additive number theory and ergodic theory." After a brief biographical sketch, the press release continues:

> Many of his discoveries carry his name. One of the most important is Szemerédi's Theorem, which shows that in any set of integers with positive density, there are arbitrarily long arithmetic progressions. Szemerédi's proof was a masterpiece of combinatorial reasoning, and was immediately recognized to be of exceptional depth and importance.

Szemerédi published a proof of his theorem in 1975. Two years later, Harry Furstenberg published another proof [181] and 24 years after that, Timothy Gowers published a third one. Actually, Gowers proved [190, Theorem 18.1] that $r_k(n) < \delta n$ if $0 < \delta \le 1/2$ and

$$2 \uparrow 2 \uparrow \delta^{-1} \uparrow 2 \uparrow 2 \uparrow (k+9) \le n.$$

Bound (6.4) is an immediate corollary of this theorem: as Erdős and Turán [159] noted,

$$r_k(n) < n/c \implies W(k,c) \le n. \tag{6.9}$$

Similarly, the upper bound (6.5) follows from the upper bound

$$r_3(n) < \frac{n}{(\log n)^{1+\delta}} \quad \text{for some positive constant } \delta.$$

Additional information on $W(k,r)$ and $r_k(n)$ can be found in [198], [61], [306].

Since 2001, further proofs of Szemerédi's theorem have been found by Vojtěch Rödl and Jozef Skokan [332, 333], by Brendan Nagle, Vojtěch Rödl, and Mathias Schacht [302], by Terence Tao [357, 358], and by others.

There is no theorem that would be to Ramsey's theorem what Szemerédi's theorem is to van der Waerden's theorem. More precisely, with $s_k(n)$ standing for the largest number of edges in an n-vertex graph that contains no k-vertex clique, we have $s_3(n) > \binom{n}{2}/2$ for all n. To see this, consider the graph with vertex-set split into two equal or nearly equal parts and two vertices adjacent if and only if they belong to distinct parts.

6.5   Ramsey Theory

DEFINITION: A *combinatorial line* in $\{1, 2, \ldots k\}^d$ is a set

$$\{x + t\Delta : t = 0, 1, \ldots k - 1\}$$

of distinct points in $\{1, 2, \ldots k\}^d$ such that $\Delta \in \{0, 1\}^d$.

For instance, in $\{1, 2, 3\}^2$ there are seven combinatorial lines:

$$\{(1,1),(1,2),(1,3)\}, \quad \{(2,1),(2,2),(2,3)\}, \quad \{(3,1),(3,2),(3,3)\} \quad \text{with } \Delta = (0,1),$$
$$\{(1,1),(2,1),(3,1)\}, \quad \{(1,2),(2,2),(3,2)\}, \quad \{(1,3),(2,3),(3,3)\} \quad \text{with } \Delta = (1,0),$$
$$\{(1,1),(2,2),(3,3)\} \qquad\qquad\qquad\qquad\qquad\qquad\qquad \text{with } \Delta = (1,1).$$

Set $\{(1,3),(2,2),(3,1)\}$ is collinear (with $\Delta = (1,-1)$) and it is not a combinatorial line.

In 1961, Alfred Hales and Robert Jewett [204] proved:

THEOREM 6.2   *For every choice of positive integers k and r there is a positive integer d such that whenever the elements of $\{1, 2, \ldots k\}^d$ are coloured by r colours, there is a combinatorial line whose elements are all of the same colour.*   □

Theorem 6.2 is stronger than van der Waerden's theorem since the bijection

$$(x_1, x_2, \ldots x_d) \; \mapsto \; 1 + \sum_{i=0}^{d-1}(x_i - 1)k^i$$

between $\{1, 2, \ldots k\}^d$ and $\{1, 2, \ldots k^d\}$ maps every combinatorial line onto an arithmetic progression.

Using ergodic theoretic techniques, Harry Furstenberg and Yitzhak Katznelson proved a theorem [182] that is to the Hales–Jewett theorem what Szemerédi's theorem is to van der Waerden's theorem. Elementary proofs of the Furstenberg–Katznelson theorem were found later by Polymath [314] and by Pandelis Dodos, Vassilis Kanellopoulos, and Konstantinos Tyros [102].

In 1970, Ron Graham and Bruce Rothschild found a far-reaching common generalization of the Hales–Jewett theorem and Ramsey's theorem. Their monumental theorem [194] on *n-parameter sets* subsumes a number of other theorems and solutions to previously unsolved conjectures. It has become the cornerstone of *Ramsey theory* [192, 196].

One fine day in the mid-1970s, Paul Erdős, Ernst Straus (1922–1983), Endre Szemerédi, and I were walking to lunch at the Stanford Faculty Club. An item much debated in the news just then was the case of a young Israeli diplomat reprimanded for allegedly engaging in a sexual act at Orly airport with a member of a foreign government. As we were crossing the Quad, Erdős reported to us the latest revelation: the young diplomat's presence at the airport overlapped with the presence of her presumed paramour by only twenty minutes, which put the accusation on a shaky ground. We all agreed that you could not trust anybody any more and that fake news was everywhere these days. "Still, there is something called a quickie," I remarked to Straus out of the corner of my mouth and he chuckled. "What's a quickie?" asked Erdős. Knowing his bashfulness concerning sex (whenever something of that ilk embarassed him, he tilted his head back and covered his face with a handkerchief), I thought it prudent to let the question pass. However, Erdős would not be denied. "What's a quickie?" he asked again and then, as I hesitated for a moment, he raised the tone of his voice and accelerated the tempo of his questions to a machine-gun speed: "What's a quickie? What's a quickie? What's a quickie?" I gave in to this onslaught and told him. "A rapid sexual intercourse," I said. The hankie came out instantly and remained on his face until we reached the steps to the faculty club.

With Paul Erdős and Endre Szemerédi at San Francisco International Airport
Courtesy of Vašek Chvátal

7 Extremal Graph Theory

7.1 Turán's Theorem

7.1.1 Two Theorems

REMINDER: The *order* of a graph is the number of its vertices. A *clique* in a graph is a set of pairwise adjacent vertices; the *clique number* $\omega(G)$ of a graph G is the number of vertices in its largest clique.

DEFINITION: A *complete k-partite graph* is a graph whose vertex-set can be split into k pairwise disjoint parts (not necessarily all of them nonempty) so that two vertices are adjacent if and only if they belong to different parts.

In 1940, while imprisoned in a labour camp in Hungary, Paul Turán proved a seminal theorem:

THEOREM 7.1 (Turán [362]) *Let n, r be integers such that $r \geq 2$. Among all the graphs of order n with clique number less than r, the unique graph with the largest number of edges is the complete $(r-1)$-partite graph whose $r-1$ parts have sizes as nearly equal as possible (meaning that every two of these sizes differ by at most one).*

(Erdős [123] later reported that Turán was informed after he finished his paper that the special case $r = 3$ had been proved in 1907 by W. Mantel and others [290].)

REMINDER: In Section 5.4, we defined the Turán function $\mathrm{ex}(F, n)$ as the largest number of hyperedges in a k-uniform hypergraph on n vertices that does not contain hypergraph F.

In the present chapter, we will consider the case of $k = 2$, when F is a graph.

NOTATION: K_r denotes the complete graph of order r.

Turán's Theorem 7.1 specifies the value of $\mathrm{ex}(K_r, n)$ in an elegant combinatorial way. This specification can be translated into an arithmetic formula, which some find less elegant:

$$\mathrm{ex}(K_r, n) = \left(1 - \frac{1}{r-1}\right)\frac{n^2}{2} - \frac{b(r-1-b)}{2(r-1)} \quad \text{where } b = n \bmod (r-1). \quad (7.1)$$

Let us verify that Theorem 7.1 implies identity (7.1).

DEFINITION: In a graph, adjacent vertices are called *neighbours*.

Given integers r and n such that $2 \le r \le n$, consider integers a, b defined by $n = a(r-1) + b$ and $0 \le b < r - 1$. In the complete $(r-1)$-partite graph of order n whose $r - 1$ parts have sizes as nearly equal as possible, b parts have size $a + 1$ and $r - 1 - b$ parts have size a; consequently, $b(a+1)$ vertices have precisely $n - (a+1)$ neighbours and $(r - 1 - b)a$ vertices have precisely $n - a$ neighbours. The total number of edges comes to

$$\frac{b(a+1)(n-a-1) + (r-1-b)a(n-a)}{2};$$

it is a routine matter to verify that this quantity equals the right-hand side of (7.1).

Thirty years later, Erdős found a beautiful refinement of Turán's theorem.

NOTATION: $d_G(v)$ denotes the *degree* of a vertex v in a graph G, defined as the number of neighbours of v in G.

THEOREM 7.2 (Erdős [118])   *Let r be an integer greater than 1. For every graph G with clique number less than r, there is a graph H such that*

(i)   G and H share their vertex-set V,
(ii) H is complete $(r-1)$-partite,
(iii) $d_G(v) \le d_H(v)$ for all v in V,
(iv) if $d_G(v) = d_H(v)$ for all v in V, then $H = G$.

To derive Theorem 7.1 from Theorem 7.2, consider any graph G of order n with clique number less than r. Theorem 7.2 guarantees the existence of a complete $(r-1)$-partite graph H such that either $H = G$ or else G has fewer edges than H. In particular, if G has the largest number of edges among all the graphs of order n with clique number less than r, then G is a complete $(r-1)$-partite graph. Finally, sizes of any two of the $r - 1$ parts of G must differ by at most 1: else moving a vertex from one of the largest parts to one of the smallest ones would increase the number of edges in G.

7.1.2   A Greedy Heuristic

Erdős's proof of Theorem 7.2 amounts to an analysis of the following algorithm:

ALGORITHM 7.3 (An attempt to find a large clique in a graph with vertex set V.)

    $V_1 = V, k = 1$;
    while $V_k \ne \emptyset$
    do   choose a vertex w_k in V_k with the largest number of neighbours in V_k;
          V_{k+1} = the set of neighbours of w_k in V_k, $k = k + 1$;
    end
    return $\{w_1, w_2, \ldots w_{k-1}\}$;

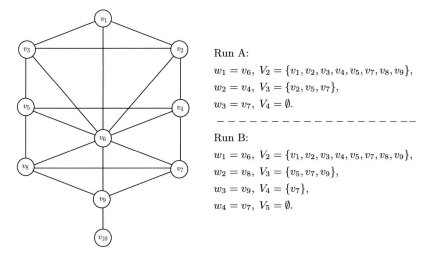

Run A:
$$w_1 = v_6, \quad V_2 = \{v_1, v_2, v_3, v_4, v_5, v_7, v_8, v_9\},$$
$$w_2 = v_4, \quad V_3 = \{v_2, v_5, v_7\},$$
$$w_3 = v_7, \quad V_4 = \emptyset.$$

- - - - - - - - - - - - - - - - - - -

Run B:
$$w_1 = v_6, \quad V_2 = \{v_1, v_2, v_3, v_4, v_5, v_7, v_8, v_9\},$$
$$w_2 = v_8, \quad V_3 = \{v_5, v_7, v_9\},$$
$$w_3 = v_9, \quad V_4 = \{v_7\},$$
$$w_4 = v_7, \quad V_5 = \emptyset.$$

Figure 7.1 Two runs of Algorithm 7.3.

The heuristic rationale for Algorithm 7.3 is that vertices of large degrees are prone to belong to large cliques. The freedom entailed in the instruction 'choose a vertex w_k in V_k with the largest number of neighbours in V_k' may influence the order of the output clique. This phenomenon is illustrated in Figure 7.1.

In each of its iterations, Algorithm 7.3 chooses a w_k that looks most promising just then and disregards all future consequences of this rash decision. Jack Edmonds dubbed another such algorithm, pertaining to a different problem, greedy,[a] and proved in [104] that this algorithm always delivers an optimal output.

TERMINOLOGY & NOTATION: $G \oplus H$ denotes the *direct sum* of G and H, which is the graph consisting of a copy of G and a copy of H that have no vertices in common; $G - H$ denotes the *join* of G and H, which is $G \oplus H$ with additional edges that join every vertex in the copy of G to every vertex in the copy of H.

Even with the most far-sighted choices of w_k in each iteration, Algorithm 7.3 may perform pitifully. For example, consider $(K_1 - \overline{K_d}) \oplus K_d$, the graph with two connected components, a vertex of degree d along with all its neighbours and a clique of order d. Given this input, Algorithm 7.3 returns a clique of order 2.

7.1.3   Proof of Theorem 7.2

Along with a clique of order $k-1$, Algorithm 7.3 can be made to deliver the complete $(k-1)$-partite graph with parts $V_1 - V_2, V_2 - V_3, \ldots, V_{k-1} - V_k (= V_{k-1})$. Let H

---

[a] Perhaps 'myopic' is a better fitting term.

denote this graph. For example, run A in Figure 7.1 delivers H with parts

$$\{v_6, v_{10}\}, \quad \{v_1, v_3, v_4, v_8, v_9\}, \quad \{v_2, v_5, v_7\}$$

and, with G the input graph, we have $d_G(v) \le d_H(v)$ for all v in V: specifically,

$$d_G(v_1) = 3, \quad d_G(v_2) = 4, \quad d_G(v_3) = 4, \quad d_G(v_4) = 4, \quad d_G(v_5) = 4,$$
$$d_H(v_1) = 5, \quad d_H(v_2) = 7, \quad d_H(v_3) = 5, \quad d_H(v_4) = 5, \quad d_H(v_5) = 7,$$

$$d_G(v_6) = 8, \quad d_G(v_7) = 4, \quad d_G(v_8) = 4, \quad d_G(v_9) = 4, \quad d_G(v_{10}) = 1.$$
$$d_H(v_6) = 8, \quad d_H(v_7) = 7, \quad d_H(v_8) = 5, \quad d_H(v_9) = 5, \quad d_H(v_{10}) = 8.$$

Now let G be an arbitrary graph with clique number less than r. We propose to prove that the graph H delivered by Algorithm 7.3 has the four properties specified in Theorem 7.2.

Proof of (i): H has this property by definition.

Proof of (ii): The **while** loop of Algorithm 7.3 maintains the invariant

vertices $w_1, w_2, \ldots w_{k-1}$ are pairwise adjacent and adjacent to all vertices in V_k

and so the set $\{w_1, w_2, \ldots w_{k-1}\}$ returned by the algorithm is a clique. It follows that $k - 1 \le \omega(G)$; since $\omega(G) < r$ by assumption, we conclude that $k \le r$. Since H is complete $(k - 1)$-partite, it is also complete $(r - 1)$-partite: just add $r - k$ empty parts if $r > k$.

Proof of (iii): We have

$$d_H(v) = (n - |V_j|) + |V_{j+1}| \quad \text{whenever } v \in V_j - V_{j+1}.$$

In the input graph G, the number of neighbours that any vertex in V_j has in V_j is at most the number of neighbours that w_j has in V_j, which is $|V_{j+1}|$; it follows that

$$d_G(v) \le |V_{j+1}| + (n - |V_j|) \quad \text{whenever } v \in V_j \tag{7.2}$$

and so $d_G(v) \le d_H(v)$ for all v in V.

Proof of (iv): If $d_G(v) = d_H(v)$ for a vertex v in $V_j - V_{j+1}$, then v satisfies (7.2) with the sign of equality, and so it must be adjacent to all the vertices outside V_j. Since $i < j \Rightarrow (V_i - V_{i+1}) \cap V_j = \emptyset$, it follows that $d_G(v) = d_H(v)$ for all v only if

$$u \in V_i - V_{i+1}, \, v \in V_j - V_{j+1}, \, i < j \quad \Rightarrow \quad u \text{ and } v \text{ are adjacent},$$

in which case every edge of H is an edge of G. This together with (iii) implies that $G = H$.

□

7.1.4 Turán's Theorem and Turán Numbers

REMINDER: The *complement* $\overline{G}$ of a graph G has the same vertices as G; two vertices are adjacent in $\overline{G}$ if and only if they are nonadjacent in G. A *stable set* in a graph is a set of pairwise nonadjacent vertices; the *stability number* $\alpha(G)$ of a graph G is the number of vertices in its largest stable set.

In Section 5.3, we defined the Turán number $T(n, r, k)$ as the smallest number of hyperedges in a k-uniform hypergraph on n vertices in which every set of r vertices contains at least one hyperedge. In particular, $T(n, r, 2)$ is the smallest number of edges in a graph G of order n such that $\alpha(G) < r$. By definition, the Turán function $\text{ex}(K_r, n)$ is the largest number of edges in a graph G of order n such that $\omega(G) < r$. Since $\alpha(G) = \omega(\overline{G})$, we conclude that

$$T(n, r, 2) = \binom{n}{2} - \text{ex}(K_r, n). \tag{7.3}$$

In Section 5.3, we remarked that Turán computed all Turán numbers $T(n, r, 2)$. This remark is justified by observation (7.3) as all $\text{ex}(K_r, n)$ are determined by Theorem 7.1.

7.2 The Erdős–Stone Theorem

NOTATION: $K_r(s)$ stands for the complete r-partite graph with precisely s vertices in each part.

On August 14, 1941, Paul Erdős and two graduate students from Princeton, Shizuo Kakutani (1911–2004) and Arthur Stone (1916–2000), were taking a stroll in Southampton on Long Island. When Kakutani took a few photographs of Erdős and Stone against the background of what turned out to be a secret radar station, a guard told them to leave and afterwards reported that "three Japanese had taken pictures of the installation and then departed in a suspicious hurry." The three mathematicians were arrested together at lunch, questioned separately by the FBI, and finally released later that night; the New York *Daily News* reported the incident the next day under the headline 3 ALIENS NABBED AT SHORT-WAVE STATION. ("Jap, Briton and Hungarian take pictures," added the *Greenville News* in South Carolina the next day and the *Courier-Journal* of Louisville, Kentucky said that "the three were carrying ten explosive films.")

Five years later, two of the three aliens published a powerful variation on the theme of Turán's theorem. This variation gives an upper bound on $\text{ex}(K_r(s), n)$; its simplified version goes as follows:

THEOREM 7.4 (Erdős and Stone [154]) *For every choice of integers r, s and a real number ε such that $r \geq 2$, $s \geq 1$, $\varepsilon > 0$, there is a positive integer $n_0(r, s, \varepsilon)$ such that*

$$n \geq n_0(r, s, \varepsilon) \quad \Rightarrow \quad \text{ex}(K_r(s), n) < \left(1 - \frac{1}{r-1} + \varepsilon\right)\binom{n}{2}.$$

The three aliens
Source: New York Daily News via Getty Images.

To prove Theorem 7.4, we will follow the line of reasoning used by Béla Bollobás and Paul Erdős [43]. The idea is to use induction on r: having found in G a large $K_{r-1}(t)$, we will proceed to find in G a $K_r(s)$ such that, for each $i = 1, 2, \ldots, r-1$, the i-th part of the $K_r(s)$ is a subset of the i-th part of the $K_{r-1}(t)$. The heart of the argument goes as follows.

LEMMA 7.5   *Let r, s, t be positive integers such that $r \geq 2$ and $s \leq t$. If a graph F contains pairwise disjoint sets $T_1, \ldots, T_r$ of vertices such that*

- *$T_1, T_2, \ldots, T_{r-1}$ are parts of a complete $(r-1)$-partite graph,*
- *$|T_1| = |T_2| = \ldots, |T_{r-1}| = t$ and $|T_r| > (s-1)\binom{t}{s}^{r-1}$,*
- *every vertex in T_r has at least $(r-2)t + s$ neighbours in $\cup_{i=1}^{r-1} T_i$,*

then F contains a $K_r(s)$ with parts $S_1, S_2, \ldots, S_r$ such that $S_i \subseteq T_i$ for all $i = 1, 2, \ldots, r$.

Proof   Since every vertex in T_r has at least $(r-2)t + s$ neighbours in $\cup_{i=1}^{r-1} T_i$, it has at least s neighbours in each of $T_1, T_2, \ldots, T_{r-1}$, and so every vertex w in T_r can be labeled by a tuple $(S_1(w), \ldots, S_{r-1}(w))$ such that each $S_i(w)$ is a set of s neighbours of w in T_i. Since there are $\binom{t}{s}^{r-1}$ possible labels, the lower bound on $|T_r|$ guarantees that at least one label appears on at least s distinct vertices; any s of these vertices may form S_r. $\square$

REMINDER: A *subgraph* of a graph G is a graph whose vertex set is a subset of the vertex set of G and where two vertices are adjacent only if they are adjacent in G. (Two vertices may be adjacent in G and nonadjacent in a subgraph of G.)

Proof of Theorem 7.4   Let us write

$$c = 1 - \frac{1}{r-1}.$$

Given r, s, ε, we have to declare a value of $n_0(r, s, \varepsilon)$; then, given any graph G whose order n is at least $n_0(r, s, \varepsilon)$ and whose number of edges is at least $(c + \varepsilon)\binom{n}{2}$, we have to find a $K_r(s)$ in G. In doing this, we may assume that

$$c + \varepsilon \le 1;$$

actually, this inequality follows from the lower bound on the number of edges of G.

We will use induction on r. In G, we will find first a complete $(r-1)$-partite graph with parts $T_1, T_2, \ldots, T_{r-1}$ of a large size t and then a set T_r that satisfies the hypothesis of Lemma 7.5. The first step is trivial when $r = 2$ and taken care of by the induction hypothesis when $r > 2$. The second step would be easier to carry out if we could assume that every vertex of G has a large degree. Unfortunately, this is not the case: the assumption that the number of edges of G is at least $(c + \varepsilon)\binom{n}{2}$ means only that the average degree of a vertex in G is at least $(c + \varepsilon)(n-1)$ and allows for individual vertices of very small degrees. Fortunately, we will be able to find in G a subgraph F of a large order m such that every vertex of F has degree larger than $(c + \varepsilon/2)(m-1)$. We will replace G by F at the very start of the proof and carry out both steps in F.

The revised outline goes as follows. Given r, s and ε, we will choose

- first a positive integer t large enough with respect to r, s, ε,
- then a positive integer m_0 large enough with respect to r, s, ε, and t,
- and finally a positive integer n_0 large enough with respect to ε and m_0.

Then we will argue in stages:

STAGE 1: As long as n_0 is large enough with respect to ε and m_0, we can find in G a subgraph F of order m such that $m \ge m_0$ and such that every vertex of F has degree larger than $(c + \varepsilon/2)(m-1)$.

STAGE 2: As long as m_0 is large enough with respect to r and t, we can find in F a complete $(r-1)$-partite graph K with parts of size t.

STAGE 3: As long as t is large enough with respect to r, s, ε and m_0 is large enough with respect to r, s, ε, and t, we can find in F more than $(s-1)\binom{t}{s}^{r-1}$ vertices such that each of them has at least $(r-2)t + s$ neighbours in K.

Now for the details. Stage 1 can be carried out by the following algorithm, where $|F|$ denotes the order of F:

```
F = G;
while  F has a vertex v of degree at most (c + ε/2)(|F| − 1)
do     remove v (and all the edges that have v for an endpoint) from F;
end
```

In the graph F produced by this algorithm, every vertex has degree larger than $(c + \varepsilon/2)(|F| - 1)$. However, it may not be immediately obvious that F has any vertices at all: as we are peeling off the deficient vertices one by one, we are making F smaller and smaller like a kitten unravelling a ball of wool. Will F not disintegrate completely and disappear in the end? The following computation shows that the answer is an emphatic "no": the total number of edges the algorithm removes from the input graph G in the process of constructing the output graph F is at most $\sum_{i=1}^{n} (c + \varepsilon/2) (i-1)$, and so F is left with at least $(\varepsilon/2) \cdot \binom{n}{2}$ edges. Writing $m = |F|$, we conclude that

$$\binom{m}{2} \geq (\varepsilon/2) \cdot \binom{n}{2} \geq (\varepsilon/2) \cdot \binom{n_0}{2},$$

and so $m \geq m_0$ as long as n_0 is large enough with respect to ε and m_0. (Here, "large enough" means $n_0 \geq 1 + (\sqrt{2/\varepsilon})m_0$.)

In Stage 2, we distinguish between two cases. In case $r = 2$ (the induction basis), insisting on $m_0 \geq t$ is enough to guarantee that F contains a set of t vertices. In case $r > 2$ (the induction step), insisting on

$$m_0 \geq n_0(r - 1, t, 1/(r - 1)(r - 2))$$

is enough to guarantee that F contains a $K_{r-1}(t)$: to see this, note that F has more than $c\binom{m}{2}$ edges and that

$$c = \left(1 - \frac{1}{r - 2}\right) + \frac{1}{(r - 1)(r - 2)}.$$

In Stage 3, let L denote the set of vertices of F that lie outside K and have at least $(r - 2)t + s$ neighbours in K. To get a lower bound on $|L|$, we will estimate in two different ways the number x of edges of F that have one endpoint in K and the other endpoint outside K. Since every vertex in K has more than $(c+\varepsilon/2)(m-1)$ neighbours in F, it has more than $(c + \varepsilon/2)(m - 1) - (|K| - 1)$ neighbours outside K, and so

$$x > |K| \cdot ((c + \varepsilon/2)(m - 1) - (|K| - 1)) > |K| \cdot ((c + \varepsilon/2)m - |K|).$$

Since every vertex in L has at most $|K|$ neighbours in K and every vertex outside $K \cup L$ has fewer than $(r - 2)t + s$ neighbours in K, we have

$$x \leq |L| \cdot |K| + |F - (K \cup L)|((r - 2)t + s) \leq |L| \cdot |K| + m((r - 2)t + s).$$

Comparing the upper bound on x with the lower bound gives

$$|L| \cdot |K| + m((r - 2)t + s) > |K|((c + \varepsilon/2)m - |K|),$$

and so

$$|L| > (c + \varepsilon/2)m - |K| - \frac{m((r - 2)t + s)}{|K|} = m\left(\frac{\varepsilon}{2} - \frac{s}{(r - 1)t}\right) - (r - 1)t.$$

As long as t is large enough to guarantee $s/(r - 1)t < \varepsilon/8$ and m is large enough to guarantee that $m\varepsilon/8 > (r - 1)t$, we can conclude that $|L| > m\varepsilon/4$. To complete the proof, note that $m\varepsilon/4 > (s-1)\binom{t}{s}^{r-1}$ as long as m is large enough with respect to r, s, ε, and t. □

Let $s(r, \varepsilon, n)$ stand for the largest nonnegative integer s such that every graph of order n with at least

$$\left(1 - \frac{1}{r-1} + \varepsilon\right)\binom{n}{2}$$

edges contains a $K_r(s)$. With this notation, the Erdős–Stone theorem shows that

$$s(r, \varepsilon, n) \to \infty \quad \text{as} \quad n \to \infty.$$

The best known bounds on $s(r, \varepsilon, n)$ are

$$(1 - \delta)\frac{\log n}{\log(1/\varepsilon)} < s(r, \varepsilon, n) < (2 + \delta)\frac{\log n}{\log(1/\varepsilon)}$$

(see [221]); for every positive δ, they hold whenever ε is small enough with respect to δ, r and n is large enough with respect to δ, r, ε.

7.3 The Erdős–Stone–Simonovits Formula

DEFINITION: The *chromatic number* $\chi(F)$ of a graph F is the smallest number of colours that can be assigned to the vertices of F in such a way that every two adjacent vertices receive distinct colours. (This definition is consistent with the definition of the chromatic number of a hypergraph given in Section 5.5.) Equivalently, $\chi(F)$ is the smallest r such that F is a subgraph of some complete r-partite graph. Graphs F with $\chi(F) \leq 2$ are called *bipartite*.

Paul Erdős and Miklós Simonovits pointed out a fundamental corollary of the Erdős–Stone theorem:

COROLLARY 7.6 (Erdős and Simonovits [152]) *For every graph F with at least one edge, we have*

$$\lim_{n \to \infty} \frac{ex(F, n)}{\binom{n}{2}} = 1 - \frac{1}{\chi(F) - 1}. \tag{7.4}$$

Proof Writing $r = \chi(F)$, $s = |F|$, we claim that

$$\left(1 - \frac{1}{r-1}\right)\binom{n-r+2}{2} \leq ex(F, n) \leq ex(K_r(s), n); \tag{7.5}$$

formula (7.4) follows from (7.5) combined with Theorem 7.4. To justify the lower bound on $ex(F, n)$ in (7.5), observe that it is a lower bound on the number of edges in the complete $(r-1)$-partite graph with parts of size $\lfloor n/(r-1) \rfloor$ and that no complete $(r-1)$-partite graph has a subgraph isomorphic to F. The upper bound is justified by observing that F is a subgraph of $K_r(s)$. □

REMINDER: When f and g are real-valued functions defined on positive integers, we write $f(n) \sim g(n)$ to mean that $\lim_{n \to \infty} f(n)/g(n) = 1$.

The Erdős–Stone–Simonovits formula (7.4) shows that

$$\text{ex}(F, n) ~\sim~ \left(1 - \frac{1}{\chi(F) - 1}\right) \binom{n}{2} \quad \text{whenever } \chi(F) \geq 3.$$

7.4   When F Is Bipartite

REMINDER: When f and g are real-valued functions defined on positive integers, we write $f(n) = o(g(n))$ to mean that $\lim_{n \to \infty} f(n)/g(n) = 0$.

When $\chi(F) = 2$, the Erdős–Stone–Simonovits formula (7.4) provides no asymptotic formula for the Turán function $\text{ex}(F, n)$: it shows only that

$$\text{ex}(F, n) ~=~ o(n^2).$$

7.4.1   An Erdős–Simonovits conjecture

Erdős and Simonovits [117, p. 119] conjectured that a simple asymptotic formula for $\text{ex}(F, n)$ exists even if $\chi(F) = 2$:

CONJECTURE 7.7   *For every bipartite graph F there are constants c and α such that $1 \leq \alpha < 2$ and*

$$\text{ex}(F, n) \sim cn^{\alpha}. \tag{7.6}$$

In [127, page 6], Erdős offered \$500 for a proof or disproof.

DEFINITIONS: A *path of length $k - 1$ between vertices u and v* is a string $w_1 w_2 \ldots w_k$ of pairwise distinct vertices such that $w_1 = u$, $w_k = v$, and each w_i with $i = 1, 2, \ldots k - 1$ is adjacent to w_{i+1}. If, in addition, w_k is adjacent to w_1, then the string $w_1 w_2 \ldots w_k w_1$ is a *cycle of length k*. A path between vertices u and v is said to *join* u and v.

Sometimes we abuse this notation a little: The *path of length $k - 1$*, denoted by P_k, may also mean the graph with vertices $w_1, \ldots w_k$ and edges $w_1 w_2, \ldots w_{k-1} w_k$. Similarly, the *cycle of length k*, denoted by C_k, may also mean the graph with vertices $w_1, \ldots w_k$ and edges $w_1 w_2, \ldots w_{k-1} w_k, w_k w_1$.

A graph is *connected* if, and only if, for every two of its vertices, u and v, there is a path (of any length) between u and v; otherwise the graph is *disconnected*. A *tree* is a connected graph which contains no cycle. A *star* is a tree where one vertex is adjacent to all the remaining vertices.

Conjecture 7.7 is known to hold true for certain special choices of F. Let us elaborate.

Erdős and Sós conjectured that every tree T of order k satisfies

$$\text{ex}(T, n) \sim \frac{k - 2}{2} n. \tag{7.7}$$

More precisely, the Erdős–Sós conjecture [113, p. 30] is:

CONJECTURE 7.8 *Every tree T of order k satisfies*

$$ex(T, n) \le \frac{k-2}{2} n. \tag{7.8}$$

Since every tree T of order k satisfies $ex(T, n) \ge (k-2)n/2$ whenever n is a multiple of $k - 1$ (to see this, consider the disjoint union of complete graphs of order $k - 1$), inequality (7.8) implies (7.7).

If T is the star of order k, then clearly $ex(T, n) = \lfloor (k-2)n/2 \rfloor$, which implies (7.8). An old result of Erdős and Gallai [137, Theorem (2.6)] asserts that the path of order k also satisfies (7.8) in place of T. For additional classes of trees which satisfy the Erdős–Sós conjecture, see [103, 162, 360]. Miklós Ajtai, János Komlós, Miklós Simonovits, and Endre Szemerédi announced that they had proved (7.8) for all trees T of order k and all n sufficiently large with respect to k.

NOTATION: $K_{r,s}$ stands for the complete bipartite graph with r vertices in one part and s vertices in the other part.

Another class of graphs known to satisfy the Erdős–Simonovits conjecture are particular complete bipartite graphs: Zoltán Füredi [178] proved that

$$ex(K_{r,2}, n) \sim \frac{\sqrt{r-1}}{2} n^{3/2} \text{ whenever } r \ge 2.$$

(The special case of $r = 2$ had been established three decades earlier by William Brown [59] and, independently and simultaneously, by Erdős, Alfréd Rényi (1921–70),[b] and Vera Sós [151].) In addition,

$$ex(K_{3,3}, n) \sim \frac{1}{2} n^{5/3}$$

has been established by Brown's lower bound in [59] and Füredi's matching upper bound in [177].

7.4.2 When *F* Is a Complete Bipartite Graph

Turán wrote his seminal paper [362] in Hungarian. Thirteen years later, he reproduced the theorem and its proof in a paper [363] written in English and there he put them in a broader context. At the same time, he co-authored with Tamás Kővári and Vera Sós another classic:

THEOREM 7.9 (Kővári, Sós, and Turán [266])

$$ex(K_{r,s}, n) \le \tfrac{1}{2}(r-1)^{1/s} n^{2-1/s} + \tfrac{1}{2}(s-1)n. \tag{7.9}$$

Proof Given a graph G of order n that contains no $K_{r,s}$, we aim to prove that the number m of its edges is at most the right-hand side of (7.9). For this purpose, we may assume that

$$m > (s-1)n/2 : \tag{7.10}$$

---

[b] Rényi was the author of the saying *A mathematician is a machine for turning coffee into theorems* that is often attributed to Erdős.

otherwise we are done. Under this assumption, consider the set $\mathbf{P}$ of all pairs (v, S) such that v is a vertex of G and S is a set of s neighbours of v. Since each v participates in precisely $\binom{d(v)}{s}$ such pairs, we have $|\mathbf{P}| = \sum_v \binom{d(v)}{s}$; since $\sum_v d(v) = 2m$, it follows from (7.10) and Lemma A.1 that

$$|\mathbf{P}| \geq n\binom{2m/n}{s}.$$

Since G contains no $K_{r,s}$, each S participates in at most $r - 1$ pairs in $\mathbf{P}$, and so

$$|\mathbf{P}| \leq \binom{n}{s}(r - 1).$$

Comparing the two bounds on $|\mathbf{P}|$, we find that

$$n\binom{2m/n}{s} \leq \binom{n}{s}(r - 1). \tag{7.11}$$

Since

$$\frac{\binom{2m/n}{s}}{\binom{n}{s}} \geq \left(\frac{\frac{2m}{n} - (s - 1)}{n}\right)^s,$$

inequality (7.11) implies that m is at most the right-hand side of (7.9).   □

REMINDER: When f and g are nonnegative real-valued functions defined on positive integers, we write $f(n) = O(g(n))$ to mean that $f(n) \leq cg(n)$ for some constant c and all sufficiently large n; we write $f(n) = \Omega(g(n))$ to mean that $f(n) \geq cg(n)$ for some positive constant c and all sufficiently large n.

For every choice of r and s, the Kővári–Sós–Turán Theorem 7.9 shows that

$$\mathrm{ex}(K_{r,s}, n) = O(n^{2-1/s}). \tag{7.12}$$

As we have noted at the end of subsection 7.4.1, this upper bound is tight when $r \geq s = 2$ and (since $\mathrm{ex}(K_{r,3}, n) \geq \mathrm{ex}(K_{3,3}, n)$ whenever $r \geq 3$) when $r \geq s = 3$. János Kollár, Lajos Rónyai, and Tibor Szabó proved that (7.12) is tight whenever r is large enough compared to s:

$$r > (s - 1)! \implies \mathrm{ex}(K_{r,s}, n) = \Omega(n^{2-1/s}) \tag{7.13}$$

(see the "Note added in proof" at the end of [256] and also [6]).

However, the best known lower bound on $\mathrm{ex}(K_{r,r}, n)$ with $r \geq 4$ is only

$$\mathrm{ex}(K_{r,r}, n) = \Omega(n^{2-2/(r+1)}),$$

which does not match the upper bound (7.12). It constitutes a special case of a more general lower bound:

THEOREM 7.10 (Erdős and Joel Spencer [153])   *If F is a graph of order s with t edges such that $t \geq 2$, then*

$$\mathrm{ex}(F, n) = \Omega(n^{2-(s-2)/(t-1)}).$$

Proof Given positive integers n and m, let $\mathbf{G}$ denote the set of all graphs with vertices $1, 2, \ldots, n$ and with m edges. Let $\mathbf{P}$ denote the set of all pairs (G, H) such that $G \in \mathbf{G}$ and H is a subgraph of G isomorphic to F. The number of one-to-one mappings from the vertex set of F to $\{1, 2, \ldots, n\}$ is $n(n-1) \cdots (n-s+1)$, and so at most n^s graphs with vertices coming from $\{1, 2, \ldots, n\}$ are isomorphic to F. Since each such graph participates in

$$\binom{\binom{n}{2} - t}{m - t}$$

pairs in $\mathbf{P}$, we have

$$|\mathbf{P}| \leq n^s \binom{\binom{n}{2} - t}{m - t}.$$

Since

$$|\mathbf{G}| = \binom{\binom{n}{2}}{m},$$

it follows that some G in $\mathbf{G}$ contains at most M subgraphs isomorphic to F, where

$$M = n^s \frac{\binom{\binom{n}{2} - t}{m - t}}{\binom{\binom{n}{2}}{m}} = n^s \frac{\binom{m}{t}}{\binom{\binom{n}{2}}{t}} \leq n^s \left(\frac{2m}{n^2}\right)^t.$$

Setting $m = \lfloor \frac{1}{8} n^{2-(s-2)/(t-1)} \rfloor$, we get $n^{s-2t}(2m)^{t-1} \leq (1/4)^{t-1}$, and so $M \leq m/2$. Removing an edge from each subgraph of G isomorphic to F, we get a graph of order n with at least $m/2$ edges which contains no subgraph isomorphic to F. $\qquad\square$

7.4.3 When Every Subgraph of F Has a Vertex of Degree at Most r

Erdős [117, p. 120] conjectured that (7.12) can be generalized:

CONJECTURE 7.11 *If F is a bipartite graph and if every subgraph of F has a vertex of degree at most s, then*

$$ex(F, n) = O(n^{2-1/s}).$$

In [134, p. 68] he attributed this conjecture as well as the following companion conjecture jointly to Simonovits and himself.

CONJECTURE 7.12 *If F is a bipartite graph with minimum degree greater than s, then*

$$ex(F, n) = \Omega(n^{2+\varepsilon-1/s})$$

For a proof or disproof of each of these conjectures he offered $500. Regarding the special case $s = 2$, he had written earlier [125, p. 14]

> Simonovits and I asked: Is it true that [Conjecture 7.11 with $s = 2$ holds true]?
> We now expect that [this] is false, but can prove nothing.

and [129, pp. 64–65]:

> I state some of our favourite conjectures with Simonovits [...] Our conjecture
> (perhaps more modestly it should be called a guess) is that $ex(F, n) = O(n^{3/2})$
> holds if any only if F is bipartite and has no subgraph each vertex of which
> has degree greater than 2. Unfortunately we could neither prove the neces-
> sity nor the sufficiency of this attractive, illuminating (but perhaps misleading)
> conjecture.

Noga Alon, Michael Krivelevich, and Benny Sudakov [5] proved the existence of a
positive constant c such that

$$ex(F, n) = O(n^{2-c/s})$$

for every bipartite graph F such that every subgraph of F has a vertex of degree at
most s.

7.4.4   When F Is a Cycle

In [114, p. 33], Erdős wrote

> I can also prove that $[ex(C_{2k}, n) = O(n^{1+1/k})]$.

and later [119, p. 78] he commented

> I never published a proof of $[ex(C_{2k}, n) = O(n^{1+1/k})]$ since my proof was
> messy and perhaps even not quite accurate and I lacked the incentive to fix
> everything up since I never could settle various related sharper conjectures –
> all these have now been proved by Bondy and Simonovits – their paper will
> soon appear.

The first published proof of this upper bound does indeed come from Adrian
Bondy and Miklós Simonovits [49]; the best currently known upper bound is Oleg
Pikhurko's [312]

$$ex(C_{2k}, n) \leq (k-1)n^{1+1/k} + 16(k-1)n.$$

Erdős and Simonovits [180, Conjecture 4.10] conjectured that $n^{1+1/k}$ is the order
of magnitude of $ex(C_{2k}, n)$ for every constant k:

CONJECTURE 7.13 $ex(C_{2k}, n) = \Omega(n^{1+1/k})$.

This conjecture is known to hold true when $k = 2$ (as we have already seen), when
$k = 3$, and when $k = 5$ (these last two lower bounds have been established by Clark
Benson [25] in a slightly different setting). When k is arbitrary, Theorem 7.10 gives

$ex(C_{2k}, n) = \Omega(n^{1+1/(2k-1)})$, and the best currently known lower bound comes from Felix Lazebnik, Vasiliy Ustimenko, and Andrew Woldar [272]:

$$ex(C_{2k}, n) = \Omega(n^{1+2/(3k-2)}).$$

Rather than excluding cycles of a single prescribed length, one may consider excluding cycles of all lengths up to a prescribed limit. Such considerations lead to a generalization of the notion of $ex(F, n)$: when $\mathcal{F}$ is a family of graphs, $ex(\mathcal{F}, n)$ denotes the largest number of edges in a graph of order n that contains no member of $\mathcal{F}$. In particular, $ex(\{C_3, C_4, \ldots, C_\ell\}, n)$ is the largest number of edges in a graph of order n where every cycle has length at least $\ell + 1$. Noga Alon, Shlomo Hoory, and Nathan Linial [4] proved that

$$ex(\{C_3, C_4, \ldots, C_{2k}\}, n) \leq \frac{1}{2}n^{1+1/k} + \frac{1}{2}n.$$

More on Turán functions $ex(F, n)$ where F is a graph and $ex(\mathcal{F}, n)$ where $\mathcal{F}$ is a family of graphs can be found, for instance, in [176], [47], and [180].

7.5   Prehistory

Here is an excerpt (with notation changed for the sake of consistency with the remainder of this chapter) from Erdős's paper [123]:

> As is well known, the theory of extremal graphs really started when Turán determined $ex(K_r, n)$ and raised several problems which showed the way to further progress. In 1935 I needed (the c's will denote positive absolute constants)
>
> $$ex(C_4, n) < c_1 n^{3/2} \qquad (7.14)$$
>
> for the following number theoretic problem [...] I proved (7.14) without much difficulty [...] I asked if (7.14) is best possible and Miss E. Klein proved
>
> $$ex(C_4, n) > c_2 n^{3/2}$$
>
> for every $c_2 > 2^{-3/2}$ and $n > n_0(c_2)$. Being struck by a curious blindness and lack of imagination, I did not at that time extend the problem from C_4 to other graphs and thus missed founding an interesting and fruitful new branch of graph theory.

When Erdős reminisced about this episode in his lectures, he liked to add [343, pp. 153–154]

> Crookes observed that leaving a photosensitive film near a cathode-ray tube causes damage to the film· it becomes exposed. He concluded that nobody should leave films near a cathode-ray tube. Röntgen observed the same phenomenon a few years later and concluded that this can be used for filming the inside of various objects. [...] It is not enough to be in the right place at the right time. You should also have an open mind at the right time.

7.6   Beyond Turán Functions

The term "extremal graph theory" denotes a wide area of results and questions where a graph parameter is maximized or minimized subject to other parameters being constrained. Here are two examples:

THEOREM 7.14 (Corollary of Theorem $1'$ in [144][c]) *Let G be a graph of order n with m edges. If k is a positive integer such that $n > 24k$ and $m \geq (2k-1)n - 2k^2 + k + 1$, then G contains k pairwise vertex-disjoint cycles.*   □

The lower bound on m in this theorem cannot be reduced. To see this, consider the graph with $2k-1$ vertices of degree $n-1$ and $n-2k+1$ vertices of degree $2k-1$.

CONJECTURE 7.15 (Case $k = 2$ of Conjecture 2 in [130]) *Every triangle-free graph of order n can be made bipartite by deletion of at most $n^2/25$ edges.*

The constant $1/25$ in this conjecture cannot be reduced. To see this, consider the graph whose vertex-set is the union of pairwise disjoint sets V_1, V_2, V_3, V_4, V_5 of equal size, where a vertex in V_i is adjacent to a vertex in V_j if and only if $|i - j|$ is 1 or 4. A weaker version of Conjecture 7.15 with the constant $1/25$ raised to $1/18$ has been proved by Erdős, Ralph Faudree (1940–2015), János Pach, and Joel Spencer [136, Theorem 2].

Several survey articles, book chapters, and entire books are devoted to extremal graph theory. These include [39], [343], [189, Chapter 10], [41, Chapter IV], [35].

The term 'anarchist' is subject to such misinterpretations (one widespread cliché holds that an anarchist is someone who aims to introduce chaos and disorder into society by violent means) that calling Erdős an anarchist could provoke a controversy. He was an anarchist. He was one in the noblest sense of the term. He lived by Louis-Auguste Blanqui's 'ni Dieu ni maître'. He acknowledged his own maxim 'property is a nuisance' as a variation on Pierre-Joseph Proudhon's theme. He accepted an invitation to speak at the 1954 International Congress of Mathematicians in Amsterdam, even though he was denied the U.S. re-entry permit: as he put it later, he "chose freedom and left the United States." The final words of his comment on this episode were "You don't let the government push you around." Blanqui himself could endorse them.

Like many fine anarchists, he was the antithesis of a snob. On one of his visits to Stanford, I proposed to him that he stay with me and my girlfriend Mari Eckstein in our one-bedroom apartment where an espresso maker was in full working order. When he accepted, Mari and I popped over to the housing office, rented a rollaway bed, and rolled it all the way to Hoskins Court. And so it came to pass that the PGOM crashed (to use the idiom of those days) in our living room just like any regular hippie would. In the first night of this arrangement, I woke up at 3AM with a start and felt a presence next to our bed. "Chvátal? Is there any orange juice?" asked the presence.

[c] The contribution that Lajos Pósa, aged thirteen at the time, made to Theorem 7.14 was an ingenious proof (which was then generalized) that every graph of order n with $3n - 5$ edges contains two vertex-disjoint cycles if $n \geq 6$: see [116, p. 4].

As the days passed, Christmas drew nearer and pulled us into a quandary. We wanted to demonstrate our affection to E.P., but what kind of a Christmas gift can you offer to a man who proclaims property to be a nuisance? Eventually, Mari hit on a brilliant idea. Contrary to defamatory reports, Erdős did not love only numbers. He could get passionate about politics and this interest of his got only nurtured by his continual travel through different countries with different systems of government. Let us get him a portable radio, said Mari, so that he can keep up with current events wherever he goes. Erdős liked his present and I offered a silent apology to his many future hosts whose sleep would be interrupted by medleys of news and classical music just like ours was for the next few nights.

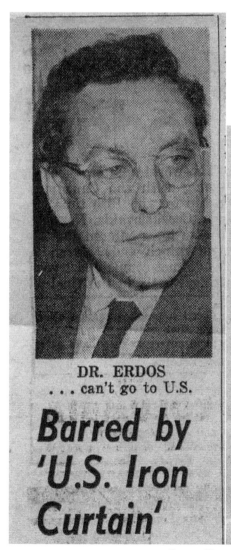

DR. ERDOS
. . . can't go to U.S.

Barred by 'U.S. Iron Curtain'

One of the world's leading mathematicians claimed in Vancouver that he has been barred from the U.S. by an "American Iron Curtain."

Dr. Paul Erdos, 45-year-old Hungarian citizen, lived in the U.S. from 1938 until he went to a mathematical conference in Amsterdam in 1954.

He said he has since been refused a visa to re-enter the U.S. as a returning resident.

Now lecturing at University of B.C. for six weeks, Dr. Erdos said he could not get a re-entry permit before he went to Amsterdam.

"I suppose I was impatient because I went anyway," he said. "I don't like that kind of thing, whether it comes from the Russians or the United States."

He said there was no justifiable reasons why he should be barred from the U.S., and he was given none.

From a Vancouver paper
Courtesy of János Pach

8   The Friendship Theorem

In the 1970s, Canadian category theorists tended to be Maoist. At one of their parties in Montreal, I overheard talk about "reactionary mathematics" and "revolutionary mathematics." I asked for the meaning of these terms and was told that reactionary mathematics obscures its truths by unintelligible exposition, whereas revolutionary mathematics makes its arguments accessible to the masses. This clarification made perfect sense to me. The fact that it came from someone inclined to explain the binomial theorem in the language of fiber bundles and diffeomorphisms of Banach analytic Lie groups only added to its charm.

In this frame of reference, it could be argued that Ryser's proof of Theorem 2.10 is a wee bit on the reactionary side: it relies on the fact that no more than m vectors in $\mathbf{R}^m$ can be linearly independent, a fact that the masses may be unacquainted with. By contrast, the original proof by de Bruijn and Erdős is undoubtedly revolutionary as it proceeds from scratch. We are going to illustrate the dichotomy on another example from Erdős's work.

8.1   The Friendship Theorem

Erdős, Rényi, and Sós proved the following theorem ([151], Theorem 6):

THEOREM 8.1   *If, in a finite graph G, every two vertices have precisely one common neighbour, then some vertex of G is adjacent to all the vertices of G except itself.*

A graph that satisfies the assumption of this theorem is shown in Figure 8.1. Theorem 8.1 is sometimes presented in the form

If, in a group of people, every two people have precisely one common friend, then the group includes a politician, someone who is a friend of everybody.

For this reason, it is known as the *Friendship Theorem*.

Proof of Theorem 8.1   Let G satisfy the hypothesis of the theorem and let V denote the set of vertices of G.

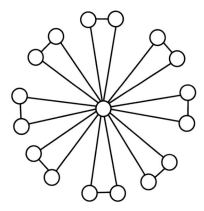

Figure 8.1 Every two vertices have precisely one common neighbour.

DEFINITIONS: We let $N(v)$ denote the set of all neighbours of vertex v; we shall write $d(v) = |N(v)|$ and refer to $d(v)$ as the *degree* of v. A graph where all vertices have the same degree is called *regular*.

CASE 1: *G is not regular.* By assumption of this case, G includes vertices x and y such that $d(x) < d(y)$. Partition V into disjoint sets S, L with $x \in S, y \in L$ by writing

$$S = \{w \in V : d(w) \le d(x)\} \quad \text{and} \quad L = \{w \in V : d(w) > d(x)\}.$$

We claim that

$$\text{all vertices in } S \text{ are adjacent to all vertices in } L. \tag{8.1}$$

To justify this claim, consider arbitrary nonadjacent vertices v, w such that $w \in L$: we will show that $d(v) \ge d(w)$, and so $v \in L$. Every neighbour u of w is distinct from v, and so it shares a unique neighbour $f(u)$ with v; since two distinct neighbours of w share no neighbour other than w, the mapping $f : N(w) \to N(v)$ is one-to-one; it follows that $d(v) \ge d(w)$.

Since $x \in S$, we have $d(x) \ge |L|$ by (8.1). Since $y \in L$, it follows that

$$|L| \le d(x) < d(y) \le |V| - 1,$$

and so $|S| = |V| - |L| > 1$, which implies that S includes a vertex distinct from x. Let x' denote this vertex. Since x and x' have precisely one common neighbour, (8.1) implies that $|L| < 2$, and so $L = \{y\}$. Now, by (8.1) again, y is adjacent to all the vertices of G except itself.

CASE 2: *G is regular.* Let d denote the common degree of vertices of G and let n denote the order of G. How many ordered triples x, y, z of vertices of G are there such that y and z are distinct neighbours of x? Choosing x first gives the answer $nd(d-1)$ and choosing x last gives the answer $n(n-1)$. Since the two answers must be equal, we conclude that

$$n = d(d-1) + 1. \tag{8.2}$$

SUBCASE 2.1: $d \leq 2$. Only two d-regular graphs G of order n such that $d \leq 2$ satisfy (8.2): the graph that consists of a single vertex and the graph that consists of three pairwise adjacent vertices. In either of these two graphs, every vertex is adjacent to all the other vertices (except itself).

SUBCASE 2.2: $d \geq 3$. We will show that this subcase cannot occur. For this purpose, note that G induces a projective plane of order $d - 1$: the vertices of G and the sets $N(v)$ with $v \in V$ have the properties of points and lines, respectively, that are specified on page 27. The mapping $v \mapsto N(v)$ preserves incidence in the sense that a point p belongs to a line L if and only if the line that p maps to includes the point that maps to L. Every such bijection between the set of points and the set of lines of a projective plane is called a *polarity* of this plane.

Reinhold Baer (1902–1979) proved [14] that

> *every polarity in a finite projective plane*
> *maps some point to a line that contains this point;*

since $v \notin N(v)$ for all v, the plane induced by G contradicts Baer's theorem.   □

How would our Maoist colleagues classify this proof? If "revolutionary" implies "self-contained," then the proof is downright reactionary. Its easy arguments dealing with Case 1 and Subcase 2.1 reduce the job to the task of proving

LEMMA 8.2   *Let G be a finite undirected regular graph of degree d. If every two vertices of G have precisely one common neighbour, then $d \leq 2$,*

which is dismissed by the reference to Baer's theorem. To make the proof revolutionary, one may extract from Baer's paper that part of his argument which suffices to prove Lemma 8.2. This is precisely what Judith Longyear (1938–95) and Tory Parsons (1941–87) seem to have done:

Proof of Lemma 8.2 from [276].

DEFINITIONS: A *walk of length k* in G is a sequence $w_0, w_1, \ldots, w_k$ of (not necessarily distinct) vertices such that each w_i with $0 \leq i < k$ is adjacent to w_{i+1}. The walk is called *open* if $w_k \neq w_0$ and it is called *closed* if $w_k = w_0$.

Let n denote the order of G and let us investigate the number c_k of closed walks of length k when k is an integer greater than 1. The number of closed walks $w_0, w_1, \ldots, w_{k-2}, w_{k-1}, w_k$ with $w_{k-2} \neq w_k$ equals $nd^{k-2} - c_{k-2}$: there are this many ways of choosing the open walk $w_0, w_1, \ldots, w_{k-2}$ and then w_{k-1} is uniquely determined. The number of closed walks $w_0, w_1, \ldots, w_{k-2}, w_{k-1}, w_k$ with $w_{k-2} = w_k$ equals $c_{k-2} \cdot d$: there are c_{k-2} ways of choosing the closed walk $w_0, w_1, \ldots, w_{k-2}$ and then there are d ways of choosing w_{k-1}. We conclude that

$$c_k = nd^{k-2} + (d - 1)c_{k-2} \quad \text{whenever } k \geq 2. \tag{8.3}$$

From (8.3) and (8.2), we find that

$$c_k \bmod (d-1) = 1 \quad \text{whenever } k \geq 2,$$

and so

$$c_k \bmod p = 1 \quad \text{whenever } k \geq 2 \text{ and } p \text{ divides } d-1;$$

in particular,

$$c_p \bmod p = 1 \quad \text{whenever } p \text{ is a prime divisor of } d-1. \tag{8.4}$$

We will complete the proof of the lemma by showing that

$$\text{every prime } p \text{ divides } c_p: \tag{8.5}$$

conjunction of (8.5) and (8.4) implies that $d-1$ has no prime divisors, and so $d \leq 2$.

A pivotal notion in the proof of (8.5) is that of rotating a closed walk $w_0, w_1,$ $\ldots w_{k-1}, w_0$, which means replacing it by $w_1, w_2, \ldots w_k, w_1$. Rotating the walk $k-1$ times, we obtain a number of closed walks; setting $w_{k+t} = w_t$ for all $t = 1, 2, \ldots, k$, we can record them neatly as

$$
\begin{array}{lll}
w_0, w_1, w_2, \ldots & & w_{k-1}, w_k \\
w_1, w_2, w_3, \ldots & & w_k, w_{k+1} \\
w_2, w_3, w_4, \ldots & & w_{k+1}, w_{k+2} \\
\cdots & & \\
w_{k-1}, w_k, w_{k+1}, \ldots & w_{2k-2}, w_{2k-1}.
\end{array} \tag{8.6}
$$

These walks may not be pairwise distinct: for example, five rotations of closed walk a, b, c, a, b, c, a produce the list

$$
\begin{array}{l}
a, b, c, a, b, c, a \\
b, c, a, b, c, a, b \\
c, a, b, c, a, b, c \\
a, b, c, a, b, c, a \\
b, c, a, b, c, a, b \\
c, a, b, c, a, b, c.
\end{array}
$$

Nevertheless, and this is crucial for the proof of (8.5),

$$\text{if } k \text{ is a prime, then list (8.6) } \textit{consists of pairwise distinct walks}. \tag{8.7}$$

To prove (8.7), consider pairs i and j such that

$$0 \leq i < j \leq k \tag{8.8}$$

and

$$\text{walks } w_i, w_{i+1}, \ldots w_{i+k} \text{ and } w_j, w_{j+1}, \ldots w_{j+k} \text{ are identical.} \tag{8.9}$$

One such pair is $i = 0, j = k$; among all pairs with properties (8.8) and (8.9), choose one that minimizes $j - i$ and set $\Delta = j - i$. If $\Delta = k$, then $i = 0, j = k$ is the only pair with properties (8.8) and (8.9), and so list (8.6) consists of pairwise distinct walks; assuming that $\Delta < k$, we will prove that k is composite.

Property (8.9) says that

walks $w_i, w_{i+1}, \ldots w_{i+k}$ and $w_{i+\Delta}, w_{i+1+\Delta}, \ldots w_{i+k+\Delta}$ are identical;

since vertices w_{i+t} with t ranging through $0, 1, \ldots, k-1$ range through $w_0, w_1, \ldots w_{k-1}$, it follows that

$$w_r = w_{r+\Delta} \quad \text{for } r = 0, 1, \ldots, k-1. \tag{8.10}$$

In turn, this implies that all of the walks

$$w_0, w_1, w_2, \ldots \qquad\qquad w_{k-1}, w_k$$
$$w_\Delta, w_{\Delta+1}, w_{\Delta+2}, \ldots \qquad w_{\Delta+k-1}, w_{\Delta+k}$$
$$w_{2\Delta}, w_{2\Delta+1}, w_{2\Delta+2}, \ldots \quad w_{2\Delta+k-1}, w_{2\Delta+k}$$
$$\cdots$$
$$w_{m\Delta}, w_{m\Delta+1}, w_{m\Delta+2}, \ldots \quad w_{m\Delta+k-1}, w_{m\Delta+k}$$

with $m = \lfloor k/\Delta \rfloor$ are identical. By definition, $k - m\Delta < \Delta$; since $m\Delta$ in place of i and k in place of j satisfy (8.9), minimality of Δ forces them to fail (8.8) in place of i and j, and so we have $m\Delta \geq k$. Since $m \leq k/\Delta$ by definition, it follows that $m = k/\Delta$, and so Δ divides k. We have $\Delta < k$ by assumption; to conclude that k is composite, observe that (8.10) guarantees $\Delta \neq 1$: since w_0 and w_1 are adjacent, they are distinct.

This concludes the proof of (8.7). To deduce (8.5), note first that every closed walk of length k belongs to precisely one of the lists (8.6); if k is a prime, then each of these lists has size k by virtue of (8.7), and so the sum of their sizes is a multiple of k; of course, this sum is c_k. □

REMINDER: $G \oplus H$ denotes the *direct sum* of G and H, which is the graph consisting of a copy of G and a copy of H that have no vertices in common; $G - H$ denotes the *join* of G and H, which is $G \oplus H$ with additional edges that join every vertex in the copy of G to every vertex in the copy of H.

With tG standing for the direct sum of t copies of G, the graph in Figure 8.1 can be recorded as $K_1 - 8K_2$. An easy corollary of Theorem 8.1 goes as follows:

COROLLARY 8.3   *In a finite graph G, every two vertices have precisely one common neighbour if and only if $G = K_1 - tK_2$ for some t.* □

8.2   Strongly Regular Graphs

In [53], Raj Chandra Bose (1901–87) introduced the following notion.

DEFINITION: A *strongly regular graph with parameters n, d, λ, μ* is an undirected graph G such that

- the order of G is n,
- G is regular of degree d,
- every two adjacent vertices of G have precisely λ common neighbours,
- every two nonadjacent vertices of G have precisely μ common neighbours.

A Necessary Condition for their Existence

Trivially, there exists a strongly regular graph with parameters $n, 0, \lambda, \mu$ if and only if $\mu = 0$ (here, the value of λ is meaningless) and there exists a strongly regular graph with parameters $n, n - 1, \lambda, \mu$ if and only if $\lambda = n - 2$ (here, the value of μ is meaningless). Let us move on to more interesting values of d.

THEOREM 8.4　*If there exists a strongly regular graph with parameters n, d, λ, μ, then*

$$(n - 1 - d)\mu = d(d - 1 - \lambda) \tag{8.11}$$

and, as long as $0 < d < n - 1$, the numbers

$$\frac{1}{2}\left(n - 1 \pm \frac{(n - 1)(\lambda - \mu) + 2d}{\sqrt{(\lambda - \mu)^2 + 4(d - \mu)}}\right) \tag{8.12}$$

are nonnegative integers.

In the special case where $\lambda = \mu = 1$, Theorem 8.4 asserts that $n = d(d - 1) + 1$ and that the numbers

$$\frac{d(d - 1)}{2} \pm \frac{d}{2\sqrt{d - 1}}$$

are nonnegative integers. Since $d(d - 1)/2$ is an integer, it follows that $d/\sqrt{d - 1}$ is an (even) integer; since

$$\frac{d^2}{d - 1} = (d + 1) + \frac{1}{d - 1},$$

it follows further that $d \leq 2$. This is another reactionary proof of Lemma 8.2; it comes from paper [375] of Herbert Wilf (1931–2012).

Proof of Theorem 8.4　Consider an arbitrary strongly regular graph G with parameters n, d, λ, μ. Identity (8.11) follows directly from counting in two different ways all open walks w_0, w_1, w_2 such that w_0 is prescribed and w_2, w_0 are nonadjacent (choosing w_2 first and w_1 second gives the left-hand side; choosing w_1 first and w_2 second gives the right-hand side). Under the assumption that $0 < d < n - 1$, we will prove that the two numbers (8.12) are nonnegative integers.

First, let us dispose of the easy case where G is disconnected. Here, $\mu = 0$, and so (8.11) combined with the assumption that $d > 0$ shows that $\lambda = d - 1$. Now the two numbers (8.12) become

$$n - \frac{n}{d+1} \quad \text{and} \quad -1 + \frac{n}{d+1}.$$

Since $\mu = 0$, the vertex-set of G partitions into pairwise disjoint cliques. Each of these cliques consists of $d+1$ vertices, and so $d+1$ divides n. It follows that the two numbers (8.12) are nonnegative integers.

From now on we may assume that G is connected.

DEFINITION: The *adjacency matrix* A of a graph with vertices $v_1, v_2, \ldots, v_n$ is defined as $A = (a_{ij})$, where

$$a_{ij} = \begin{cases} 1 & \text{if } v_i, v_j \text{ are adjacent,} \\ 0 & \text{if } v_i, v_j \text{ are nonadjacent.} \end{cases}$$

If G is a strongly regular graph with parameters n, d, λ, μ, then its adjacency matrix A has size $n \times n$ and satifies the equation

$$A^2 = dI + \lambda A + \mu(J - I - A)$$

where I is the $n \times n$ identity matrix and J is the $n \times n$ all-ones matrix: the entry in the i-th row and the j-th column of A^2 counts the number of common neighbours of v_i and v_j.

The rest of the argument resorts to the following (not quite revolutionary) tools of linear algebra.

DEFINITIONS: The *trace* of a square matrix is the sum of its diagonal entries. An *eigenvalue* (or *proper value* or *characteristic root* or *latent root*) of a square matrix M is a number r such that $Mx = rx$ for some nonzero vector x.

The *Principal Axis Theorem* states that for every $n \times n$ real symmetric matrix M there are pairwise orthogonal nonzero real vectors $x^1, x^2, \ldots x^n$ and real numbers $r_1, r_2, \ldots r_n$ such that $Mx^i = r_i x^i$ for all $i = 1, 2, \ldots, n$ and such that every eigenvalue of M appears on the list $r_1, r_2, \ldots r_n$.

A related theorem states that $\sum_{i=1}^{n} r_i$ equals the trace of M.

The eigenvalues $r_1, r_2, \ldots r_n$ in the Principal Axis Theorem are not necessarily distinct.

DEFINITION: The number of times an eigenvalue appears on the list $r_1, r_2, \ldots r_n$ of the Principal Axis Theorem is called its *multiplicity*.

We will apply these tools to the adjacency matrix A of G. Every real vector x such that $Ax = dx$ represents an assignment of real weights to the vertices of G such that the weight of each vertex multiplied by d equals the sum of the weights of its neighbours. Since G is regular of degree d, the all-ones vector e satisfies $Ae = de$; since G is connected, every vector x such that $Ax = dx$ must be a multiple of e (consider any vertex with the largest weight and observe that all of its neighbours must also have the largest weight). It follows that d is an eigenvalue of A and that its multiplicity is one. Now let r be any other eigenvalue of A. There is a nonzero vector x orthogonal to e and such that $Ax = rx$. Since $e^T x = 0$, we have $Jx = 0$, and so

$$(A^2 + (\mu - \lambda)A + (\mu - d)I)x = \mu Jx = 0;$$

since $Ax = rx$, we have

$$(A^2 + (\mu - \lambda)A + (\mu - d)I)x = (r^2 + (\mu - \lambda)r + (\mu - d))x;$$

since $x \neq 0$, it follows that $r^2 + (\mu - \lambda)r + (\mu - d) = 0$, and so

$$r = \frac{1}{2}\left((\lambda - \mu) \pm \sqrt{(\lambda - \mu)^2 + 4(d - \mu)} \right).$$

These are two distinct numbers (since $d < n - 1$, we have $\mu \leq d$; since $d > 0$, we have $\lambda \leq d - 1$); when m^+ and m^-, respectively, denote their multiplicities as eigenvalues of A, we have

$$1 + m^+ + m^- = n$$

and, since the trace of A equals zero (in fact, all n diagonal entries of A are zeros),

$$d + \frac{m^+}{2}\left((\lambda - \mu) + \sqrt{(\lambda - \mu)^2 + 4(d - \mu)} \right)$$
$$+ \frac{m^-}{2}\left((\lambda - \mu) - \sqrt{(\lambda - \mu)^2 + 4(d - \mu)} \right) = 0.$$

The solution of this system of two equations in variables m^+ and m^- is the pair of numbers (8.12). $\qquad \square$

Conditions of Theorem 8.4, necessary for the existence of a strongly regular graph with parameters n, d, λ, μ, are not sufficient: for instance, there is no strongly regular graph with parameters $21, 10, 4, 5$. (No easily verifiable necessary and sufficient conditions for the existence of strongly regular graphs with prescribed parameters are known: for instance, it is not known whether there is a strongly regular graph with parameters $65, 32, 15, 16$ or not.) The method used in the proof of Theorem 8.4 can be traced back to paper [87] by W. S. Connor and Willard H. Clatworthy (1915–2010); its special case $\lambda = 0$, $\mu = 1$ was used in [215] by Alan Hoffman (1924–2021) and R. R. Singleton (see Theorem 8.6) and its special case $\lambda = \mu = 1$ was used in [375] by Herbert Wilf (see the remark after Theorem 8.4).

Moore Graphs of Diameter Two

How many vertices can a regular graph of degree d have if every two of its nonadjacent vertices have at least one common neighbour? Consider any such graph; let n denote its order and let $N(w)$ denote the set of neighbours of any of its vertices w. For every vertex u, precisely $n - 1 - d$ vertices are distinct from u and nonadjacent to u; by assumption, each of these $n - 1 - d$ vertices is a neighbour of a neighbour of u, and so

$$n - 1 - d = \left| \bigcup_{v \in N(u)} (N(v) - (\{u\} \cup N(u))) \right|.$$

Since

$$\left| \bigcup_{v \in N(u)} (N(v) - (\{u\} \cup N(u))) \right| \le$$

$$\sum_{v \in N(u)} |N(v) - (\{u\} \cup N(u))| \le d(d-1), \quad (8.13)$$

it follows that $n \le d^2 + 1$. Graphs that attain this bound have a name:

DEFINITION: A *Moore graph of diameter two* is a graph with the following properties:

- it is regular of some degree d such that $d \ge 2$,
- every two of its nonadjacent vertices have at least one common neighbour,
- its order is $d^2 + 1$.

PROPOSITION 8.5   *A graph is a Moore graph of diameter two if and only if it is strongly regular with parameters n, d, λ, μ such that $n = d^2 + 1$, $d \ge 2$, $\lambda = 0$, $\mu = 1$.*

Proof   The "if" part is evident. To prove the "only if" part, consider an arbitrary Moore graph G of diameter two and degree d; let n denote its order and let $N(w)$ denote the set of neighbours of any of its vertices w. Since $n = d^2 + 1$, both inequalities in (8.13) with an arbitrarily chosen vertex u must hold as equations. Since the first inequality in (8.13) holds as equation, the sets $N(v) - (\{u\} \cup N(u))$ with $v \in N(u)$ are pairwise disjoint, and so every vertex nonadjacent to u shares precisely one neighbour with u. Since the second inequality in (8.13) holds as equation, we have $|N(v) - (\{u\} \cup N(u))| = d - 1$ whenever $v \in N(u)$, and so no vertex adjacent to u shares a neighbour with u. Since vertex u can be chosen arbitrarily, it follows that G is a strongly regular graph with $\lambda = 0$, $\mu = 1$.   □

The cycle C_5 is a Moore graph of diameter two and degree 2. The *Petersen graph* shown in Figure 8.2 is a Moore graph of diameter two and degree 3. (In terms introduced on page 135, the Petersen graph is the *Kneser graph $KG_{2,1}$*: its ten vertices are the $\binom{5}{2}$ two-point subsets of a fixed five-point set and two of these vertices are adjacent if and only if, as two-point sets, they are disjoint.)

THEOREM 8.6 (Hoffman and Singleton [215])   *If there exists a Moore graph of diameter two and degree d, then d is one of the numbers $2, 3, 7, 57$.*

Hoffman and Singleton [215] constructed a Moore graph of diameter two and degree 7 (which is now called the *Hoffman-Singleton graph*); existence of a Moore graph of diameter two and degree 57 is unknown.

Proof of Theorem 8.6   As in Theorem 8.4, the numbers $(d^2 \pm \Delta)/2$ with

$$\Delta = \frac{d(d-2)}{\sqrt{4d-3}} \quad (8.14)$$

must be integers; since $(d^2 + \Delta)/2$ is an integer, Δ is an integer. This implies that the numerator $d(d-2)$ in (8.14) is zero or else the denominator $\sqrt{4d-3}$ is rational.

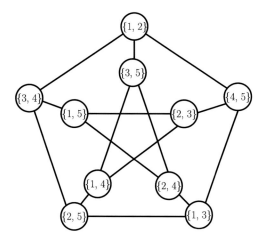

Figure 8.2 The Petersen graph.

If the numerator is zero, then $d = 0$ or $d = 2$; the option $d = 0$ is excluded by the assumption that $d \geq 2$. If the denominator is rational, then (since the square root of a positive integer is rational only if it is an integer) $4d - 3 = s^2$ for some positive integer s. Substituting $(s^2 + 3)/4$ for d in (8.14), we find that

$$s\left(s^3 - 2s - 16\Delta\right) = 15,$$

and so s divides 15; this means that s is one of the numbers $1, 3, 5, 15$, and so d is one of the numbers $1, 3, 7, 57$; the option $d = 1$ is excluded by the assumption that $d \geq 2$. □

Additional information on strongly regular graphs can be found in [58] and elsewhere.

A man who loved only numbers? The title of that book is a clear libel. Anyone who has seen him play table tennis can testify to this. When a ping-pong paddle appeared in his hand, the serenity of his above-the-fray position was all gone and a look of ferocious concentration came over his face. In this context, he did love competition. And he was justifiably proud of the speed of his reflexes.

In the early 1970s, I drove Paul Erdős and Ernst Straus from Waterloo to York University. As we were zipping along the 401, the PGOM took a sudden interest in the dense and fast traffic. "I have never learned to drive, you know. But I think I would be good at it," he announced. "My reflexes are quite quick, you see," he added dreamily. "Perhaps you could stop on the shoulder and let me take over for a while?" I suggested that there might be a good reason why student drivers, their ping-pong prowess notwithstanding, practise first in empty parking lots rather than on busy highways, but he would not have any of such nonsense. It took the combined wiles of Professor Straus and myself (Oh no! Look at the time! I do hope we will make it to York in time for the lecture!) to talk him into abandoning the project.

Courtesy of Gábor Simonyi

Photos by George Csicsery for his film "N is a Number" (1993). All rights reserved

9 Chromatic Number

9.1 The Chromatic Number

In Section 7.3, we defined the chromatic number $\chi(F)$ of a graph F as the smallest number of colours that can be assigned to the vertices of F in such a way that every two adjacent vertices receive distinct colours. This concept sneaked into graph theory through a puzzle with a recreational mathematics flavour, the celebrated *Four Colour Conjecture*. The first written reference to this conjecture occurs in a letter written on 23 October 1852 by Augustus De Morgan (1806–71) to William Rowan Hamilton (1805–65):

> A student of mine asked me today to give him a reason for a fact which I did not know was a fact – and do not yet. He says that if a figure be anyhow divided and the compartments differently coloured so that figures with any portion of common boundary line are differently coloured – four colours may be wanted, but not more – the following is the case in which four colours are wanted. Query cannot a necessity for five or more be invented.

In present-day terms, the Four Colour Conjecture asserts that the chromatic number of every *planar graph* (meaning a graph that can be drawn in the plane with no two edges intersecting except possibly at their endpoints) is at most four. (Equivalently, the Four Colour Conjecture asserts that the answer to De Morgan's query is negative.) Its history from 1852 to 1936 is explored in Chapters 6 and 9 of [30]. The Four Colour Conjecture was proved in 1976 by Kenneth Appel (1932–2013) and Wolfgang Haken with assistance of John Koch [11, 12]; they used hundreds of hours of computer time to work through the details of their proof. A simplified computer-aided proof [328] was found in 1994 by Neil Robertson, Daniel Sanders, Paul Seymour, and Robin Thomas (1962–2020).

The Four Colour Theorem combines the discrete notion of chromatic number with the continuous notion of planarity. Rigorous definition of a planar drawing is not simple: it involves the definition of a continuous mapping from an interval into the plane, which in turn requires a definition of real numbers, and even that starting point is already intricate. A theorem of István Fáry (1922–84) shows that all these complications can be bypassed: it asserts that a graph has a planar drawing if and only if it has a planar drawing with each edge drawn as a straight line segment [164]. We may assume (by shifting each vertex imperceptibly if need be) that all vertices in such drawings have rational coordinates or even (by magnifying the scale of the drawing) integer coordinates, and so planar drawings are reduced to mapping each vertex onto an ordered pair of integers subject to easily verifiable arithmetic conditions.

A celebrated theorem of Kazimierz Kuratowski (1896–1980) provides another characterization of planarity in purely discrete terms.

DEFINITION: *A subdivision of a graph* is any graph obtained from the original graph by repeating the following operation any finite number of times: introduce a new vertex x, choose an edge uv, and replace this edge by the pair of edges ux, xv. (Since the "any finite number" here can mean zero, every graph is a subdivision of itself.)

Kuratowski's theorem [268] asserts that a graph is planar if and only if it contains no subdivision of K_5 and no subdivision of $K_{3,3}$. (One corollary of this theorem is an algorithm that, given a graph G, decides whether or not G is planar. John Hopcroft and Robert Tarjan [219] designed an algorithm for doing this in time linear in the order of G; for more recent and simpler algorithms of this kind, see [55].)

In a sharp contrast with these mixed and tangled origins of χ, the Erdős–Stone–Simonovits formula

$$\lim_{n \to \infty} \frac{\mathrm{ex}(F,n)}{\binom{n}{2}} = 1 - \frac{1}{\chi(F) - 1}$$

shines with a stark simplicity and puts χ directly in the center of extremal graph theory.

9.2   The Unbearable Weakness of the Bound $\chi \geq \omega$

Every graph G of order n satisfies

$$\chi(G) \geq \omega(G) \tag{9.1}$$

since every two vertices in a clique must get distinct colours. In this section, we will show (Corollary 9.2) that this inequality is next to useless for most graphs.

Erdős's lower bound $r(k,k) > 2^{k/2}$ on Ramsey numbers shows that for every positive integer k there is a graph G of order $\lfloor 2^{k/2} \rfloor$ such that $\alpha(G) < k$ and $\omega(G) < k$. To turn it around, the lower bound shows the existence of graphs G of arbitrarily large order n such that $\alpha(G) < 2 \lg n$ and $\omega(G) < 2 \lg n$. Erdős's proof shows not just that such graphs exist: it shows that they constitute a large proportion of all graphs of order n, a proportion that gets more and more overwhelming as n gets larger and larger. Let us elucidate this point.

DEFINITION: Saying that *almost all graphs* have some prescribed property has the following meaning: when $f(n)$ denotes the number of graphs with vertices $1, 2, \ldots, n$ that have the property, we have

$$\lim_{n \to \infty} \frac{f(n)}{2^{n(n-1)/2}} = 1.$$

Here, the denominator counts all graphs with vertices $1, 2, \ldots, n$ (whether they have the property or not).

REMINDER: We let $\lg x$ stand for the binary logarithm $\log_2 x$.

THEOREM 9.1   *Almost all graphs G of order n have $\alpha(G) < 2\lg n$ and $\omega(G) < 2\lg n$.*

DEFINITION: An *induced subgraph* of graph G is a graph whose vertex set is a subset of the vertex set of G and where two vertices are adjacent if and only if they are adjacent in G. (All graphs of order at most n are subgraphs of the complete graph K_n, but only complete graphs are its induced subgraphs.)

Proof of Theorem 9.1   Writing $s = \lceil 2\lg n \rceil$, let $\mathbf{P}$ denote the set of all pairs (G, F), where G is a graph with vertices $1, 2, \ldots, n$ and F is an induced subgraph of G of order s with either all $\binom{s}{2}$ edges or else no edges at all. The theorem asserts that almost all graphs appear as G in none of these pairs. In the standard notation used on page 93, where

$$f(n) = o(g(n)) \text{ means that } \lim_{n \to \infty} f(n)/g(n) = 0,$$

the assertion is that the number of graphs appearing as G in the pairs in $\mathbf{P}$ is $o(2^{n(n-1)/2})$.

Let us prove this. The number of graphs appearing as G in the pairs in $\mathbf{P}$ is at most $|\mathbf{P}|$. Since there are precisely

$$2\binom{n}{s}$$

choices of F and since each F appears in precisely

$$2^{\binom{n}{2} - \binom{s}{2}}$$

pairs in $\mathbf{P}$, we have

$$|\mathbf{P}| \;=\; 2\binom{n}{s} \cdot 2^{\binom{n}{2} - \binom{s}{2}}.$$

The proof is completed by showing that

$$2\binom{n}{s} \cdot 2^{\binom{n}{2} - \binom{s}{2}} \;=\; o\left(2^{\binom{n}{2}}\right).$$

To do this, just observe that

$$\frac{2\binom{n}{s}}{2^{s(s-1)/2}} \;\leq\; \frac{2n^s/s!}{2^{s(s-1)/2}} \;\leq\; \frac{2}{s!} \cdot 2^{s/2} \;=\; o(1). \qquad \square$$

Theorem 9.1 implies that bound (9.1) is very weak for an overwhelming majority of all graphs:

COROLLARY 9.2   *Almost all graphs of order n have*

$$\chi(G) \;\geq\; \frac{n}{4\lg^2 n} \cdot \omega(G).$$

Proof Every graph G of order n satisfies

$$\chi(G) \geq n/\alpha(G)$$

since at most $\alpha(G)$ of the n vertices of G can get the same colour. Combined with this bound, Theorem 9.1 shows that almost all graphs of order n have

$$\chi(G) \geq \frac{n}{2\lg n} \geq \frac{n}{4\lg^2 n} \cdot \omega(G). \qquad \Box$$

Nevertheless, if all degrees are large, then the bound $\chi(G) \geq \omega(G)$ is tight:

THEOREM 9.3 (Special case of Theorem 1.1 in [9]) *If a graph G of order n has at least one edge and if each of its vertices has degree greater than*

$$\left(1 - \frac{3}{3\omega(G) - 1}\right)n,$$

then $\chi(G) = \omega(G)$. $\qquad \Box$

By the way, the bound $\chi(G) \geq n/2\lg n$ used in Corollary 9.2 is asymptotically tight:

$$\text{almost all graphs } G \text{ of order } n \text{ have } \chi(G) = (1 + o(1))\frac{n}{2\lg n}. \qquad (9.2)$$

This formula follows from a much finer result of Béla Bollobás [38].

9.3   The End of Hajós's Conjecture

The Four Colour Theorem and Kuratowski's theorem together show that

if G contains no subdivision of K_5 and no subdivision of $K_{3,3}$,

then $\chi(G) \leq 4$.

György Hajós (1912–72) conjectured that this assertion may be strengthened to the more elegant

if G contains no subdivision of K_5, then $\chi(G) \leq 4$

and that, more generally,

if G contains no subdivision of K_{t+1}, then $\chi(G) \leq t$. $\qquad (9.3)$

A part of the appeal of Hajós's conjecture (9.3) comes from the fact that it constitutes a weak converse of the bound $\chi(G) \geq \omega(G)$, which may be stated as

if $\chi(G) \leq t$, then G contains no K_{t+1}.

To put it differently, let $\omega^*(G)$ denote the largest s such that G contains a subdivision of K_s: in this notation, Hajós's conjecture is $\chi(G) \leq \omega^*(G)$ and we have $\omega^*(G) \geq \omega(G)$.

It seems that [99], published in 1952, is the first written reference to (9.3); an attribution of (9.3) to Hajós has appeared some thirteen years later in [101]. It is easy to see that the special cases $t = 1$ and $t = 2$ of (9.3) are valid (if $\omega^*(G) = 1$, then G contains no edge, and so $\chi(G) \leq 1$; if $\omega^*(G) = 2$, then G contains no cycle, and so $\chi(G) \leq 2$). The next special case, $t = 3$, was confirmed [203] by Hugo Hadwiger (1908–81) already in 1942; unaware of this work, Gabriel Andrew Dirac (1925–84) duplicated it [99] some nine years later. In 1979, Paul Catlin (1948–95) disproved [65] all cases where $t > 5$: for every positive integer t, he constructed a graph G such that $\omega^*(G) \leq t$ and $\chi(G) \geq 5(t - 1)/4$. The special cases $t = 4$ (the conjectured strengthening of the Four Colour Theorem) and $t = 5$ of Hajós's conjecture remain open.

REMINDER: A *path of length $k-1$ between vertices u and v* is a string $w_1 w_2 \ldots w_k$ of pairwise distinct vertices such that $w_1 = u$, $w_k = v$, and each w_i with $i = 1, 2, \ldots k - 1$ is adjacent to w_{i+1}. A path between vertices u and v is said to *join* u and v.

Catlin's construction is simple. If $t = 2s + 1$ with a positive integer s, then take pairwise disjoint cliques C_0, C_1, C_2, C_3, C_4 such that $|C_i| = s$ for all i and, with subscript arithmetic modulo 5, add all the edges between C_i and C_{i+1}; if $t = 2s$ with a positive integer s, then take the same graph and remove one vertex from each of C_2 and C_3. These graphs G have $\alpha(G) = 2$, and so the lower bound $\chi(G) \geq 5(t - 1)/4$ follows from $\chi \geq n/\alpha$. To check that $\omega^*(G) \leq t$, we note first that $\omega(G) \leq t$, then that every two nonadjacent vertices of G are separated by some set of $t - 1$ vertices (where, as usual, vertices u and v are said to be separated by a set C of vertices if, and only if, every path between u and v has at least one interior vertex in C), and finally that

> every subdivision of K_{t+1} contains a set S of $t + 1$ vertices such that no two vertices of S are separated by a set of $t - 1$ vertices.

Hajós's conjecture stated that the ratio χ/ω^* never exceeds 1; in Catlin's counterexamples, this ratio gets arbitrarily close to $5/4$. About a year after Catlin submitted his paper for publication, Paul Erdős and Siemion Fajtlowicz [135] showed that far more impressive counterexamples are provided by almost all graphs: in almost all graphs of order n, the same ratio χ/ω^* is at least $c\sqrt{n}/\lg n$ for some positive constant c. Their argument uses the following observation:

LEMMA 9.4 *If a graph of order n contains a subdivision of K_s, then it contains a subgraph of order s with at least $\binom{s}{2} - (n - s)$ edges.*

Proof By assumption, the graph contains a set S of s vertices and a family $\mathcal{F}$ of $\binom{s}{2}$ paths, pairwise disjoint except for the vertices at their ends, such that each pair of distinct vertices in S is joined by one of the paths in $\mathcal{F}$. For each pair of distinct nonadjacent vertices u, v in S, choose an interior vertex $f(\{u, v\})$ of the path in $\mathcal{F}$ that joins u and v. This mapping f is one-to-one and its range is disjoint from S; it follows that there are at most $n - s$ unordered pairs of distinct non-adjacent vertices in S, and so the graph induced by S has at least $\binom{s}{2} - (n - s)$ edges. $\qquad\square$

From this lemma and from Turán's Theorem 7.1, Erdős and Fajtlowicz deduce an upper bound on $\omega^*(G)$:

LEMMA 9.5   *Every graph G of order n satisfies*

$$\omega^*(G) \le \sqrt{2\omega(G)n}.$$

Proof   Given an arbitrary graph G of order n, write $s = \omega^*(G)$ and $r = \omega(G)$. Lemma 9.4 guarantees that G contains a subgraph of order s with at least $\binom{s}{2} - (n-s)$ edges, and so

$$\binom{s}{2} - (n - s) \le \text{ex}(K_{r+1}, s).$$

Turán's theorem guarantees that

$$\text{ex}(K_{r+1}, s) \le \frac{s^2}{2}\left(1 - \frac{1}{r}\right).$$

It follows that

$$\binom{s}{2} - (n - s) \le \frac{s^2}{2}\left(1 - \frac{1}{r}\right),$$

and so $s^2/2r \le n - s/2 < n$. $\qquad\square$

Erdős and Fajtlowicz note that, by Theorem 9.1 and Lemma 9.5, almost all graphs G of order n have

$$\chi(G) \ge \frac{n}{2\lg n} \quad \text{and} \quad \omega^*(G) \le 2\sqrt{n\lg n},$$

and so [135, Theorem 2]

$$\frac{\chi(G)}{\omega^*(G)} \ge \frac{\sqrt{n}}{4(\lg n)^{3/2}}.$$

Then they go on to show that a little extra work shaves a factor $\sqrt{\frac{1}{2}\lg n}$ off the upper bound on $\omega^*(G)$. Most of this extra work goes into proving that in almost all graphs of order n, every subgraph of order $\lceil\sqrt{8n}\,\rceil$ has fewer than

$$\frac{3}{4}\binom{\lceil\sqrt{8n}\,\rceil}{2}$$

edges. (For a proof, see Section A.5.) This result combined with Lemma 9.4 implies that almost all graphs G of order n have $\omega^*(G) < \lceil\sqrt{8n}\,\rceil$, and so $\omega^*(G) < \sqrt{8n}$. Here is the punchline [135, Theorem 3]:

THEOREM 9.6   *Almost all graphs G of order n have*

$$\frac{\chi(G)}{\omega^*(G)} = \Omega\left(\frac{\sqrt{n}}{\log n}\right). \qquad\square$$

Hadwiger's Conjecture

Hajós conjecture turned out to be false, but a similar conjecture that also subsumes the Four Colour Theorem may turn out to be true. This conjecture involves the following notion.

DEFINITION: A graph G has a K_{t+1} *minor* if its vertex set contains $t+1$ nonempty pairwise disjoint subsets such that each of these subsets induces a connected subgraph of G and every two of these subsets are linked by at least one edge of G (in the sense that each of the two subsets includes one endpoint of the edge).

CONJECTURE 9.7   *If G does not have a K_{t+1} minor, then $\chi(G) \leq t$.*

Conjecture 9.7 was proposed [203] in 1942 by Hugo Hadwiger (1908–81). Obviously, if a graph contains a subdivision of K_{t+1}, then it has a K_{t+1} minor; obviously, the converse is true for $t = 1$, $t = 2$, and $t = 3$. However, the graph in Figure 9.1. has a K_5 minor and contains no subdivision of K_5 (only four of its vertices have degree at least 4). Therefore Hadwiger's conjecture is weaker than Hajós's conjecture. Nevertheless, even Hadwiger's conjecture is strong enough to subsume the Four Colour Theorem: it is easy to show that every graph with a K_5 minor must contain a subdivision of K_5 or a subdivision of $K_{3,3}$.

Validity of Hadwiger's conjecture for $t = 1$, $t = 2$, and $t = 3$ follows from validity of Hajós conjecture for these values of t.

Validity of Hadwiger's conjecture for $t = 4$ follows from the Four Colour Theorem: Four decades before the Four Colour Conjecture was resolved and six years before Hadwiger announced his conjecture, Klaus Wagner (1910–2000) proved in [373] that all planar graphs are 4-colourable if and only if all graphs that do not have a K_5 minor are 4-colourable.

Validity of Hadwiger's conjecture for $t = 5$ also follows from the Four Colour Theorem: In 1993, Robertson, Seymour, and Thomas [329] proved that every minimal counterexample G to Hadwiger's conjecture for $t = 5$ must be *apex*, meaning that it includes a vertex v such that G with v deleted is planar. However, every apex graph is 5-colourable by the Four Colour Theorem.

All other instances of Hadwiger's conjecture remain unresolved, but progress toward the conjecture has been made in other ways.

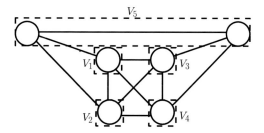

Figure 9.1 A graph and its K_5 minor.

REMINDER: When f and g are nonnegative real-valued functions defined on positive integers, we write $f(n) = O(g(n))$ to mean that $f(n) \leq cg(n)$ for some constant c and all sufficiently large n.

In 1963, Wagner [374] proved the existence of a function f such that

$$\text{if } G \text{ does not have a } K_{t+1} \text{ minor, then } \chi(G) \leq f(t). \qquad (9.4)$$

Actually, he proved that $f(t+1) \leq 2f(t)$ for all positive t, which implies $f(t) = O(2^t)$. The latest improvement of this upper bound comes from Alexandr Kostochka [262, 263] and, independently, Andrew Thomason [359]: $f(t) = O(t\sqrt{\log t})$.

In 1979, Bollobás, Catlin, and Erdős proved [42] that Hadwiger's conjecture is true for almost all graphs: they proved that almost all graphs of order n have a K_t minor such that

$$t \geq \frac{n}{\sqrt{\lg n} + 4}$$

and appealed to the previously known result [44, Theorem 4] that almost all graphs of order n have chromatic number at most $(1 + o(1))n/\lg n$. (This upper bound was superseded by (9.2) only later.)

More on Hadwiger's conjecture can be found in [340].

9.4   Graphs with a Large Chromatic Number and No Triangles

Theorem 9.1 implies that there is no constant upper bound on the ratio $\chi(G)/\omega(G)$ for all graphs G, but it does not preclude an analogue of (9.4), the existence of a function f such that $\chi(G) \leq f(\omega(G))$ for all graphs G. Nonexistence of such a function is implied by the following theorem.

THEOREM 9.8   *For every positive integer k there is a graph G such that $\chi(G) > k$ and $\omega(G) = 2$.*

Proofs   Theorem 9.8 was proved by several people independently of each other.

9.4.1   Zykov

The first seems to have been Alexander Alexandrovich Zykov in 1949 (Theorem 8 of [378]). He exhibited a way of transforming any graph F with $\omega(F) = 2$ into a graph G with $\chi(G) > \chi(F)$ and $\omega(G) = 2$. His construction goes as follows:

Let n stand for the order of F and write $k = \chi(F)$. Begin with k pairwise disjoint copies $F_1, F_2, \ldots, F_k$ of F and a set S of n^k additional pairwise nonadjacent vertices; put the vertices in S in a one-to-one correspondence with ordered k-tuples $(v_1, v_2, \ldots, v_k)$ where each v_i is a vertex of F_i and make each vertex in S adjacent to the k vertices in the corresponding k-tuple. The resulting graph G contains no triangle.

To see that $\chi(G) > k$, consider an arbitrary assignment of colours $1, 2, \ldots, k$ to the vertices of G: we will find a pair of adjacent vertices that have the same colour. If one of the k colours is missing from some F_i, then we are done as $\chi(F_i) = k$; else every F_i includes a vertex of any prescribed colour. In the latter case, let v_i denote a vertex of F_i whose colour is i. Now the vertex in S that is adjacent to all of $v_1, v_2, \ldots, v_k$ has the same colour as one of them.

9.4.2   Tutte

Next were several people in North America. In 1953, Peter Ungar [367] proposed proving Theorem 9.8 as one of the Advanced Problems in the American Mathematical Monthly. Its solution by Blanche Descartes [96], published in 1954, used a different way of transforming any graph F with $\omega(F) = 2$ into a graph G with $\chi(G) > \chi(F)$ and $\omega(G) = 2$. This construction goes as follows:

Let n stand for the order of F and write $k = \chi(F)$, $s = k(n-1) + 1$. Begin with $\binom{s}{n}$ pairwise disjoint copies of F a set S of s additional pairwise nonadjacent vertices; put the copies of F in a one-to-one correspondence with sets of n vertices in S and join the n vertices of each copy of F to the corresponding n vertices in S by a set of n pairwise disjoint edges. The resulting graph G contains no triangle.

To see that $\chi(G) > k$, consider an arbitrary assignment of colours $1, 2, \ldots, k$ to the vertices of G: we will find a pair of adjacent vertices that have the same colour. Since $|S| > k(n-1)$, one of the colours must appear on n distinct vertices in S. If this colour is missing from the corresponding copy of F, then we are done as $\chi(F) = k$; else one of the vertices in this copy and its neighbour in S have the same colour.

Actually, Blanche Descartes was a pen name used by William T. Tutte (1917–2002) and his friends R. Leonard Brooks (1916–93), Cedric A. B. Smith (1917–2002), and Arthur H. Stone [344]. On this particular occasion, Tutte alone was using it; under this pseudonym, he also discussed the special case $\chi(G) = 4$ as early as 1947 [95]. Ungar's Advanced Problem 4526 was also solved by John B. Kelly (whose solution [243] was identical with Tutte's) and by Ungar himself (whose solution [368] was identical with Zykov's).

9.4.3   Mycielski

A year later, yet another proof of Theorem 9.8 was published by Jan Mycielski [300]. His way of transforming any graph F with $\omega(F) = 2$ into a graph G with $\chi(G) > \chi(F)$ and $\omega(G) = 2$ goes as follows:

Label the vertices of F as $u_1, u_2, \ldots, u_n$; take additional pairwise nonadjacent vertices $v_1, v_2, \ldots, v_n$ and make each v_i adjacent to all the neighbours of u_i; finally, add one more vertex w and make it adjacent to all of $v_1, v_2, \ldots, v_n$. The resulting graph G contains no triangle (in particular, a triangle $v_i u_j u_k$ in G would force a triangle $u_i u_j u_k$ in F).

DEFINITION: A *proper colouring* of a graph is an assignment of colours to its vertices such that adjacent vertices always receive distinct colours.

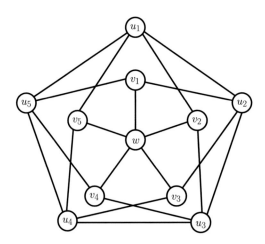

Figure 9.2 Mycielski's graph with $\omega = 2$ and $\chi = 4$.

To prove that $\chi(F) < \chi(G)$, we will transform an arbitrary proper colouring g of G by $\chi(G)$ colours into a proper colouring f of F by $\chi(G) - 1$ colours. Let us show that this can be done by setting, for all $i = 1, 2, \ldots, n$,

$$f(u_i) = \begin{cases} g(u_i) & \text{if } g(u_i) \neq g(w), \\ g(v_i) & \text{if } g(u_i) = g(w). \end{cases}$$

As colour $g(w)$ is missing from the range of f, we only need prove that f is a proper colouring of F. For this purpose, consider arbitrary adjacent vertices u_i and u_j: our task is proving that $f(u_i) \neq f(u_j)$. Since g is a proper colouring of G, at least one of $g(u_i)$ and $g(u_j)$ is different from $g(w)$; switching subscripts if necessary, we may assume that $g(u_i) \neq g(w)$, and so $f(u_i) = g(u_i)$. Now the desired conclusion $f(u_i) \neq f(u_j)$ follows from observing that $f(u_j)$ is $g(u_j)$ or $g(v_j)$, that u_i is adjacent to both u_j and v_j, and that g is a proper colouring of G.

Mycielski's construction transforms K_2 into C_5 and it transforms C_5 into the graph in Figure 9.2.

9.4.4   Erdős and Hajnal

Paul Erdős and András Hajnal [139, Theorem 6] found a way of deducing Theorem 9.8 from Theorem 3.9 without resorting to induction on k:

Take a positive integer n large enough to ensure that $n \rightarrow (3)^2_k$, let the vertices of G be all the ordered pairs (a, b) of integers such that $0 \leq a < b < n$, and let G have all the edges of the form $(a, b)(b, c)$. Verifying that $\omega(G) = 2$ and $\chi(G) > k$ is a straightforward exercise.

It is a part of the graph-theoretic lore (see, for instance, Problem 9.26 of [279]) that the assumption $n \rightarrow (3)^2_k$ in the Erdős–Hajnal construction can be replaced by the weaker $n > 2^k$. To see this, consider an arbitrary assignment of k colours to the vertices of G and, for each integer a such that $0 \le a < n$, let $S(a)$ denote the set of colours that appear on all the vertices (a, x). If $n > 2^k$, then $S(a) = S(b)$ for some integers a, b such that $0 \le a < b < n$, in which case vertex (a, b) has the same colour as some vertex (b, c).

Here, the condition $n > 2^k$ is not only sufficient, but also necessary to guarantee $\chi(G) > k$. To prove this, we have to exhibit a proper colouring of the vertices of G by k colours when $n = 2^k$. Each vertex of G is an ordered pair (a, b) of integers such that $0 \le a < b < n$; in the binary expansions $a = \sum_{i=0}^{k-1} a_i 2^i$ and $b = \sum_{i=0}^{k-1} b_i 2^i$, there is at least one subscript i such that $a_i = 0$ and $b_i = 1$; colour the vertex by any such i.

9.4.5 Lovász

Another simple construction of graphs with a large chromatic number and no triangles involves a notion introduced by Martin Kneser (1928–2004).

DEFINITION: When s and d are positive integers, the vertices of the *Kneser graph*, denoted $KG_{s,d}$ in [278], are s-point subsets of $\{1, 2, \ldots 2s + d\}$ and two of them are adjacent if and only if they are disjoint.

We have $\chi(KG_{s,d}) \le d + 2$. (To see this, consider an arbitrary s-point subset S of $\{1, 2, \ldots 2s + d\}$ and the smallest integer i appearing in S. If $i \le d + 1$, then give S colour i; else give S colour $d + 2$.) In 1955, Kneser conjectured that this upper bound is best possible [250]. Twenty-two years later, László Lovász proved Kneser's conjecture [278]:

THEOREM 9.9 $\quad \chi(KG_{s,d}) = d + 2$. $\qquad\qquad\qquad\qquad\qquad\qquad$ □

This breakthrough introduced topological methods in graph theory. Like the proverbial sledgehammer cracking a walnut, Theorem 9.9 provides another proof of Theorem 9.8: $KG_{s,d}$ contains no triangle when when $s > d$. We shall return to the sledgehammer at the end of next section. $\qquad\qquad\qquad\qquad\qquad\qquad\qquad$ □

Theorem 9.8 allows us to define $n(k)$ as the smallest order of a graph G such that $\omega(G) = 2$ and $\chi(G) = k$. It is known ([78, 187, 227]) that

$$n(2) = 2, \ n(3) = 5, \ n(4) = 11, \ n(5) = 22, \ 32 \le n(6) \le 40$$

and that ([247])

$$c_1 k^2 \log k \ \le \ n(k) \ \le \ c_2 k^2 \log k$$

with some positive constants c_1, c_2. (Since the values of c_1, c_2 are unspecified, $\log k$ to any fixed base will do here.)

9.5   Graphs with a Large Chromatic Number and No Short Cycles

REMINDER: A *cycle of length k* is a string $w_1 w_2 \ldots w_k$ of pairwise distinct vertices such that each w_i with $i = 1, 2, \ldots k - 1$ is adjacent to w_{i+1} and w_k is adjacent to w_1.

Theorem 9.8 asserts that there are graphs of arbitrarily large chromatic numbers without cycles of length three; its Tutte–Kelly proof shows that the "three" can be replaced by "at most five"; John B. Kelly and L. M. Kelly ([243], p. 786) conjectured that it can be further replaced by "at most ℓ" for every integer ℓ. About three years later, this conjecture was proved by Erdős [110]:

THEOREM 9.10   *For every choice of positive integers k and ℓ, there is a graph G of order n with $\alpha(G) < n/k$ and without cycles of length at most ℓ.*

Erdős proved Theorem 9.10 without actually constructing the relevant graphs. Nine years later, Lovász [277] found a *constructive* proof of the Kelly-Kelly conjecture; eleven years after that, Jaroslav Nešetřil and Vojtěch Rödl [304] found a simpler constructive proof of the conjecture.

Erdős's proof of Theorem 9.10 uses again the technique featured in the proof of Theorem 9.1, but it adds a couple of new twists. The first of them comes from the observation that expecting almost all graphs to have the property of Theorem 9.10 would be hopeless: Theorem 9.1 shows that almost all graphs contain triangles (and even much larger cliques). To get anywhere with counting arguments, we must restrict our attention to graphs that are *sparse* in the sense of having a small number of edges relative to their order.

DEFINITION: Consider any function m that assigns to every positive integer n a positive integer $m(n)$ and consider any property that every graph has or has not. Saying that *almost all graphs with n vertices and m(n) edges* have this property has the following meaning (introduced by Erdős and Rényi in 1960 [148]): When $f(n)$ denotes the number of graphs with vertices $1, 2, \ldots, n$ and with $m(n)$ edges that have the property, we have

$$\lim_{n \to \infty} \frac{f(n)}{\binom{\binom{n}{2}}{m(n)}} = 1.$$

Here, the denominator counts all graphs with vertices $1, 2, \ldots, n$ and $m(n)$ edges (whether they have the property or not).

One could set out to prove Theorem 9.10 by proving that almost all graphs G of order n with some number $m(n)$ of edges (depending not only on n, but also on the prescribed constants k and ℓ) have $\alpha(G) < n/k$ and no cycles of length at most ℓ. However, this plan cannot be implemented: if $m(n) \leq n/2$, then all graphs G with n vertices and $m(n)$ edges have $\alpha(G) \geq n/2$ (this follows from Turán's theorem) and if $m(n) \geq n/2$, then almost all graphs G with n vertices and $m(n)$ edges contain a cycle (this is Theorem 5b of [148]). Erdős deals with this difficulty by relaxing the requirement that G contain no cycles of length at most ℓ and strengthening the requirement that $\alpha(G) < n/k$. He proves that, for a suitable choice of the function m,

almost all graphs G with n vertices and $m(n)$ edges
contain fewer than n cycles of length at most ℓ
and have more than n edges in every induced subgraph of order $\lceil n/k \rceil$.

Then he gets graphs with the properties of Theorem 9.10 by choosing a graph with n vertices and $m(n)$ edges at random and removing an edge from each of its cycles of length at most ℓ.

Actually, his argument proves the following refinement of Theorem 9.10:

THEOREM 9.11 *For every choice of an integer ℓ such that $\ell \geq 3$ and a positive δ such that $\delta < 1/3\ell$, there is a positive integer n_0 with the following property: For every integer n greater than n_0, there is a graph G of order n without cycles of length at most ℓ and with $\alpha(G) < \lfloor n^{1-\delta} \rfloor$.*

We will treat two parts of this argument as separate lemmas.

LEMMA 9.12 *Let ℓ be an integer such that $\ell \geq 3$ and let δ be a number such that $0 < \delta < 1/3\ell$. Then almost all graphs with n vertices and $\lfloor n^{1+3\delta} \rfloor$ edges contain fewer than n cycles of length at most ℓ.*

Proof With $m(n) = \lfloor n^{1+3\delta} \rfloor$, let $\mathbf{P}$ denote the set of all pairs (G, F), where G is a graph with vertices $1, 2, \ldots, n$ and with $m(n)$ edges and F is a cycle of length at most ℓ in G. To reduce the clutter in the formulas that follow, we shall write simply m for $m(n)$.

The lemma asserts that the number of graphs appearing as G in at least n of the pairs in $\mathbf{P}$ is

$$o\left(\binom{\binom{n}{2}}{m}\right).$$

To prove this, note first that the number of graphs appearing as G in at least n pairs in $\mathbf{P}$ is at most $|\mathbf{P}|/n$. For each $i = 3, 4, \ldots, \ell$, there are fewer than n^i choices of a cycle F with i vertices (actually, the exact number of these choices is $\binom{n}{i}\frac{1}{2}(i-1)!$, but we do not need this degree of precision) and every such F appears in precisely

$$\binom{\binom{n}{2} - i}{m - i}$$

pairs in $\mathbf{P}$. It follows that

$$|\mathbf{P}| \leq \sum_{i=3}^{\ell} n^i \binom{\binom{n}{2} - i}{m - i}.$$

The bulk of the proof is a computation showing that

$$\frac{1}{n} \sum_{i=3}^{\ell} n^i \binom{\binom{n}{2} - i}{m - i} \binom{\binom{n}{2}}{m}^{-1} = o(1).$$

We have

$$\sum_{i=3}^{\ell} n^i \left(\frac{\binom{n}{2} - i}{m - i} \right) \left(\frac{\binom{n}{2}}{m} \right)^{-1} = \sum_{i=3}^{\ell} n^i \binom{m}{i} \binom{\binom{n}{2}}{i}^{-1}$$

$$\leq \sum_{i=3}^{\ell} n^i \left(\frac{m}{\binom{n}{2}} \right)^i = \sum_{i=3}^{\ell} \left(\frac{2m}{n-1} \right)^i ;$$

since $\sum_{i=3}^{\ell} x^i < 2x^{\ell}$ whenever $x \geq 2$ (this can be verified by easy induction on ℓ or directly from the formula $\sum_{i=3}^{\ell} x^i = x^3(x^{\ell-2} - 1)/(x-1)$), we conclude that

$$\frac{1}{n} \sum_{i=3}^{\ell} \left(\frac{2m}{n-1} \right)^i < \frac{2}{n} \left(\frac{2m}{n-1} \right)^{\ell} = O \left(\frac{m^{\ell}}{n^{\ell+1}} \right) = O \left(\frac{n^{3\delta\ell}}{n} \right) = o(1). \quad \square$$

REMINDER: Both $f(n) \sim g(n)$ and $f(n) = (1 + o(1))g(n)$ mean that $\lim_{n \to \infty} f(n)/g(n) = 1$.

LEMMA 9.13   *For every constant δ such that $0 < \delta < 1/3$, almost all graphs with n vertices and $\lfloor n^{1+3\delta} \rfloor$ edges have more than n edges in every subgraph induced by $\lfloor n^{1-\delta} \rfloor$ vertices.*

Proof   With $m(n) = \lfloor n^{1+3\delta} \rfloor$ and $s(n) = \lfloor n^{1-\delta} \rfloor$, let **P** denote the set of all pairs (G, F), where G is a graph with vertices $1, 2, \ldots, n$ and with $m(n)$ edges and F is an induced subgraph of G with $s(n)$ vertices and with at most n edges. To reduce the clutter in the formulas that follow, we shall write simply m and s for $m(n)$ and $s(n)$, respectively.

The lemma asserts that the number of graphs appearing as G in the pairs in **P** is

$$o \left(\binom{\binom{n}{2}}{m} \right).$$

To prove this, note first that the number of graphs appearing as G in the pairs in **P** is at most $|\mathbf{P}|$. For each nonnegative integer i, there are precisely

$$\binom{n}{s} \binom{\binom{s}{2}}{i}$$

choices of a graph F with s vertices coming from $\{1, 2, \ldots, n\}$ and with i edges; every such F appears in precisely

$$\binom{\binom{n}{2} - \binom{s}{2}}{m - i}$$

pairs in **P**; it follows that

$$|\mathbf{P}| = \sum_{i=0}^{n} \binom{n}{s} \binom{\binom{s}{2}}{i} \binom{\binom{n}{2} - \binom{s}{2}}{m - i}.$$

The bulk of the proof is a computation showing that

$$\binom{n}{s} \sum_{i=0}^{n} \binom{\binom{s}{2}}{i} \binom{\binom{n}{2} - \binom{s}{2}}{m - i} \binom{\binom{n}{2}}{m}^{-1} = o(1).$$

Here are its ingredients:

- Since $s \leq n$, we have $\binom{n}{s} \leq n^s \leq s^n$ whenever $s \geq 3$.
- For all sufficiently large n, we have $3 \leq n \leq \frac{1}{2}\binom{s}{2}$, which implies

$$\sum_{i=0}^{n} \binom{\binom{s}{2}}{i} \leq (n+1)\binom{\binom{s}{2}}{n} \leq \binom{s}{2}^n \leq s^{2n}.$$

- For all sufficiently large n, we have $m \leq \frac{1}{2}\left(\binom{n}{2} - \binom{s}{2}\right)$, which implies

$$\binom{\binom{n}{2} - \binom{s}{2}}{m - i} \leq \binom{\binom{n}{2} - \binom{s}{2}}{m}$$

for all $i = 0, 1, \ldots, m$; since $1 + x \leq e^x$ for all x, it follows that

$$\binom{\binom{n}{2} - \binom{s}{2}}{m}\binom{\binom{n}{2}}{m}^{-1} \leq \left(\frac{\binom{n}{2} - \binom{s}{2}}{\binom{n}{2}}\right)^m = \left(1 - \frac{s(s-1)}{n(n-1)}\right)^m$$

$$\leq \exp\left(-\frac{ms(s-1)}{n(n-1)}\right) \leq \exp\left(-0.99\,\frac{ms^2}{n^2}\right).$$

Combining these ingredients, we find that, for all sufficiently large n,

$$\binom{n}{s}\sum_{i=0}^{n} \binom{\binom{s}{2}}{i}\binom{\binom{n}{2} - \binom{s}{2}}{m - i}\binom{\binom{n}{2}}{m}^{-1} \leq s^{3n}\exp\left(-0.99\,ms^2/n^2\right)$$

$$\leq \exp\left(n(3\ln s - 0.98\,n^\delta)\right) = o(1). \quad \square$$

Proof of Theorem 9.11 Together, Lemma 9.13 and Lemma 9.12 guarantee that almost all graphs with n vertices and $\lfloor n^{1+3\delta} \rfloor$ edges contain fewer than n cycles of length at most ℓ and have more than n edges in every subgraph induced by $\lfloor n^{1-\delta} \rfloor$ vertices. It follows that there is a positive integer n_0 with the following property:

> For every integer n greater than n_0, there is a graph H of order n
> which contains fewer than n cycles of length at most ℓ
> and has more than n edges in every subgraph induced by $\lfloor n^{1-\delta} \rfloor$ vertices.

Removing an edge from each cycle of length at most ℓ, we transform H into a graph G with no cycles of length at most ℓ and with $\alpha(G) < \lfloor n^{1-\delta} \rfloor$. $\quad \square$

Erdős stated [110, page 37] that by using a little more care he can prove by the same method the following result:

> *For every choice of a positive integer k, there are positive constants c and n_0 such that for every integer n greater than n_0 there is a graph G of order n without cycles of length less than $c \log n$ and with $\chi(G) \geq k$.*

Béla Bollobás worked out an elegant version of the proof [35, p. 256, Theorem 4.1]:

THEOREM 9.14   *For every choice of integers k and n such that $n \geq k \geq 2$, there is a graph G of order n without cycles of length less than $\lfloor \log n/4 \log k \rfloor$ and with $\chi(G) \geq k$.*

Sketch of proof First, let us dismiss the easy cases. If $k = 2$, then the graph with n vertices and a single edge has the properties required of G. If $k = 3$, then the graph consisting of a cycle on n vertices (when n is odd) or a cycle on $n - 1$ vertices and a vertex of degree zero (when n is even) has the properties required of G since $\log_3 n \leq 4(n - 1)$ whenever $n \geq 3$. If $n < k^{16}$, then $\lfloor \log n/4 \log k \rfloor \leq 3$, and so the complete graph on k vertices has the properties required of G. Having made these observations, we may assume that $k \geq 4$ and $n \geq k^{16}$.

Given positive integers k and n such that $k \geq 4$ and $n \geq k^{16}$, Bollobás sets

$$m = k^3 n, \quad t = \frac{3n^{7/8}}{\lfloor \log n/4 \log k \rfloor},$$

and proves that

(i) at least two thirds of all graphs with n vertices and m edges have fewer than t cycles of length less than $\lfloor \log n/4 \log k \rfloor$,

(ii) at least two thirds of all graphs with n vertices and m edges have more than t edges in every subgraph induced by $\lfloor n/(k - 1) \rfloor$ vertices.

It follows that at least one third of all graphs with n vertices and m edges have both fewer than t cycles of length less than $\lfloor \log n/4 \log k \rfloor$ and more than t edges in every subgraph induced by $\lfloor n/(k - 1) \rfloor$ vertices. Removing one edge from each cycle of length less than $\lfloor \log n/4 \log k \rfloor$ in any such graph produces a G without cycles of length less than $\lfloor \log n/4 \log k \rfloor$ and with $\alpha(G) < n/(k - 1)$. Finally, $\chi(G) \geq k$ follows from $\chi(G) \geq n/\alpha(G) > k - 1$. □

An Erdős–Gallai Conjecture and Its Resolution

NOTATION: Let $\lambda(n, k)$ denote the smallest ℓ such that every graph of order n and chromatic number at least k must contain an *odd* cycle of length at most ℓ.

Theorem 9.14 implies that

$$\lambda(n, k) \geq \lfloor \log n/4 \log k \rfloor.$$

Erdős conjectured that $\lambda(n, 4)$ grows like $\log n$, but Gallai, unaware of this conjecture, constructed graphs showing that $\lambda(n, 4) \geq n^{1/2}$. By 1962, Erdős and Gallai conjectured [111, page 171] that $\lambda(n, k)$ grows like $n^{1/(k-2)}$. Some fifteen years later, Lovász proved that it grows at least this fast. He never published this result, but Erdős reported it in [126, page 154].

Lovász's proof of the Erdős–Gallai conjecture is related to Kneser's conjecture. He submitted his proof of the latter conjecture for publication on March 4, 1977. Eleven weeks later [17], Imre Bárány gave a quick derivation of Lovász's's Theorem 9.9 from a theorem [183] of David Gale (1921–2008),

the d-dimensional sphere contains a set of $2n + d$ points
that meets each open hemisphere in at least n points,

combined with a theorem [51] of Karol Borsuk (1905–82),

if the d-dimensional sphere is covered by $d + 1$ open subsets,
then one of these subsets contains a pair of antipodal points.

On July 5–9, 1977, University of Waterloo held a conference in honour of Professor W. T. Tutte on the occasion of his sixtieth birthday. It was during this conference that Lovász proved the Erdős–Gallai conjecture. The vertex set of his graph is a maximal subset of the unit $(k - 2)$-dimensional sphere such that, for a fixed positive ε, the minimum distance between two vertices is at least ε; two vertices are adjacent if and only if the distance between them is at least $2 - 2\varepsilon$. Borsuk's theorem guarantees that the chromatic number of this graph is at least k (given an assignment of colours $1, 2, \ldots, k - 1$ to the vertices, let the i-th open subset of the sphere consist of all its points that have distance less than ε from at least one vertex coloured i). When $\varepsilon = cn^{-1/(k-2)}$ with a constant c, the order of this graph is $O(n)$ and all of its odd cycles have length $\Omega(n^{1/(k-2)})$.

About nine months later, Alexander (Lex) Schrijver combined Bárány's method and an explicit construction of the set featured in Gale's theorem to prove that $KG_{s,d}$ (defined on page 135) owes its chromatic number to its far smaller subgraph. A corollary of his result is another proof of the Erdős–Gallai conjecture. We are going to outline its details.

DEFINITION: When s and d are positive integers, $SG_{s,d}$ denotes the graph whose vertices are stable sets of size s in the cycle $w_1 w_2 \ldots w_{2s+d}$ and two of them are adjacent if and only if they are disjoint.

THEOREM 9.15   *[337, Theorem 2]* $\chi(SG_{s,d}) = d + 2$.   □

We will not reproduce Schrijver's proof of this theorem, but will note that $SG_{s,d}$ is indeed far smaller than $KG_{s,d}$:

PROPOSITION 9.16   *The number of stable sets of size s in C_{2s+d} equals* $\frac{2s+d}{s}\binom{s+d-1}{s-1}$.

Proof Let M denote the number of stable sets of size s in C_{2s+d} and let N denote the number of pairs (S, w) such that S is a stable set of size s in C_{2s+d} which includes vertex w. By this definition, $N = Ms$. Symmetry shows that there is a number x (depending on s and d) such that every vertex of the cycle C_{2s+d} belongs to precisely x stable sets of size s. Since $N = (2s + d)x$, we have $M = (2s + d)x/s$. By definition, x is the number of stable sets S of size s in C_{2s+d} which include w_1. Arranging each $\{i : w_i \in S\}$ into an increasing sequence $i_1, i_2, \ldots, i_s$, we find that x equals the number of integer sequences $i_2, \ldots, i_s$ such that $i_2 \geq 3$, $i_3 \geq i_2 + 2$, $i_4 \geq i_3 + 2$, $\ldots i_s \geq i_{s-1} + 2$ and $i_s \leq 2s + d - 1$. To conclude that $x = \binom{s+d-1}{s-1}$, let us point out a bijection between our sequences $i_2, \ldots, i_s$ and increasing integer sequences $j_2, \ldots, j_s$ such that $j_2 \geq 1$ and $j_s \leq s + d - 1$. This bijection is given by $j_t = i_t - t$ for all $t = 2, \ldots s$.   □

As Lovász remarked in [278],

$$\text{every odd cycle in } KG_{s,d} \text{ has length at least } 1 + 2s/d. \qquad (9.5)$$

To see this, consider an arbitrary odd cycle $S_1 S_2 \ldots S_\ell$ in $KG_{s,d}$. Since every S_{i+1} with $1 \le i \le \ell - 2$ is disjoint from $S_i \cup S_{i+2}$, we have $|S_i \cup S_{i+2}| \le s + d$, and so $|S_{i+2} - S_i| \le d$ for all $i = 1, 2, \ldots \ell - 2$. As $|S_{i+2} - S_1| \le |S_{i+2} - S_i| + |S_i - S_1|$, induction on t shows that $|S_{1+2t} - S_1| \le td$ for all $t = 1, 2, \ldots \lfloor (\ell - 1)/2 \rfloor$. In particular, if ℓ is odd, then $|S_\ell - S_1| \le (\ell - 1)d/2$; since S_ℓ and S_1 are disjoint, we conclude that $(\ell - 1)d/2 \ge s$, which is (9.5).

Now let us deduce

$$\lambda(n, k) \ge \tfrac{1}{7} n^{1/(k-2)} \tag{9.6}$$

from Theorem 9.15, Proposition 9.16, and observation (9.5). Given any pair of positive integers n and k, we have to find a graph of order n and chromatic number at least k in which every odd cycle has length at least $\tfrac{1}{7} n^{1/(k-2)}$. For this purpose, set $d = k - 2$ and let s be the largest integer such that $\frac{2s+d}{s} \binom{s+d-1}{s-1} \le n$. Consider the graph that consists of $SG_{s,d}$ and an additional $n - \frac{2s+d}{s} \binom{s+d-1}{s-1}$ vertices of degree 0. This graph has order n, chromatic number k, and each of its odd cycles has length at least $1 + 2s/d$; to complete the proof of (9.6), we only need verify that

$$1 + 2s/d \ge \tfrac{1}{7} n^{1/d}.$$

To do this, note that maximality of s guarantees

$$n < \frac{2s+2+d}{s+1} \binom{s+d}{s} = \frac{2s+2+d}{s+1} \binom{s+d}{d}.$$

We have

$$\frac{2s+2+d}{s+1} = 2 + \frac{d}{s+1} \le 2 + \frac{d}{2} \le 2.5d;$$

since $k! > (k/e)^k$ for all positive integers k, we have

$$\binom{s+d}{d} < \left(\frac{e(s+d)}{d} \right)^d;$$

it follows that

$$n^{1/d} < 2.5e \left(1 + \frac{s}{d} \right) < 2.5e \left(1 + \frac{2s}{d} \right).$$

Hal Kierstead, Endre Szemerédi, and Tom Trotter [246] complemented (9.6) by proving that

$$\lambda(n, k) \le 4(k-2)n^{1/(k-2)} + 1. \tag{9.7}$$

9.6 An Upper Bound on the Chromatic Number

Erdős and Hajnal [140] defined the *colouring number* of finite and infinite graphs.

DEFINITION: When G is a finite graph, its colouring number $\text{Col}(G)$ is the smallest integer k for which the vertices of G can be enumerated in a sequence $v_1, v_2, \ldots, v_n$ in such a way that each v_j has fewer than k neighbours v_i with $i < j$.

Section §3 of [140] begins with the sentence

It is well known that $\chi(G) \leq \text{Col}(G)$ for every graph G.

This upper bound is easy to justify: if the vertices of G are enumerated as $v_1, v_2, \ldots, v_n$ in such a way that each v_j has fewer than $\text{Col}(G)$ neighbours v_i with $i < j$, then a proper colouring of G by colours $1, 2, \ldots, \text{Col}(G)$ can be constructed by scanning the sequence from v_1 to v_n and colouring each v_j with the smallest positive integer that appears as a colour on none of its neighbours v_i with $i < j$.

George Szekeres and Herbert Wilf established the following upper bound on the chromatic number:

THEOREM 9.17 ([354])

$$\chi(G) \leq \max_F \min_v d_F(v) + 1,$$

where F runs through all subgraphs of G and $d_F(v)$ stands for the degree of vertex v in graph F.

Proof We will prove the theorem by showing that

$$\text{Col}(G) \leq \max_F \min_v d_F(v) + 1. \tag{9.8}$$

For this purpose, we will use induction on the order of G and we will write $SW(G)$ for the right-hand side of (9.8). For the induction basis, note that the one-vertex graph G has $\text{Col}(G) = SW(G) = 1$. In the induction step consider an arbitrary graph H of order $n + 1$, let w be the vertex of H that minimizes $d_H(v)$, and let G denote H with vertex w (and all the edges incident to w) removed. By the induction hypothesis, the vertices of G can be enumerated in a sequence $v_1, v_2, \ldots, v_n$ in such a way that each v_j has fewer than $SW(G)$ neighbours v_i with $i < j$; if we set $v_{n+1} = w$, then v_{n+1} has precisely $d_H(w)$ neighbours v_i with $i < n + 1$; since $SW(G) \leq SW(H)$ and $d_H(w) = \min_v d_H(v) < SW(H)$, we conclude that each v_j with $1 \leq j \leq n + 1$ has fewer than $SW(H)$ neighbours v_i with $i < j$. $\qquad\square$

Actually, we have

$$\text{Col}(G) = SW(G)$$

for all graphs G. To verify that $\text{Col}(G) \geq SW(G)$, consider an arbitrary ordering $v_1, v_2, \ldots, v_n$ of vertices of G; let F be the subgraph of G that maximizes $\min_v d_F(v)$ and let j be the largest subscript such that v_j belongs to F. Since v_j has $d_F(v_j)$ neighbours v_i in F and since all of these neighbours v_i have $i < j$, we have $\text{Col}(G) > d_F(v_j) \geq \min_v d_F(v) = SW(G) - 1$.

Unlike the chromatic number, the colouring number is easy to compute: implicit in our inductive proof of $\text{Col}(G) \leq SW(G)$ is an efficient recursive algorithm that, given any graph G of order n, enumerates the vertices of G in a sequence $v_1, v_2, \ldots, v_n$ so that each v_j has fewer than $SW(G)$ neighbours v_i with $i < j$.

The upper bound in Theorem 9.17 is next to useless for almost all graphs: the typical ratio $\mathrm{Col}(G)/\chi(G)$ grows logarithmically with the number of vertices. More precisely, almost all graphs G of order n have

$$\chi(G) \sim \frac{n}{2 \lg n} \quad \text{and} \quad \mathrm{Col}(G) \sim \frac{n}{2}.$$

The asymptotic formula for $\chi(G)$ is (9.2) and the asymptotic formula for $\mathrm{Col}(G)$ follows from another property of almost all graphs G: the degrees of all their n vertices are $(1 + o(1))n/2$. We will prove this in Subsection A.5; a much finer result of Erdős and Rényi is Theorem 10 in [148].

Still, Theorem 9.17 is often useful. In particular, we will use it in a proof that the bound in Theorem 9.14 cannot be improved beyond a constant factor.

THEOREM 9.18 (Theorem 1 of [111]) *Every graph G of order n with $\chi(G) \geq k \geq 4$ contains a cycle of length at most $2\lceil \log n / \log(k - 2) \rceil$.*

A major part of the proof of this theorem consists of the following lemma:

LEMMA 9.19 *If d, r are integers such that $d \geq 3$, $r \geq 2$ and if F is a graph such that $\min_v d_F(v) = d$ and F contains no cycle of length at most $2r$, then F has more than $(d - 1)^r$ vertices.*

Proof Choose a vertex s of F and let V_i denote the set of all vertices v of F such that i is the length of a shortest path between s and v. (In particular, $V_0 = \{s\}$ and V_1 consists of all the neighbours of s.)

Now consider arbitrary distinct vertices v and w in V_i. There are a path of length i between s and v and a path of length i between s and w; the union of these two paths contains a path P of length at most $2i$ between v and w. Let us make a note of this:

there is a path P_{vw} of length at most $2i$ between any two vertices v, w in V_i.

When $v, w \in V_i$, adding edge vw to P_{vw} would create a cycle of length at most $2i + 1$, and so our assumption that F contains no cycle of length at most $2r$ implies that

(P1) if $1 \leq i \leq r - 1$, then no two vertices in V_i are adjacent.

When $v, w \in V_{i-1}$, adding a path of length two to P_{vw} (whose length is at most $2i - 2$) would create a cycle of length at most $2i$, and so our assumption that F contains no cycle of length at most $2r$ implies that

(P2) if $1 \leq i \leq r$, then every vertex in V_i has only one neighbour in V_{i-1}.

Now consider an arbitrary integer i in the range $1 \leq i \leq r - 1$. By definition, the neighbours of every vertex in V_i come from $V_{i-1} \cup V_i \cup V_{i+1}$; this fact and (P1) and (P2) together imply that every vertex in V_i has at least $d - 1$ neighbours in V_{i+1}; it follows that there are at least $(d - 1)|V_i|$ edges with one endpoint in V_i and the other endpoint in V_{i+1}. By (P2), every vertex in V_{i+1} has at most one neighbour in V_i, and so there are at most $|V_{i+1}|$ edges with one endpoint in V_i and the other endpoint in V_{i+1}. We conclude that

$$|V_{i+1}| \geq (d-1)|V_i| \quad \text{for all } i = 1, 2, \ldots r - 1.$$

Since $|V_1| \geq d$, it follows that $|V_i| \geq d(d-1)^{i-1}$ for all $i = 1, 2, \ldots r$, and so

$$\sum_{i=0}^{r} |V_i| \geq 1 + \sum_{i=1}^{r} d(d-1)^{i-1} > (d-1)^r. \qquad \square$$

Proof of Theorem 9.18   Setting $r = \lceil \log n / \log(k-2) \rceil$, note that $(k-2)^r \geq n$. Theorem 9.17 guarantees that G contains a subgraph F such that $\min_v d_F(v) \geq k - 1$. In turn, Lemma 9.19 guarantees that F contains a cycle of length at most $2r$. $\square$

We will use Theorem 9.17 again in the next section.

9.7   Small Subgraphs Do Not Determine Chromatic Number

Since $\chi(F) \geq 3$ if and only if F contains an odd cycle, inequality (9.6) highlights the global character of the chromatic number: there are arbitrarily large graphs such that $\chi(G) > k$ and yet $\chi(F) \leq 2$ for every subgraph F of G with fewer than $\frac{1}{7} n^{1/(k-2)}$ vertices. Inequality (9.7) shows that the upper bound on the order of F cannot be raised by a factor greater than $28(k-2)$; Erdős proved that it can be raised dramatically as soon as the upper bound on $\chi(F)$ is incremented by 1:

THEOREM 9.20 (Theorem 2 of [111])   *For every positive integer k, there are a positive constant δ and graphs G with arbitrarily large orders n such that $\chi(G) \geq k$ and yet $\chi(F) \leq 3$ for every subgraph F of G with at most δn vertices.*

We will treat two parts of the proof of this theorem as separate lemmas.

LEMMA 9.21   *For every positive integer c, there are positive constants n_0 and δ with the following property: If $n \geq n_0$, then in at least 99% of all graphs with n vertices and cn edges, every subgraph with s vertices such that $0 < s \leq \delta n$ has fewer than $3s/2$ edges.*

Proof   Given a positive integer c, set

$$n_0 = 6c + 3 \quad \text{and} \quad \delta = \frac{1}{64e^5 c^3};$$

given an integer n, let **P** denote the set of all pairs (G, F), where G is a graph with vertices $1, 2, \ldots, n$ and cn edges and F is an induced subgraph of G such that the number s of vertices of F is positive and at most δn and the number j of edges of F is at least $3s/2$.

We aim to prove that, when $n \geq n_0$, at least 99% of all graphs with vertices $1, 2, \ldots, n$ vertices and cn edges appear as G in no pair in **P**, which means that the number of graphs appearing as G in these pairs is at most

$$\frac{1}{100} \binom{\binom{n}{2}}{cn}.$$

To prove this, note first that the number of graphs appearing as G in the pairs in $\mathbf{P}$ is at most $|\mathbf{P}|$. If a subgraph F of G has s vertices and at least $3s/2$ edges, then $s \geq 4$; the number of such subgraphs F with precisely j edges is

$$\binom{n}{s}\binom{\binom{s}{2}}{j}$$

and each of them appears in precisely

$$\binom{\binom{n}{2} - \binom{s}{2}}{cn - j}$$

pairs in $\mathbf{P}$. It follows that

$$|\mathbf{P}| = \sum_{4 \leq s \leq \delta n} \sum_{j \geq 3s/2} \binom{n}{s}\binom{\binom{s}{2}}{j}\binom{\binom{n}{2} - \binom{s}{2}}{cn - j}.$$

The bulk of the proof is a computation showing that

$$\sum_{4 \leq s \leq \delta n} \sum_{j \geq 3s/2} \binom{n}{s}\binom{\binom{s}{2}}{j}\binom{\binom{n}{2} - \binom{s}{2}}{cn - j}\binom{\binom{n}{2}}{cn}^{-1} \leq \frac{1}{100}.$$

Here are its ingredients with $j \geq 3s/2$ assumed throughout:

- By (A.3), we have

$$\binom{n}{s} \leq \left(\frac{en}{s}\right)^s \leq \left(\frac{en}{s}\right)^{2j/3}.$$

- By (A.2) with j in place of k, we have

$$\binom{\binom{s}{2}}{j} \leq \frac{\binom{s}{2}^j}{j!} \leq \left(\frac{es(s-1)}{2j}\right)^j \leq \left(\frac{es}{3}\right)^j.$$

- When $n \geq n_0 = 6c + 3$, we have $\binom{n}{2} - cn \geq n^2/3$, and so

$$\binom{\binom{n}{2} - \binom{s}{2}}{cn - j}\binom{\binom{n}{2}}{cn}^{-1} \leq \binom{\binom{n}{2}}{cn - j}\binom{\binom{n}{2}}{cn}^{-1}$$

$$= \frac{(\binom{n}{2} - cn)!}{(\binom{n}{2} - cn + j)!} \cdot \frac{(cn)!}{(cn-j)!} \leq \left(\frac{cn}{\binom{n}{2} - cn}\right)^j \leq \left(\frac{3cn}{n^2}\right)^j = \left(\frac{3c}{n}\right)^j.$$

Combining these ingredients, we find that, whenever $n \geq n_0$ and $s \leq \delta n$, we have

$$\binom{n}{s}\binom{\binom{s}{2}}{j}\binom{\binom{n}{2} - \binom{s}{2}}{cn - j}\binom{\binom{n}{2}}{cn}^{-1} \leq \left(\frac{en}{s}\right)^{2j/3} \cdot \left(\frac{es}{3}\right)^j \cdot \left(\frac{3c}{n}\right)^j$$

$$= \left(\frac{ce^{5/3}s^{1/3}}{n^{1/3}}\right)^j \leq \left(ce^{5/3}\delta^{1/3}\right)^j = \left(\frac{1}{4}\right)^j$$

and so

$$\sum_{j \geq 3s/2} \binom{n}{s}\binom{\binom{s}{2}}{j}\binom{\binom{n}{2} - \binom{s}{2}}{cn - j}\binom{\binom{n}{2}}{cn}^{-1} \leq \sum_{j \geq 3s/2} \left(\frac{1}{4}\right)^j$$

$$= \left(\frac{1}{4}\right)^{\lceil 3s/2 \rceil} \sum_{i=0}^{\infty} \left(\frac{1}{4}\right)^i \le \left(\frac{1}{4}\right)^{3s/2} \sum_{i=0}^{\infty} \left(\frac{1}{4}\right)^i = \frac{4}{3}\left(\frac{1}{4}\right)^{3s/2} = \frac{4}{3}\left(\frac{1}{8}\right)^s,$$

and so

$$\sum_{4 \le s \le \delta n} \sum_{j \ge 3s/2} \binom{n}{s}\binom{\binom{s}{2}}{j}\binom{\binom{n}{2}-\binom{s}{2}}{cn-j}\binom{\binom{n}{2}}{cn}^{-1} \le$$

$$\frac{4}{3} \sum_{4 \le s \le \delta n} \left(\frac{1}{8}\right)^s \le \frac{4}{3} \sum_{s=4}^{\infty} \left(\frac{1}{8}\right)^s = \frac{4}{3}\left(\frac{1}{8}\right)^4 \frac{8}{7} = \frac{1}{2688}. \qquad \square$$

REMINDER: A *stable set* in a graph is a set of pairwise nonadjacent vertices; the *stability number* $\alpha(G)$ of a graph G is the number of vertices in its largest stable set.

LEMMA 9.22 *If m is an integer-valued function of n such that $e^3 n \le m(n) \le \binom{n}{2}$ for all sufficiently large n, then for every positive ε almost all graphs G with n vertices and $m(n)$ edges have*

$$\alpha(G) < (1+\varepsilon)\frac{n^2}{m(n)} \ln \frac{m(n)}{n}.$$

Proof With

$$s(n) = \left\lceil (1+\varepsilon)\frac{n^2}{m(n)} \ln \frac{m(n)}{n} \right\rceil,$$

let $\mathbf{P}$ denote the set of all pairs (G, F), where G is a graph with vertices $1, 2, \ldots, n$ and with $m(n)$ edges and F is an induced subgraph of G with $s(n)$ vertices and with no edges at all. To reduce the clutter in the formulas that follow, we shall write simply m, s for $m(n), s(n)$, respectively.

The lemma asserts that almost all graphs with n vertices and m edges appear as G in no pair in $\mathbf{P}$, which means that the number of graphs appearing as G in these pairs is

$$o\left(\binom{\binom{n}{2}}{m}\right).$$

To prove this, note first that the number of graphs appearing as G in the pairs in $\mathbf{P}$ is at most $|\mathbf{P}|$. There are precisely

$$\binom{n}{s}$$

choices of F and every F appears in precisely

$$\binom{\binom{n}{2}-\binom{s}{2}}{m}$$

pairs in $\mathbf{P}$; it follows that

$$|\mathbf{P}| = \binom{n}{s}\binom{\binom{n}{2}-\binom{s}{2}}{m}.$$

The bulk of the proof is a computation showing that

$$\binom{n}{s}\left(\frac{\binom{n}{2}-\binom{s}{2}}{m}\right)\binom{\binom{n}{2}}{m}^{-1} = o(1).$$

Here are its ingredients :

- By (A.2), we have

$$\binom{n}{s} \le \left(\frac{en}{s}\right)^s.$$

- Since $1+x \le e^x$ for all x, we have

$$\left(\frac{\binom{n}{2}-\binom{s}{2}}{m}\right)\binom{\binom{n}{2}}{m}^{-1} \le \left(\frac{\binom{n}{2}-\binom{s}{2}}{\binom{n}{2}}\right)^m$$

$$= \left(1 - \frac{\binom{s}{2}}{\binom{n}{2}}\right)^m \le e^{-ms(s-1)/n(n-1)}.$$

- For all sufficiently large n we have

$$s \ge \ln n$$

(if $m \le n^{3/2}$, then $s \ge n^{1/2}\ln(m/n) \ge 3n^{1/2}$; if $m \ge n^{3/2}$, then $s \ge 2\ln(m/n) \ge \ln n$), and so $s \ge 1+1/\varepsilon$, and so

$$\frac{m(s-1)}{n(n-1)} \ge \frac{m(s-1)}{n^2} \ge \frac{ms}{(1+\varepsilon)n^2} \ge \ln\frac{m}{n}.$$

Combining these ingredients, we find that

$$\binom{n}{s}\left(\frac{\binom{n}{2}-\binom{s}{2}}{m}\right)\binom{\binom{n}{2}}{m}^{-1} \le \left(\frac{en}{s}\right)^s e^{-ms(s-1)/n(n-1)}$$

$$\le \left(\frac{en}{s}\right)^s\left(\frac{n}{m}\right)^s \le \left(\frac{e}{\ln(m/n)}\right)^s \le \left(\frac{e}{3}\right)^s \le \left(\frac{e}{3}\right)^{\ln n} = o(1). \quad\square$$

The upper bound of Lemma 9.22 is asymptotically tight in the sense that

$$\alpha(G) = (1+o(1))\frac{n^2}{m(n)}\ln\frac{m(n)}{n}$$

for almost all graphs G with n vertices and $m(n)$ edges such that $m(n)=o(n^2)$. A far finer asymptotic formula for $\alpha(G)$ has been established by Alan Frieze [174].

Proof of Theorem 9.20   Given a positive integer k, choose a positive integer c so that $c/\ln c > 2k$ and $c > e^3$. By Lemma 9.22, almost all graphs G with n vertices and cn edges have

(i) $\alpha(G) < n/k$;

by Lemma 9.21, there is a positive constant δ such that in at least 99% of all sufficiently large graphs G with n vertices and cn edges,

(ii) every subgraph F of G with s vertices such that $0 < s \leq \delta n$ has fewer than $3s/2$ edges.

It follows that, for all sufficently large n, at least 98% of all graphs G with n vertices and cn edges have both properties (i) and (ii). Property (i) implies that $\chi(G) > n/k$; property (ii) implies that

(iii) every subgraph F of G with at most δn vertices has $\min_v d_F(v) \leq 2$;

by Theorem 9.17, property (iii) implies that every subgraph of G with at most δn vertices has chromatic number at most 3.   □

The paradigm of the arguments pioneered by Erdős in his proofs of Theorems 9.1, 9.11, and 9.20 is simple, even though the requisite computations may be intricate. Here it goes:

LEMMA 9.23 (Counting pairs in two different ways)   *Let Ω and Δ be finite sets, let* **P** *be a subset of $\Omega \times \Delta$, and let s be a positive integer. For every element F of Δ, let $t(F)$ denote the number of pairs in* **P** *that contain F. Then at most $\frac{1}{s}\sum_{F\in\Delta} t(F)$ elements of Ω appear in at least s pairs in* **P**.

Proof   With r standing for the number of elements of Ω that appear in at least s pairs in **P**, we have

$$|\mathbf{P}| = \sum_{G\in\Omega}\ \sum_{(G,F)\in\mathbf{P}} 1 \ \geq\ rs,$$

$$|\mathbf{P}| = \sum_{F\in\Delta}\ \sum_{(G,F)\in\mathbf{P}} 1 \ =\ \sum_{F\in\Delta} t(F),$$

and so $rs \leq \sum_{F\in\Delta} t(F)$.   □

For instance, in Lemma 9.12, Ω is the set of all graphs with vertices $1, 2, \ldots, n$ and with m edges; Δ is the set of all cycles of length at most ℓ with vertices coming from $\{1, 2, \ldots, n\}$; a pair (G, F) in $\Omega \times \Delta$ belongs to **P** if and only if F is a subgraph of G. Here, $s = n$ and the proof of the lemma amounts to establishing that $\frac{1}{n}\sum_{F\in\Delta} t(F) = o\left(|\Omega|\right)$.

In the proofs of Theorem 9.1, Lemma 9.13, Lemma 9.22, and Lemma 9.21, Lemma 9.23 is used in its rudimentary form, where $s = 1$ and $t(y)$ does not depend on y.

We shall come back to Lemma 9.23 in Section 10.4.

In the fall of 1974, having just returned to Stanford after a two-year hiatus, I bought another second-hand Mustang and once again fell under the spell of a fast car in a way that would befit an adolescent. With Neal Cassady never far from my thoughts, I kept the gas pedal down to the floor whenever feasible. I was given a $120 fine for driving at 110 mph on highway 5 to Los Angeles. My passengers often pleaded with me to slow down and occasionally they elected to return home by public transport.

Just as it happened earlier in Ontario, I slipped into the role of Erdős's personal driver in California. My first assignment in this new phase of my career was to take him from Stanford to Santa Barbara. At first, I kept to a sedate five miles above the speed limit: the frequent frightened reactions of my fellow travellers were vivid in my mind. Eventually though, I began stealthily accelerating until ours was the fastest car on this stretch of the 101. Soft slapping of tires on the tarmac and the intermittent *whoosh* of our overtaking cars in the slower lane were like a soothing background music.

After an hour or so of this an automobile appeared in my rearview mirror and when it closed the gap between us, I moved to the right lane to let it pass. When Erdős noticed the car to our left inching its way past us, he gave me a critical look. "Why are you going so slowly?" he asked.

10   Thresholds of Graph Properties

Paul Erdős and his friend Alfréd Rényi laid the foundations of the theory of random graphs in [147] and [148]. Here is its key notion.

DEFINITION: The *probability that a random graph with n vertices and m edges has some property* (such as the property of being connected or the property of containing a triangle) is defined as the ratio between the number of graphs with vertices $1, 2, \ldots, n$ and with m edges that have this property and the total number

$$\binom{\binom{n}{2}}{m}$$

of all graphs with vertices $1, 2, \ldots, n$ and with m edges.

As noted on page 136, Erdős and Rényi introduced in [148] the expression *almost all graphs with n vertices and m(n) edges have a certain property,* meaning that the probability of a random graph with n vertices and $m(n)$ edges having this property tends to 1 as n tends to infinity. Their study of random graphs revealed that a number of graph properties have *threshold functions* θ, meaning that

- if $\lim_{n \to \infty} m(n)/\theta(n) = 0$,
  then almost all graphs with n vertices and $m(n)$ edges lack the property,
- if $\lim_{n \to \infty} m(n)/\theta(n) = \infty$,
  then almost all graphs with n vertices and $m(n)$ edges have the property.

In [148], and even more emphatically in its extended abstract [149], Erdős and Rényi proposed to interpret their random graphs with vertices $1, 2, \ldots, n$ as evolving step by step through adding edges one at a time, beginning with no edges at all at time 0. Each of the $\binom{n}{2}$ candidates is equally likely to be introduced into the graph at time 1, then each of the remaining $\binom{n}{2} - 1$ candidates is equally likely to be added at time 2, and so on: at time t, the graph has acquired t edges and each of the missing $\binom{n}{2} - t$ edges is equally likely to be added to it. Clearly, all of the

$$\binom{\binom{n}{2}}{t}$$

graphs with vertices $1, 2, \ldots, n$ and t edges are equally likely to appear at time t.

Alfred Rényi and Paul Erdős in 1957

Courtesy of János Pach

10.1   Connectivity

REMINDER: When f, g, d are real-valued functions defined on positive integers, we write $f(n) = o(g(n))$ to mean that $\lim_{n \to \infty} f(n)/g(n) = 0$ and we write $f(n) = g(n) + o(d(n))$ to mean that $f(n) - g(n) = o(d(n))$.

The first theorem on random graphs that Erdős and Rényi proved implies that $n \log n$ is the threshold function for the property of being connected. Actually, this theorem is far more precise:

THEOREM 10.1   *([147, Theorem 1]) Let $p(n, m)$ denote the probability that a random graph with n vertices and m edges is connected. If c is a real number and if $m : N \to N$ is a function such that $m(n) = \frac{1}{2} n \ln n + cn + o(n)$, then*

$$\lim_{n \to \infty} p(n, m(n)) = e^{-e^{-2c}}.$$

Since

$$e^{-e^{1.528}} < 0.01 \quad \text{and} \quad e^{-e^{-4.602}} > 0.99,$$

this theorem guarantees that more than 99% of the random graphs that evolve step by step through adding edges one at a time switch from being disconnected to being connected within the time window that extends from $t = \frac{1}{2}n \ln n - \lceil 0.764n \rceil$ to $t = \frac{1}{2}n \ln n + \lceil 2.301n \rceil$.

We will prove Theorem 10.1 in Section 10.1.4. The proof uses tools that are applicable in more general settings; we will present these tools in Section 10.1.1.

10.1.1 The Inclusion-Exclusion Principle and Bonferroni Inequalities

Of the 28 governors of a certain university, 11 drink and gamble, 21 drink (and possibly also gamble), and 17 gamble (and possibly also drink). How many of them neither drink nor gamble? To answer this question, let us write

$x(\{1, 2\})$	for the number of governors who drink and gamble,
$x(\{1\})$	for the number of governors who drink but do not gamble,
$x(\{2\})$	for the number of governors who gamble but do not drink,
$x(\emptyset)$	for the number of governors who neither drink nor gamble.

We are told that

$$
\begin{aligned}
x(\{1,2\}) &&&&&= 11 \\
x(\{1,2\}) &+x(\{1\}) &&&&= 21 \\
x(\{1,2\}) &&+x(\{2\}) &&&= 17 \\
x(\{1,2\}) &+x(\{1\}) &+x(\{2\}) &+x(\emptyset) &&= 28;
\end{aligned}
$$

solving this system of equations, we find that $x(\{1\}) = 10$, $x(\{2\}) = 6$, and $x(\emptyset) = 1$.

More generally, consider any finite set S and any two subsets S_1, S_2 of S. These two subsets partition S into four pairwise disjoint *atoms*

$$S_1 \cap S_2, \quad S_1 - S_2, \quad S_2 - S_1, \quad S - (S_1 \cup S_2),$$

some of which may be empty. Let $x(\{1, 2\})$, $x(\{1\})$, $x(\{2\})$, $x(\emptyset)$ denote the sizes of these atoms: let us write

$x(\{1, 2\})$	for the number of elements in both S_1 and S_2,
$x(\{1\})$	for the number of elements in S_1 and outside S_2,
$x(\{2\})$	for the number of elements in S_2 and outside S_1,
$x(\emptyset)$	for the number of elements outside both S_1 and S_2.

Furthermore, let us write

$y(\{1, 2\})$	for the number $x(\{1, 2\})$ of elements in both S_1 and S_2,
$y(\{1\})$	for the number of elements in S_1,
$y(\{2\})$	for the number of elements in S_2,
$y(\emptyset)$	for the total number of elements in S.

By definition, we have

$$
\begin{aligned}
x(\{1,2\}) &&&&&= y(\{1,2\}) \\
x(\{1,2\}) &+x(\{1\}) &&&&= y(\{1\}) \\
x(\{1,2\}) &&+x(\{2\}) &&&= y(\{2\}) \\
x(\{1,2\}) &+x(\{1\}) &+x(\{2\}) &+x(\emptyset) &&= y(\emptyset);
\end{aligned}
$$

solving this system of equations, we find that

$$
\begin{aligned}
x(\{1,2\}) &= y(\{1,2\}) \\
x(\{1\}) &= -y(\{1,2\}) +y(\{1\}) \\
x(\{2\}) &= -y(\{1,2\}) +y(\{2\}) \\
x(\emptyset) &= y(\{1,2\}) -y(\{1\}) -y(\{2\}) +y(\emptyset).
\end{aligned}
$$

The formula for $x(\emptyset)$ is a rudimentary case of so-called *inclusion-exclusion principle*. To explain why this term is being used, let us write out the derivation of the formula as

$$
\begin{aligned}
x(\emptyset) &= y(\emptyset) - \Big(x(\{1\}) + x(\{2\}) + x(\{1,2\}) \Big) \\
&= y(\emptyset) - \Big(y(\{1\}) + y(\{2\}) \Big) + x(\{1,2\}) \\
&= y(\emptyset) - \Big(y(\{1\}) + y(\{2\}) \Big) + y(\{1,2\}).
\end{aligned}
$$

To calculate the number $x(\emptyset)$ of governors who neither drink nor gamble, we begin by including all $y(\emptyset)$ governors; this is an upper bound on $x(\emptyset)$, and so next we exclude all $y(\{1\})$ governors who drink (and possibly also gamble) as well as all $y(\{2\})$ governors who gamble (and possibly also drink); but then the $y(\{1,2\})$ governors who both drink and gamble have been excluded twice in the resulting count $y(\emptyset) - (y(\{1\}) + y(\{2\}))$ and we compensate for the double exclusion by including them back.

To take this principle to its full generality, consider any finite set S and any n subsets $S_1, S_2, \ldots, S_n$ of S. These n subsets partition S into 2^n pairwise disjoint atoms, some of which may be empty. Let $x(I)$ with I running through subsets of $\{1, 2, \ldots, n\}$ denote the sizes of these atoms, so that $x(I)$ is the number of elements that belong to an S_k if and only if $k \in I$. Furthermore, for each subset J of $\{1, 2, \ldots, n\}$, let $y(J)$ denote the number of elements that belong to an S_k if (but not necessarily only if) $k \in J$. By definition, we have

$$
\sum_{I \supseteq J} x(I) = y(J) \qquad \text{for all } J. \tag{10.1}
$$

THEOREM 10.2   *(The inclusion-exclusion principle.) Every solution of (10.1) satisfies*

$$
x(\emptyset) = \sum_J (-1)^{|J|} y(J).
$$

Proof   If i is a positive integer, then $\sum_{j=0}^{i}(-1)^j \binom{i}{j} = (-1+1)^i = 0$ by the binomial formula; if $i = 0$, then $\sum_{j=0}^{i}(-1)^j \binom{i}{j} = (-1)^0 \binom{0}{0} = 1$. It follows that every solution of (10.1) satisfies

$$
\sum_J (-1)^{|J|} y(J) = \sum_J (-1)^{|J|} \left(\sum_{I \supseteq J} x(I) \right) = \sum_I \left(\sum_{J \subseteq I} (-1)^{|J|} \right) x(I) =
$$

$$
\sum_I \left(\sum_{j=0}^{|I|} (-1)^j \binom{|I|}{j} \right) x(I) = x(\emptyset). \qquad \square
$$

Our justification of the term "inclusion-exclusion" hints at the following inequalities.

THEOREM 10.3   *(Bonferroni inequalities.) Every solution of (10.1) satisfies*

$$x(\emptyset) \le \sum_{|J| \le k} (-1)^{|J|} y(J) \quad \textit{for every nonnegative even integer } k,$$

$$x(\emptyset) \ge \sum_{|J| \le k} (-1)^{|J|} y(J) \quad \textit{for every nonnegative odd integer } k.$$

Proof If k and i are positive integers, then $\binom{i-1}{k+1} + \binom{i-1}{k} = \binom{i}{k+1}$, and so straightforward induction on k shows that $\sum_{j=0}^{k} (-1)^j \binom{i}{j} = (-1)^k \binom{i-1}{k}$. It follows that every solution of (10.1) satisfies, for every nonnegative integer k,

$$\sum_{|J| \le k} (-1)^{|J|} y(J) = \sum_{|J| \le k} (-1)^{|J|} \left(\sum_{I \supseteq J} x(I) \right) = \sum_{I} \left(\sum_{\substack{J \subseteq I \\ |J| \le k}} (-1)^{|J|} \right) x(I)$$

$$= \sum_{I} \left(\sum_{j=0}^{k} (-1)^j \binom{|I|}{j} \right) x(I) = x(\emptyset) + (-1)^k \sum_{|I| \ge 1} \binom{|I|-1}{k} x(I). \quad \square$$

Theorem 10.3 is implicit in the proof of the inclusion-exclusion principle published by Károly Jordán (1871–1959) in 1927 [230]. Its inequalities are named *Bonferroni inequalities* after Carlo Emilio Bonferroni (1892–1960), who, having proved them later [50], used them again and again in statistics.

10.1.2   Lemma on Isolated Vertices

DEFINITION: A vertex in a graph is called *isolated* if its degree is zero.

NOTATION: We let **N** stand for the set of all positive integers.

REMINDER: When f, g, d are real-valued functions defined on positive integers, we write $f(n) = O(g(n))$ to mean that $|f(n)| \le cg(n)$ for some positive constant c and all sufficiently large n and we write $f(n) = g(n) + O(d(n))$ to mean that $f(n) - g(n) = O(d(n))$.

LEMMA 10.4   *Let $f(n, m)$ denote the probability that a random graph with n vertices and m edges has no isolated vertices. If c is a real number and if $m : N \to N$ is a function such that $m(n) = \frac{1}{2} n \ln n + cn + o(n)$, then*

$$\lim_{n \to \infty} f(n, m(n)) = e^{-e^{-2c}}.$$

Proof To reduce the clutter in the formulas that follow, let us write simply m for $m(n)$. Given a positive integer n, consider the set S of all graphs with vertices $1, 2, \ldots, n$ and with m edges. Letting S_j denote the set of all graphs in S where vertex j is isolated produces the situation of Section 10.1.1 with $x(\emptyset)$ the number of graphs in S that have no isolated vertices and with $y(J)$ the number of graphs in S where all the vertices in J (and possibly some other vertices, too) are isolated. Since

$$x(\emptyset) = f(n,m)\binom{\binom{n}{2}}{m} \quad \text{and} \quad y(J) = \binom{\binom{n-|J|}{2}}{m},$$

Theorem 10.3 guarantees that

$$f(n,m) \le \sum_{|J| \le k} (-1)^{|J|} \binom{\binom{n-|J|}{2}}{m} \binom{\binom{n}{2}}{m}^{-1} \quad \text{for every nonnegative even integer } k,$$

$$f(n,m) \ge \sum_{|J| \le k} (-1)^{|J|} \binom{\binom{n-|J|}{2}}{m} \binom{\binom{n}{2}}{m}^{-1} \quad \text{for every nonnegative odd integer } k,$$

which means that

$$\sum_{j=0}^{2k+1} (-1)^j \binom{n}{j} \binom{\binom{n-j}{2}}{m} \binom{\binom{n}{2}}{m}^{-1}$$

$$\le f(n,m) \le$$

$$\sum_{j=0}^{2k} (-1)^j \binom{n}{j} \binom{\binom{n-j}{2}}{m} \binom{\binom{n}{2}}{m}^{-1} \quad (10.2)$$

for every nonnegative integer k. We will prove that, for every nonnegative integer j,

$$\lim_{n \to \infty} \binom{n}{j} \binom{\binom{n-j}{2}}{m} \binom{\binom{n}{2}}{m}^{-1} = \frac{e^{-2cj}}{j!}; \quad (10.3)$$

substituting into (10.2), we get

$$\sum_{j=0}^{2k+1} (-1)^j \frac{(e^{-2c})^j}{j!} \le \liminf_{n \to \infty} f(n,m) \le \limsup_{n \to \infty} f(n,m) \le \sum_{j=0}^{2k} (-1)^j \frac{(e^{-2c})^j}{j!}$$

and the proof of the lemma is concluded by recalling that

$$\lim_{k \to \infty} \sum_{j=0}^{2k+1} (-1)^j \frac{x^j}{j!} = \lim_{k \to \infty} \sum_{j=0}^{2k} (-1)^j \frac{x^j}{j!} = e^{-x}.$$

Proof of (10.3)

Note that

$$\ln\left(\binom{\binom{n-j}{2}}{m} \binom{\binom{n}{2}}{m}^{-1}\right) = \ln \prod_{i=0}^{m-1} \left(\frac{\binom{n-j}{2} - i}{\binom{n}{2} - i} \right) = \sum_{i=0}^{m-1} \ln\left(1 - \frac{\binom{n}{2} - \binom{n-j}{2}}{\binom{n}{2} - i} \right),$$

and so

$$m \ln\left(1 - \frac{\binom{n}{2} - \binom{n-j}{2}}{\binom{n}{2} - m + 1} \right) \le \ln\left(\binom{\binom{n-j}{2}}{m} \binom{\binom{n}{2}}{m}^{-1} \right) \le m \ln\left(1 - \frac{\binom{n}{2} - \binom{n-j}{2}}{\binom{n}{2}} \right).$$

Since

$$\frac{\binom{n}{2} - \binom{n-j}{2}}{\binom{n}{2} - m + 1} = \frac{\binom{n}{2} - \binom{n-j}{2}}{\binom{n}{2}} \cdot \frac{\binom{n}{2}}{\binom{n}{2} - m + 1}$$

and

$$\frac{\binom{n}{2}}{\binom{n}{2} - m + 1} = 1 + \frac{m-1}{\binom{n}{2} - m + 1} = 1 + O(n^{-1} \ln n),$$

it follows that

$$\ln\left(\binom{\binom{n-j}{2}}{m}\binom{\binom{n}{2}}{m}^{-1}\right) = m \ln\left(1 - \left(1 + O(n^{-1} \ln n)\right)\frac{\binom{n}{2} - \binom{n-j}{2}}{\binom{n}{2}}\right).$$

This and

$$\frac{\binom{n}{2} - \binom{n-j}{2}}{\binom{n}{2}} = \frac{j(n-j) + \binom{j}{2}}{\binom{n}{2}} = \frac{2j}{n} + O(n^{-2}),$$

show that

$$\ln\left(\binom{\binom{n-j}{2}}{m}\binom{\binom{n}{2}}{m}^{-1}\right) = m \ln\left(1 - \frac{2j}{n} + O(n^{-2} \ln n)\right).$$

Finally, since $x - x^2 \le \ln(1 + x) \le x$ whenever $x \ge -0.5$ (an easy exercise in calculus), we conclude that

$$\ln\left(1 - \frac{2j}{n} + O(n^{-2} \ln n)\right) = -\frac{2j}{n} + O(n^{-2} \ln n),$$

and so

$$\ln\left(\binom{\binom{n-j}{2}}{m}\binom{\binom{n}{2}}{m}^{-1}\right) = \left(\tfrac{1}{2}n \ln n + cn + o(n)\right) \cdot \left(-\frac{2j}{n} + O(n^{-2} \ln n)\right)$$

$$= -j \ln n - 2jc + o(1),$$

which means that

$$\binom{\binom{n-j}{2}}{m}\binom{\binom{n}{2}}{m}^{-1} = (1 + o(1))n^{-j} e^{-2cj}.$$

Since $\binom{n}{j} = (1 + o(1))n^j/j!$, this implies (10.3). □

10.1.3 Lemma on a Single Nontrivial Component

LEMMA 10.5 *If $m : N \to N$ is a function such that $m(n) = \frac{1}{2}n \ln n + O(n)$, then almost all graphs with n vertices and m(n) edges have only one component with more than one vertex.*

Proof Let us call a component of a graph *nontrivial* if it has more than one vertex. If a graph of order n has two or more nontrivial components, then the smallest of these components has at most $n/2$ vertices. Therefore we can prove the lemma by proving that almost all graphs with n vertices and $m(n)$ edges have no nontrivial component with at most $n/2$ vertices. For this purpose, let **P** denote the set of all pairs (G, F), where G is a graph with vertices $1, 2, \ldots, n$ and with $m(n)$ edges and F is a component of G with at least 2 and at most $n/2$ vertices. Again, let us write simply m for $m(n)$. We aim to prove that almost all graphs with n vertices and m edges appear as G in no pair in $\mathbf{P}$, which means that the number of graphs appearing as G in these pairs is

$$o\left(\binom{\binom{n}{2}}{m}\right).$$

To prove this, note first that the number of graphs appearing as G in the pairs in $\mathbf{P}$ is at most $|\mathbf{P}|$. Every pair (G, F) in $\mathbf{P}$ can be manufactured by first choosing the number s of vertices of F (which is at least 2 and at most $\lfloor n/2 \rfloor$), then choosing the actual set S of these vertices, then choosing the number t of edges of F (which is at least 1 and at most $s(s-1)/2$; we will rely only on the lower bound), and finally choosing the set of edges of G. Since each edge of G has either both endpoints in S or both endpoints outside S, the set of edges of G can be chosen in at most

$$\sum_{t=1}^{m} \binom{\binom{s}{2}}{t}\binom{\binom{n-s}{2}}{m-t}$$

ways; it follows that

$$|\mathbf{P}| \le \sum_{s=2}^{\lfloor n/2 \rfloor} \binom{n}{s} \sum_{t=1}^{m} \binom{\binom{s}{2}}{t}\binom{\binom{n-s}{2}}{m-t}.$$

The remainder of the proof is a computation showing that

$$\sum_{s=2}^{\lfloor n/2 \rfloor} \binom{n}{s} \sum_{t=1}^{m} \binom{\binom{s}{2}}{t}\binom{\binom{n-s}{2}}{m-t}\binom{\binom{n}{2}}{m}^{-1} = o(1). \tag{10.4}$$

To prove (10.4), choose a function f that maps integers greater than 1 to positive integers in such a way that

$$\lim_{n \to \infty} f(n) = \infty \text{ but } f(n) = o(\ln n).$$

We are going to prove separately that

$$\sum_{s=f(n)}^{\lfloor n/2 \rfloor} \binom{n}{s} \sum_{t=1}^{m} \binom{\binom{s}{2}}{t}\binom{\binom{n-s}{2}}{m-t}\binom{\binom{n}{2}}{m}^{-1} = o(1) \tag{10.5}$$

and that

$$\sum_{s=2}^{f(n)} \binom{n}{s} \sum_{t=1}^{m} \binom{\binom{s}{2}}{t}\binom{\binom{n-s}{2}}{m-t}\binom{\binom{n}{2}}{m}^{-1} = o(1). \tag{10.6}$$

Both proofs rely on the fact that assumption $m = \frac{1}{2}n \ln n + O(n)$ means the existence of a positive constant c such that

$$\frac{1}{2}n \ln n - cn \le m \le \frac{1}{2}n \ln n + cn \text{ for all } n. \tag{10.7}$$

Proof of (10.5)

Here, the lower bound $t \geq 1$ is superfluous: we will prove that for all sufficiently large s we have

$$s \leq n/2 \implies \binom{n}{s} \sum_{t=0}^{m} \binom{\binom{s}{2}}{t} \binom{\binom{n-s}{2}}{m-t} \binom{\binom{n}{2}}{m}^{-1} \leq (1/2)^s. \qquad (10.8)$$

This bound implies that

$$\sum_{s=f(n)}^{\lfloor n/2 \rfloor} \binom{n}{s} \sum_{t=0}^{m} \binom{\binom{s}{2}}{t} \binom{\binom{n-s}{2}}{m-t} \binom{\binom{n}{2}}{m}^{-1} \leq$$

$$\sum_{s=f(n)}^{\lfloor n/2 \rfloor} \left(\frac{1}{2}\right)^s \leq \sum_{s=f(n)}^{\infty} \left(\frac{1}{2}\right)^s = 2 \left(\frac{1}{2}\right)^{f(n)} = o(1).$$

To prove (10.8), we argue as follows. Since

$$\sum_{t=0}^{m} \binom{\binom{s}{2}}{t} \binom{\binom{n-s}{2}}{m-t} \binom{\binom{n}{2}}{m}^{-1}$$

$$= \binom{\binom{s}{2} + \binom{n-s}{2}}{m} \binom{\binom{n}{2}}{m}^{-1}$$

$$= \binom{\binom{n}{2} - s(n-s)}{m} \binom{\binom{n}{2}}{m}^{-1}$$

$$\leq \left(\frac{\binom{n}{2} - s(n-s)}{\binom{n}{2}}\right)^{m}$$

$$= \left(1 - \frac{s(n-s)}{\binom{n}{2}}\right)^{m}$$

$$\leq \exp\left(-s(n-s)\frac{2m}{n^2-n}\right)$$

$$\leq \exp\left(-s(n-s)\frac{2m}{n^2}\right),$$

we have

$$\sum_{t=0}^{m} \binom{\binom{s}{2}}{t} \binom{\binom{n-s}{2}}{m-t} \binom{\binom{n}{2}}{m}^{-1} \leq \exp\left((n-s)\frac{2m}{n^2}\right)^{-s};$$

by (10.7), we have

$$(n-s)\frac{2m}{n^2} \geq \frac{(n-s)\ln n - 2c(n-s)}{n} \geq \ln n - s\frac{\ln n}{n} - 2c;$$

by (A.3), we have $\binom{n}{s} < (en/s)^s$. It follows that

$$\binom{n}{s}\sum_{t=0}^{m}\binom{\binom{s}{2}}{t}\binom{\binom{n-s}{2}}{m-t}\binom{\binom{n}{2}}{m}^{-1} \leq \exp\left(-1+\ln s - s\frac{\ln n}{n} - 2c\right)^{-s}.$$

Now it only remains to be shown that for all sufficiently large s we have

$$n \geq 2s \implies -1 + \ln s - s\frac{\ln n}{n} - 2c \geq \ln 2. \tag{10.9}$$

Since $n^{-1}\ln n$ decreases as n increases away from $n = 3$, we have

$$n \geq 2s \implies \frac{\ln n}{n} \leq \frac{\ln(2s)}{2s} = \frac{\ln 2}{2s} + \frac{\ln s}{2s}$$

and so

$$n \geq 2s \implies -1 + \ln s - s\frac{\ln n}{n} - 2c \geq -1 + \frac{\ln s}{2} - \frac{\ln 2}{2} - 2c,$$

which proves (10.9) for all sufficiently large s.

Proof of (10.6)

Here, bound (10.8) is useless and the lower bound $t \geq 1$ is indispensable. We will prove that for all sufficiently large n we have

$$s \leq f(n) \implies \binom{n}{s}\sum_{t=1}^{m}\binom{\binom{s}{2}}{t}\binom{\binom{n-s}{2}}{m-t}\binom{\binom{n}{2}}{m}^{-1} \leq \frac{1}{n^{1/2}}, \tag{10.10}$$

which evidently implies (10.6).

Our proof of (10.10) relies on the inequality

$$\sum_{t=1}^{m}\binom{\binom{s}{2}}{t}\binom{\binom{n-s}{2}}{m-t} \leq \binom{s}{2}\binom{\binom{s}{2}+\binom{n-s}{2}-1}{m-1}$$

which can be understood by interpreting both sides in terms of disjoint sets A and B such that $|A| = \binom{s}{2}$ and $|B| = \binom{n-s}{2}$: the left-hand side counts the number of m-point subsets E of $A \cup B$ such that $|E \cap A| \geq 1$ whereas the right-hand side counts the number of pairs (E, e) such that E is an m-point subset of $A \cup B$ and $e \in E \cap A$ (by choosing first an e in A and then an E that contains e).

To prove (10.10), we argue as follows. Since

$$\sum_{t=1}^{m}\binom{\binom{s}{2}}{t}\binom{\binom{n-s}{2}}{m-t}\binom{\binom{n}{2}}{m}^{-1}$$

$$\leq \binom{s}{2}\binom{\binom{s}{2}+\binom{n-s}{2}-1}{m-1}\binom{\binom{n}{2}}{m}^{-1}$$

$$= \binom{s}{2}\binom{\binom{n}{2}-s(n-s)-1}{m-1}\binom{\binom{n}{2}}{m}^{-1}$$

$$= \frac{ms(s-1)}{n(n-1)} \left(\frac{\binom{n}{2} - s(n-s) - 1}{m-1} \right) \binom{\binom{n}{2} - 1}{m-1}^{-1}$$

$$\leq \frac{ms(s-1)}{n(n-1)} \left(\frac{\binom{n}{2} - s(n-s) - 1}{\binom{n}{2} - 1} \right)^{m-1}$$

$$= \frac{ms(s-1)}{n(n-1)} \left(1 - \frac{s(n-s)}{\binom{n}{2} - 1} \right)^{m-1}$$

$$\leq \frac{ms(s-1)}{n(n-1)} \exp\left(-s(n-s) \frac{2m-2}{n^2 - n - 2} \right)$$

$$\leq \frac{ms(s-1)}{n(n-1)} \exp\left(-s(n-s) \frac{2m-2}{n^2} \right)$$

$$\leq \frac{ms(s-1)}{n(n-1)} \exp\left(-s(n-s) \frac{2m}{n^2} + \frac{2s}{n} \right),$$

we have

$$\sum_{t=1}^{m} \binom{\binom{s}{2}}{t} \binom{\binom{n-s}{2}}{m-t} \binom{\binom{n}{2}}{m}^{-1} \leq \frac{ms^2}{n^2} \exp\left(-s(n-s) \frac{2m}{n^2} + \frac{2s}{n} \right);$$

by (10.7), we have

$$\frac{2m}{n^2} \leq \frac{\ln n + 2c}{n}$$

and

$$s(n-s) \frac{2m}{n^2} \geq \frac{s(n-s)\ln n - 2cs(n-s)}{n} \geq s\ln n - \frac{s^2 \ln n}{n} - 2cs;$$

trivially $\binom{n}{s} < n^s$. It follows that

$$\binom{n}{s} \sum_{t=1}^{m} \binom{\binom{s}{2}}{t} \binom{\binom{n-s}{2}}{m-t} \binom{\binom{n}{2}}{m}^{-1} \leq \frac{s^2(\ln n + 2c)}{2n} \exp\left(\frac{s^2 \ln n}{n} + 2cs + \frac{2s}{n} \right)$$

$$= \frac{s^2(\ln n + 2c) \cdot n^{s^2/n} \cdot e^{2s/n} \cdot e^{2cs}}{2n}.$$

Now it only remains to be shown that for all sufficiently large n we have

$$s \leq f(n) \implies s^2(\ln n + 2c) \cdot n^{s^2/n} \cdot e^{2s/n} \cdot e^{2cs} \leq 2n^{1/2}. \tag{10.11}$$

Assumption $f(n) = o(\ln n)$ guarantees that for all sufficiently large n we have

$$f(n)^2(\ln n + 2c) \leq n^{1/8}, \quad n^{f(n)^2/n} \leq n^{1/8}, \quad e^{2f(n)/n} \leq n^{1/8}, \quad e^{2cf(n)} \leq n^{1/8}.$$

(The last of these bounds is the bottleneck which makes the assumption $f(n) = o(\ln n)$ necessary; the first three bounds can be guaranteed by weaker assumptions.) This proves (10.11). $\square$

10.1.4   Proof of Theorem 10.1

With $m : \mathbf{N} \to \mathbf{N}$ a function such that $m(n) = \frac{1}{2}n \ln n + cn + o(n)$, let us write

$\mathcal{A}_n$ for the set of all graphs with vertices $1, 2, \ldots, n$ and with $m(n)$ edges,
$\mathcal{B}_n$ for the set of all graphs in $\mathcal{A}_n$ that have only one nontrivial component,
$\mathcal{C}_n$ for the set of all graphs in $\mathcal{A}_n$ that have no isolated vertices.

Since a graph is connected if and only if it has only one nontrivial component and no isolated vertices, we have

$$p(n, m(n)) = \frac{|\mathcal{B}_n \cap \mathcal{C}_n|}{|\mathcal{A}_n|} = \frac{|\mathcal{C}_n|}{|\mathcal{A}_n|} - \frac{|\mathcal{C}_n - \mathcal{B}_n|}{|\mathcal{A}_n|};$$

Lemma 10.4 asserts that

$$\lim_{n \to \infty} \frac{|\mathcal{C}_n|}{|\mathcal{A}_n|} = e^{-e^{-2c}};$$

Lemma 10.5 asserts that

$$\lim_{n \to \infty} \frac{|\mathcal{B}_n|}{|\mathcal{A}_n|} = 1,$$

which implies

$$\lim_{n \to \infty} \frac{|\mathcal{C}_n - \mathcal{B}_n|}{|\mathcal{A}_n|} = 0. \qquad \square \qquad (10.12)$$

Limiting formula (10.12) shows that almost all graphs with n vertices and $m(n)$ edges either are connected or have isolated vertices. In this sense, isolated vertices are the last barrier to being connected. A theorem of Bollobás and Thomason [45, a special case ($k = 1$) of Theorem 4] makes the preceding statement even more explicit:

THEOREM 10.6   *Consider the random graph of order n that evolves step by step through adding edges one at a time. The probability that this graph becomes connected at the moment when all of its vertices have acquired at least one neighbour tends to 1 as n tends to infinity.* $\qquad \square$

10.2   Subgraphs

Turán's Theorem 7.1 shows that $\mathrm{ex}(K_r, n)$, the largest number of edges in a graph of order n that has no subgraph isomorphic to the complete graph K_r of order r, equals

$$\frac{r - 2}{2(r - 1)} \cdot n^2 + O(1).$$

In particular, if $r \geq 3$, then the smallest number $m(n)$ of edges that forces the presence of a K_r in *all* graphs with n vertices and $m(n)$ edges grows quadratically with n. Erdős and Rényi have shown that a much smaller $m(n)$ forces the presence of a K_r in *almost all* graphs with n vertices and $m(n)$ edges: $n^{2-2/(r-1)}$ is the threshold function for the property of containing a K_r. This assertion is a special case of our next theorem.

DEFINITION: A graph is called *balanced* if none of its subgraphs has a larger ratio of the number of edges to the number of vertices than the graph itself.

THEOREM 10.7 *Let F be a connected balanced graph with r vertices and s edges. If*

$$\lim_{n\to\infty} \frac{m(n)}{n^{2-r/s}} = 0,$$

then almost all graphs with n vertices and m(n) edges contain no subgraph isomorphic to F; if

$$\lim_{n\to\infty} \frac{m(n)}{n^{2-r/s}} = \infty, \tag{10.13}$$

then almost all graphs with n vertices and m(n) edges contain a subgraph isomorphic to F.

Other noteworthy special cases of Theorem 10.7 assert that n is the threshold function for the property of containing a prescribed cycle and that $n^{1-1/k}$ is the threshold function for the property of containing a prescribed tree with k edges. (In turn, Theorem 10.7 is a special case of Theorem 1 in [148], which concerns the threshold for the occurrence of at least one of a prescribed family of balanced graphs.) The assumption that F is balanced cannot be dropped: to see this, consider a triangle plus an isolated vertex (or, if you insist on a connected F, then a K_4 with a one-edge tree glued to one of its vertices).

10.2.1   A Lemma

Let us state the central part of the proof of Theorem 10.7 as a separate lemma.

LEMMA 10.8 *If X(G), with G running through a finite set Ω, are nonnegative integers and if at least one of these integers is positive, then*

$$\frac{(\sum_{G\in\Omega} X(G))^2}{\sum_{G\in\Omega} X(G)^2} \le |\{G \in \Omega : X(G) > 0\}| \le \sum_{G\in\Omega} X(G).$$

Proof Writing $A = \{G : X(G) > 0\}$, note that $\sum_{G\in\Omega} X(G) = \sum_{G\in A} X(G)$ and $\sum_{G\in\Omega} X(G)^2 = \sum_{G\in A} X(G)^2$, and so the Lemma reduces to

$$\frac{(\sum_{G\in A} X(G))^2}{\sum_{G\in A} X(G)^2} \le |A| \le \sum_{G\in A} X(G).$$

The upper bound on $|A|$ follows from observing that $|A| = \sum_{G\in A} 1$ and $1 \le X(G)$ for all G in A. The lower bound comes from setting $a_G = X(G)$ and $b_G = 1$ in the classic Cauchy–Bunyakovsky–Schwarz inequality

$$\left(\sum_{G\in A} a_G b_G\right)^2 \le \left(\sum_{G\in A} a_G^2\right)\left(\sum_{G\in A} b_G^2\right),$$

which holds for all choices of real numbers a_G, b_G with G running through A (see Subsection A.1.2). $\qquad\square$

10.2.2   Proof of Theorem 10.7

Given positive integers n and m, let $\Omega(n, m)$ denote the set of all graphs with vertices $1, 2, \ldots, n$ and with m edges; let $\Phi(n)$ denote the set of all graphs isomorphic to F and with vertices coming from $\{1, 2, \ldots, n\}$. For every G in $\Omega(n, m)$ and every H in $\Phi(n)$, write

$$\chi(G, H) = \begin{cases} 1 & \text{if } H \text{ is a subgraph of } G, \\ 0 & \text{otherwise} \end{cases}$$

and write $X(G) = \sum_{H \in \Phi(n)} \chi(G, H)$. In this notation, $X(G) > 0$ means that G has a subgraph isomorphic to F; we are going to prove that

$$\lim_{n \to \infty} \frac{m(n)}{n^{2-r/s}} = 0 \quad \Rightarrow \quad \lim_{n \to \infty} \frac{\sum_{G \in \Omega(n, m(n))} X(G)}{\binom{\binom{n}{2}}{m(n)}} = 0; \tag{10.14}$$

and

$$\lim_{n \to \infty} \frac{m(n)}{n^{2-r/s}} = \infty \quad \Rightarrow \quad \lim_{n \to \infty} \frac{\left(\sum_{G \in \Omega(n, m(n))} X(G)\right)^2}{\binom{\binom{n}{2}}{m(n)} \sum_{G \in \Omega(n, m(n))} X(G)^2} = 1. \tag{10.15}$$

Lemma 10.8 guarantees that the theorem follows from these two implications.

Proof of (10.14)

Observe that

$$\sum_{G \in \Omega(n, m)} X(G) = \sum_{G \in \Omega(n, m)} \sum_{H \in \Phi(n)} \chi(G, H)$$

$$= \sum_{H \in \Phi(n)} \sum_{G \in \Omega(n, m)} \chi(G, H)$$

$$= \sum_{H \in \Phi(n)} \binom{\binom{n}{2} - s}{m - s}$$

$$= |\Phi(n)| \binom{\binom{n}{2} - s}{m - s}$$

and so

$$\frac{\sum_{G \in \Omega(n, m)} X(G)}{\binom{\binom{n}{2}}{m}} = |\Phi(n)| \frac{\binom{m}{s}}{\binom{\binom{n}{2}}{s}}. \tag{10.16}$$

Now (10.14) follows since $|\Phi(n)| \leq n^r$ and

$$n^r \binom{m}{s} \binom{\binom{n}{2}}{s}^{-1} \leq n^r \left(\frac{m}{\binom{n}{2}}\right)^s = \left(\frac{2n}{n-1} \cdot \frac{m}{n^{2-r/s}}\right)^s = (1 + o(1)) \left(\frac{2m}{n^{2-r/s}}\right)^s.$$

Proof of (10.15)

We will prove (10.15) by proving that $\lim_{n \to \infty} m(n) n^{-2+r/s} = \infty$ implies

$$\sum_{G \in \Omega(n,m(n))} X(G) = (1 + o(1)) |\Phi(n)| \binom{\binom{n}{2}}{m(n)} \left(\frac{2m(n)}{n^2}\right)^s \qquad (10.17)$$

and

$$\sum_{G \in \Omega(n,m(n))} X(G)^2 = (1 + o(1)) |\Phi(n)|^2 \binom{\binom{n}{2}}{m(n)} \left(\frac{2m(n)}{n^2}\right)^{2s}. \qquad (10.18)$$

Proof of (10.17)

Consider an arbitrary function $m : \mathbf{N} \to \mathbf{N}$ that satisfies (10.13). By assumption, $s \geq r/2$ (since G is balanced, none of its vertices is isolated), and so $\lim_{n \to \infty} m(n) = \infty$. Therefore

$$\binom{m(n)}{s} \binom{\binom{n}{2}}{s}^{-1} = (1 + o(1)) \left(\frac{2m(n)}{n^2}\right)^s$$

and (10.17) follows from (10.16).

Proof of (10.18)

To prove (10.18), observe first that

$$
\begin{aligned}
\sum_{G \in \Omega(n,m(n))} X(G)^2 &= \sum_{G \in \Omega(n,m(n))} \Big(\sum_{H \in \Phi(n)} \chi(G,H)\Big)^2 \\
&= \sum_{G \in \Omega(n,m(n))} \Big(\sum_{A \in \Phi(n)} \chi(G,A)\Big)\Big(\sum_{B \in \Phi(n)} \chi(G,B)\Big) \\
&= \sum_{G \in \Omega(n,m(n))} \sum_{A \in \Phi(n)} \sum_{B \in \Phi(n)} \chi(G,A) \chi(G,B) \\
&= \sum_{A \in \Phi(n)} \sum_{B \in \Phi(n)} \sum_{G \in \Omega(n,m(n))} \chi(G,A) \chi(G,B),
\end{aligned}
$$

and so (10.18) can be proved by proving that, for all A in $\Phi(n)$,

$$\sum_{B \in \Phi(n)} \sum_{G \in \Omega(n,m(n))} \chi(G,A) \chi(G,B)$$

$$= (1 + o(1)) |\Phi(n)| \binom{\binom{n}{2}}{m(n)} \left(\frac{2m(n)}{n^2}\right)^{2s}. \qquad (10.19)$$

Now consider a fixed A in $\Phi(n)$ and let $\Psi(n)$ denote the set of all graphs in $\Phi(n)$ that share no edge with A. We will prove (10.19) by proving that

$$\sum_{B \in \Psi(n)} \sum_{G \in \Omega(n,m(n))} \chi(G,A) \chi(G,B)$$

$$= (1 + o(1)) |\Phi(n)| \binom{\binom{n}{2}}{m(n)} \left(\frac{2m(n)}{n^2}\right)^{2s} \qquad (10.20)$$

and that

$$\sum_{B\in\Phi(n)-\Psi(n)}\sum_{G\in\Omega(n,m(n))}\chi(G,A)\chi(G,B)$$

$$= o(1)|\Phi(n)|\binom{\binom{n}{2}}{m(n)}\left(\frac{2m(n)}{n^2}\right)^{2s}. \quad (10.21)$$

Proof of (10.20)

Consider an arbitrary B in $\Phi(n)$ and let j denote the number of edges shared by A and B (so that $B \in \Psi(n)$ if and only if $j = 0$). Since the union of A and B has $2s-j$ edges, we have

$$\sum_{G\in\Omega(n,m(n))}\chi(G,A)\chi(G,B) = \binom{\binom{n}{2}-(2s-j)}{m(n)-(2s-j)};$$

since

$$\binom{\binom{n}{2}-(2s-j)}{m(n)-(2s-j)} = \binom{\binom{n}{2}}{m(n)}\binom{m(n)}{2s-j}\binom{\binom{n}{2}}{2s-j}^{-1}$$

$$= (1+o(1))\binom{\binom{n}{2}}{m(n)}\left(\frac{2m(n)}{n^2}\right)^{2s-j},$$

we conclude that

$$\sum_{G\in\Omega(n,m(n))}\chi(G,A)\chi(G,B)$$

$$= (1+o(1))\binom{\binom{n}{2}}{m(n)}\left(\frac{2m(n)}{n^2}\right)^{2s}\left(\frac{n^2}{2m(n)}\right)^{j}. \quad (10.22)$$

In particular,

$$\sum_{B\in\Psi(n)}\sum_{G\in\Omega(n,m(n))}\chi(G,A)\chi(G,B)$$

$$= (1+o(1))|\Psi(n)|\binom{\binom{n}{2}}{m(n)}\left(\frac{2m(n)}{n^2}\right)^{2s}. \quad (10.23)$$

For all $i = 0, 1, \ldots, r$, let $\Phi(n, i)$ denote the set of graphs in $\Phi(n)$ that share precisely i vertices with A. Since the order of A is r, we have

$$\Phi(n) = \bigcup_{i=0}^{r}\Phi(n, i).$$

The number of graphs in $\Phi(n)$ with a prescribed vertex set V does not change when V ranges over all r-point subsets of $\{1, 2, \ldots, n\}$, and so it equals

$$\frac{|\Phi(n)|}{\binom{n}{r}};$$

it follows that

$$|\Phi(n, i)| = \binom{r}{i}\binom{n-r}{r-i}\frac{|\Phi(n)|}{\binom{n}{r}} \quad \text{for all } i = 0, 1, \ldots, r. \quad (10.24)$$

In particular, $|\Phi(n, 0)| = (1 + o(1))|\Phi(n)|$; since $\Phi(n, 0) \subseteq \Psi(n) \subseteq \Phi(n)$, we conclude that

$$|\Psi(n)| = (1 + o(1))|\Phi(n)|.$$

Substituting into (10.23) we get (10.20).

Proof of (10.21)

Consider an arbitrary B in $\Phi(n) - \Psi(n)$, let i denote the number of vertices shared by A and B, and let j denote the number of edges shared by A and B. Since G is balanced (apart from our earlier derivation of $\lim_{n\to\infty} m(n) = \infty$ from (10.13), this assumption is used only here), we have

$$j/i \le s/r.$$

By assumption (10.13), we have $n^2/m = o(n^{r/s})$, and so (10.22) implies

$$\sum_{G \in \Omega(n,m(n))} \chi(G, A)\chi(G, B) = (1 + o(1)) \binom{\binom{n}{2}}{m(n)} \left(\frac{2m(n)}{n^2}\right)^{2s} (o(1) \cdot (n^{r/s}))^j;$$

since $B \notin \Psi(n)$, we have $j > 0$, and so we can replace the $(o(1) \cdot (n^{r/s}))^j$ by $o(1) \cdot (n^{jr/s})$. It follows that

$$\sum_{G \in \Omega(n,m(n))} \chi(G, A)\chi(G, B) = o(1) \binom{\binom{n}{2}}{m(n)} \left(\frac{2m(n)}{n^2}\right)^{2s} n^i.$$

The proof of (10.21) is completed by observing that

$$|\Phi(n, i)| = O(n^{-i})|\Phi(n)| \quad \text{for all } i = 0, 1, \ldots, r$$

follows directly from (10.24). $\qquad\qquad\qquad\qquad\qquad\qquad\qquad\qquad\qquad\qquad\qquad\square$

10.3 Evolution of Random Graphs and the Double Jump

The threshold function of a graph property is the approximate time around which almost all of the evolving graphs switch from not having the property to having it. This switch is often referred to as a *phase transition,* which is the term originally reserved for relatively abrupt structural changes in thermodynamic systems.

The most striking results of [148] reveal the phase transition that random graphs with n vertices and $(1 + o(1))cn$ edges go through as the proportionality constant c moves across its critical point at $1/2$: the structure of the graph changes dramatically when c reaches the critical point and then immediately again when it rises above this point. As c increases through all positive real numbers, almost all graphs with n vertices and $(1 + o(1))cn$ edges see the number of vertices in their largest component jump from order $\log n$ in the region $0 < c < 1/2$ to about $n^{2/3}$ at $c = 1/2$ and then again to order n in the region $1/2 < c$. Erdős and Rényi dubbed this phenomenon a *double jump* and they dubbed the component with the linear number of vertices (which is provably unique when $c > 1/2$) the *giant component* [148, p. 52]. Because of this phase transition, evolution of random graphs is classified into three phases:

> - the *subcritical phase,* before time cn with $c < 1/2$,
> - the *critical phase,* at time around $n/2$,
> - the *supercritical phase,* after time cn with $c > 1/2$.

In the subcritical phase, the largest component in almost all graphs with n vertices and $(1 + o(1))cn$ edges has

$$\frac{1 + o(1)}{2c - 1 - \ln 2c} \left(\ln n - \frac{5}{2} \ln \ln n \right)$$

vertices [148, §7].

In the critical phase, the largest component in almost all graphs has a number of vertices that, for every sequence ω_n such that $\lim_{n \to \infty} \omega_n = \infty$, is sandwiched between $n^{3/2}/\omega_n$ and $n^{3/2}\omega_n$ [148, Theorem 7c].

In the supercritical phase, the largest component in almost all of graphs with n vertices and $(1 + o(1))cn$ edges has $(1 + o(1))\gamma(c)n$ vertices, where $\gamma : (1/2, \infty) \to (0, 1)$ is an increasing continuous function such that $\lim_{c \to 1/2} \gamma(c) = 0$ and $\lim_{c \to \infty} \gamma(c) = 1$. Specifically, $\gamma(c) = 1 - x$, where x is the unique solution of $0 < x < 1$, $\ln x = 2c(x - 1)$ [148, Theorem 9b].

In addition, the second largest component in almost all of graphs with n vertices and $(1 + o(1))cn$ edges in the supercritical phase has

$$\frac{1 + o(1)}{2c - 1 - \ln 2c} \left(\ln n - \frac{5}{2} \ln \ln n \right)$$

vertices [148, Theorem 7a]. Since $2c - \ln 2c$ increases when $c \geq 1/2$, the second largest component keeps getting smaller: from time to time, the new added edge happens to join the giant component with the second largest component, in which case the second largest component gets absorbed into the giant and the third largest component gets promoted to the vacant second position.

As the number of edges grows in the supercritical phase, the number of vertices in the second largest component eventually drops all the way to 1 and then the graph consists of its giant component plus some isolated vertices; a special case of Theorem 2c of [148] implies that this happens at time barely past $\frac{1}{4}n \ln n$, which is halfway to the point where the giant absorbs all the remaining isolated vertices, and so the graph becomes connected (Theorem 10.1).

The Critical Window

In a seminal paper [36], Béla Bollobás (see also [40, Theorem 6.9]) subjected the critical phase to a more precise study. He proved that, when $s(n) \geq 2(\ln n)^{1/2}n^{2/3}$ but $s(n) = o(n)$, almost all graphs with n vertices and $n/2 + s(n)$ edges have a component with $(4 + o(1))s(n)$ vertices and their second largest component is a tree with $o(s(n))$ vertices.

Next, Tomasz Łuczak [281, Theorem 3] proved that these conclusions hold even when the lower bound on $s(n)$ is relaxed to $\lim_{n \to \infty} s(n)/n^{2/3} = \infty$ and that the largest component in almost all graphs with n vertices and $n/2 - s(n)$ edges has $o(s(n))$ vertices. Therefore the giant component materializes while time passes through the *critical window* $n/2 + \lambda n^{2/3}$ with parameter λ sliding through all real numbers.

A conclusive step in the search for the number of vertices of the emerging giant component came from Boris Pittel [313]. For every real number λ and for every positive a, the probability that the largest component of a random graph with n vertices and $n/2 + (\lambda + o(1))n^{2/3}$ edges has at most $an^{2/3}$ vertices tends to a limit $p(\lambda, a)$ as n tends to infinity; we have

$$\lim_{\lambda \to -\infty} p(\lambda, a) = 1, \quad \lim_{\lambda \to \infty} p(\lambda, a) = 0 \quad \text{for all } a,$$
$$\lim_{a \to 0+} p(\lambda, a) = 0, \quad \lim_{a \to \infty} p(\lambda, a) = 1 \quad \text{for all } \lambda.$$

Pittel found an explicit (and complicated) formula for $p(\lambda, a)$. More recent results in this direction are contained in [301].

DEFINITION: A connected component of a graph is called *complex* if it has more edges than vertices (which is another way of saying that it contains at least two cycles).

We have already noted that almost all graphs with n vertices and $(1 + o(1))cn$ edges such that $c < 1/2$ have no complex components; Kolchin [255] strengthened this by proving that almost all graphs with n vertices and $m(n)$ edges such that $\lim_{n \to \infty} (m(n) - n/2)/n^{2/3} = -\infty$ have no complex components; Łuczak [281] proved that almost all graphs with n vertices and $m(n)$ edges such that $\lim_{n \to \infty} (m(n) - n/2)/n^{2/3} = +\infty$ have precisely one complex component. Between these two opposite ends of the critical window, it is not too rare to see any number of complex components: a special case of a far more general theorem of Łuczak, Pittel, and Wierman [283, Theorem 3] asserts that for every real number λ and for every nonnegative integer r, the probability that a graph with n vertices and $n/2 + (\lambda + o(1))n^{2/3}$ edges has precisely r complex components tends to a positive limit as n tends to infinity. Janson, Knuth, Łuczak, and Pittel proved [224, Theorem 15] that the evolving graph has at most one complex component throughout all stages of its development with probability tending to $5\pi/18$ (which is about 0.873) as the number of its vertices tends to infinity.

Łuczak, Pittel, and Wierman also described the structure of the complex components of graphs evolving through the critical window.

DEFINITION: The *2-core* of a graph is its maximal subgraph in which every vertex has degree at least 2.

Another way of describing the 2-core of a graph goes as follows: Given the graph, first remove all of its vertices that have degree at most 1; this removal may expose new vertices of degree at most 1; remove these as well and continue recursively until all the vertices in the remaining graph (which may turn out to have no vertices at all) have degree at least 2. This remaining graph is the 2-core of the input graph and reconstructing the input amounts to gluing a tree to each vertex of the core (some or all of these trees may consist of a single vertex and gluing such a tree to a vertex of the core means doing nothing).

Theorem 4 of [283] asserts that, for every sequence ω_n such that $\lim_{n\to\infty}\omega_n = \infty$, all complex components of almost all graphs in the critical window are such that

- in the 2-core, all vertices have degree at most 3 and at most ω_n of them have degree 3,
- in the 2-core, each path between two vertices of degree 3 has at least $n^{1/3}/\omega_n$ edges and at most $n^{1/3}\omega_n$ edges,
- in the 2-core, each cycle that passes through at least one vertex of degree 3 has at least $n^{1/3}/\omega_n$ edges and at most $n^{1/3}\omega_n$ edges,
- the largest tree glued to a vertex of the 2-core has at least $n^{2/3}/\omega_n$ vertices and at most $n^{2/3}\omega_n$ vertices.

Since each component of almost all graphs in the subcritical phase has at most one cycle, almost all graphs in this phase are planar; Erdős and Rényi [148, Theorem 8b] proved that almost all graphs in the supercritical phase are nonplanar. Łuczak, Pittel, and Wierman [283, Theorem 5] refined this by proving that for every real number λ, the probability that a random graph with n vertices and $n/2 + (\lambda + o(1))n^{2/3}$ edges is planar tends to a limit $p(\lambda)$ as n tends to infinity and that $\lim_{\lambda\to-\infty}p(\lambda) = 1$, $\lim_{\lambda\to\infty}p(\lambda) = 0$: the switch from planarity to nonplanarity also takes place in the critical window.

Additional information on random graphs can be found, for instance, in the monographs [40, 175, 225] and in Chapter 10 of [7] and in the endearing (even if occasionally erroneous) introduction [311].

10.4   Finite Probability Theory

The material of the preceding three sections is usually presented in terms of probability theory. In the present section, we shall sketch this general context.

When a coin is flipped, it comes up heads or tails (we disregard the possibility of its landing on its edge); the coin is called *unbiased* or *fair* if the probability of a head equals the probability of a tail (and so both equal $1/2$). We have used the term "probability" just now. What does this term mean here?

One way of answering this question is to consider not just a single flip, but an infinite sequence of flips; the *relative frequency* of heads in the first n flips is defined as the number of heads in these first n flips divided by n; now the coin is said to be unbiased if, and only if, these relative frequencies tend to their limit $1/2$ as n increases. This *frequentist* approach to building up a theory of probability was championed by John Venn (1834–1923), Richard von Mises (1883–1953), Hans Reichenbach (1891–1953), and others. Eventually, it gave way to the *axiomatic* approach of Andrey Nikolaevich Kolmogorov (1903–87), where certain functions are called *probability distributions* without fretting over their interpretation by coins, dice, cards, and such.

From the axiomatic point of view, an unbiased coin is simply the probability distribution $p : \{\text{head}, \text{tail}\} \to [0, 1]$ with $p(\text{head}) = p(\text{tail}) = 1/2$: this definition reflects the intuition that the head and the tail are equally likely to come out from a flip of the coin. More generally, if some experiment has finitely many possible outcomes and if there is no reason to favour any of these outcomes over the others, then all of the outcomes are equally likely. This is reflected in the notion of a *uniform probability distribution, p* : $\Omega \to [0, 1]$ with Ω a finite set and

$$p(\omega) = \frac{1}{|\Omega|} \quad \text{for all } \omega \text{ in } \Omega.$$

As probability theory originated in problems involving games of chance [188], its initial focus was on uniform probability distributions.

DEFINITIONS: When Ω is a finite set and $p : \Omega \to [0, 1]$ is a mapping such that

$$\sum_{\omega \in \Omega} p(\omega) = 1,$$

the ordered pair (Ω, p) is called a (finite) *probability space.* Its Ω is called its *sample space* and its p is called its *probability distribution.*

In the preceding three sections, we dealt exclusively with probability spaces where the sample space is the set of all graphs with vertices $1, 2, \ldots, n$ and with m edges for some fixed n and m and where the probability distribution is uniform. In the present section, we will deal with arbitrary finite probability spaces, but we will avoid infinite probability spaces since their axioms are more complicated.

DEFINITIONS & NOTATION: A *random variable* is a real-valued function defined on the sample space Ω of a probability space. The *expectation* E(X) of a random variable X, also known as its *mean,* is defined by

$$E(X) = \sum_{\omega \in \Omega} p(\omega)X(\omega)$$

and its *variance* Var(X) is defined by

$$\text{Var}(X) = E\left((X - E(X))^2\right).$$

It follows directly from the definition of expectation that

$$E(X + Y) = E(X) + E(Y)$$

for every two random variables X, Y; this fact is known as *linearity of expectation.* In particular, we have

$$\begin{aligned}
\text{Var}(X) &= E((X - E(X))^2) \\
&= E(X^2 - 2E(X)X + E(X)^2) \\
&= E(X^2) - 2E(X)E(X) + E(X)^2 \\
&= E(X^2) - E(X)^2.
\end{aligned}$$

DEFINITIONS & NOTATION: Subsets of a sample space are called *events* and the *probability* Prob(A) *of event A* is defined by

$$\text{Prob}(A) = \sum_{\omega \in A} p(\omega).$$

In recording probabilities of events, we will use the standard shorthand, where

$$\text{Prob}(X \geq t)$$

stands for $\text{Prob}(\{\omega : X(\omega) \geq t\})$ and so on.

The *probability distribution of a random variable X* is defined as

$$p_X : \Omega_X \to [0, 1],$$

where Ω_X stands for the set of values of X and $p_X(t)$ is defined as $\text{Prob}(X = t)$.

Many properties of a random variable can be deduced from its probability distribution alone: for instance,

$$E(X) = \sum_{t \in \Omega_X} t p_X(t),$$

$$\text{Var}(X) = \sum_{t \in \Omega_X} t^2 p_X(t) - \left(\sum_{t \in \Omega_X} t p_X(t) \right)^2.$$

Here are generalizations of the two bounds in our Lemma 10.8:

LEMMA 10.9 (Markov's inequality)   *If X is a nonnegative random variable and t is positive, then*

$$\text{Prob}(X \geq t) \leq \frac{E(X)}{t}.$$

Proof $E(X) \geq \text{Prob}(X \geq t) \cdot t.$ □

LEMMA 10.10   *Every random variable X satisfies*

$$E(X^2) > 0 \quad \Rightarrow \quad \text{Prob}(X \neq 0) \geq \frac{E(X)^2}{E(X^2)}.$$

Proof   Setting $A = \{\omega : X(\omega) \neq 0\})$ and $a_\omega = p(\omega)^{1/2} X(\omega)$, $b_\omega = p(\omega)^{1/2}$ in the Cauchy–Bunyakovsky–Schwarz inequality

$$\left(\sum_{\omega \in A} a_\omega b_\omega \right)^2 \leq \left(\sum_{\omega \in A} a_\omega^2 \right) \left(\sum_{\omega \in A} b_\omega^2 \right),$$

note that

$$\sum_{\omega \in A} a_\omega b_\omega = E(X), \quad \sum_{\omega \in A} a_\omega^2 = E(X^2), \quad \sum_{\omega \in A} b_\omega^2 = \text{Prob}(X \neq 0). \quad □$$

Markov's inequality with $(X - E(X))^2$ in place of X and $t^2 \text{Var}(X)$ in place of t reads

$$t > 0, \text{Var}(X) > 0 \quad \Rightarrow \quad \text{Prob}\left((X - E(X))^2 \geq t^2 \text{Var}(X) \right) \leq \frac{1}{t^2},$$

which can be written as

$$t > 0, \text{Var}(X) > 0 \quad \Rightarrow \quad \text{Prob}\left(|X - E(X)| \geq t \cdot \text{Var}(X)^{1/2} \right) \leq \frac{1}{t^2}. \quad (10.25)$$

Inequality (10.25) is known as *Chebyshev's inequality*.

Chebyshev's inequality with $t = |E(X)| \cdot \text{Var}(X)^{-1/2}$ reads

$$E(X) \neq 0, \text{Var}(X) > 0 \implies \text{Prob}\left(|X - E(X)| \geq |E(X)|\right) \leq \frac{\text{Var}(X)}{E(X)^2}.$$

As $X = 0 \implies |X - E(X)| \geq |E(X)|$, this implies

$$E(X) \neq 0, \text{Var}(X) > 0 \implies \text{Prob}(X = 0) \leq \frac{\text{Var}(X)}{E(X)^2},$$

which can be presented as

$$E(X) \neq 0, \text{Var}(X) > 0 \implies \text{Prob}(X \neq 0) \geq 2 - \frac{E(X^2)}{E(X)^2}. \tag{10.26}$$

Let us point out that (10.26) is not as good as

$$E(X^2) > 0 \implies \text{Prob}(X \neq 0) \geq \frac{E(X)^2}{E(X^2)} \tag{10.27}$$

of Lemma 10.10: If the hypothesis of (10.26) is satisfied, then the hypothesis of (10.27) is satisfied since $E(X^2) = \text{Var}(X) + E(X)^2$. Furthermore, since $1/r > 2 - r$ whenever $r > 1$, the lower bound in (10.27) is strictly greater than the lower bound in (10.26) except when $E(X^2) = E(X)^2$, in which case the two bounds are equal.

DEFINITIONS: The *k-th moment* of a random variable X means $E(X^k)$. Using Lemma 10.9 (in particular, using it to prove the existence of an ω such that $X(\omega) = 0$) is called *the first moment method*; using (10.27) or just (10.26) (in particular, using one of them to prove the existence of an ω such that $X(\omega) \neq 0$) is called *the second moment method*.

Paul Erdős with his mother Anna (1880–1971)
Source:
www.geni.com/people/Johanna-Anna-Erdos-Englander/6000000013005264173

Erdős was strongly attached to his mother and from 1964 the two of them were inseparable. In 1971, she died at the age of ninety and he went into a deep depression. To chase at least some of the blues away, he immersed himself in mathematics to an extent surpassing even his earlier standard. While his mother was alive, he occasionally resorted to Benzedrine and Ritalin to sharpen his concentration and accelerate his thought processes; now, in order to sustain nineteen-hour workdays, he made these stimulants the staple of his diet.

In 1979, Ron Graham became so worried about his addiction that he decided to fight it: he bet Erdős $500 that he could not stay off his drugs for a month. Erdős did abstain for this length of time and doggedly put up with withdrawal symptoms such as sluggishness. As he collected his wager, he reproached Graham for setting mathematics back a month and, having declared himself not an addict, immediately resumed his daily intake of stimulant pills.

In the spring of 1971, I discovered marijuana and, like many other enthusiastic neophytes, I was ready to convert the world to my new creed. In the fall, Erdős came to McGill and on the first day of his visit the two of us had a dinner at a Moroccan restaurant. In this congenial ambiance I offered to share a bit of my supply with him. He pricked up his ears and inquired about the effects of cannabis on mathematical proficiency. When I told him that, at least in my own experience, they were negligible, he lost interest.

However, one thing leading to another, he offered to let me try Ritalin. It would be particularly beneficial, he said, if I struggled with a difficult problem and needed just an extra nudge to get me over the last hurdle. As I had no pressing need of that extra nudge just then, I politely declined.

Such was the evening where each of us had a shot at pushing drugs and both of us failed. Apparently neither of us was destined for this particular vocation.

11 Hamilton Cycles

11.1   A Theorem That Involves Degrees of Vertices

During his visit to University of Waterloo in the fall of 1970, Erdős gave a lecture for its Department of Combinatorics and Optimization and presented there a proof of his fresh refinement of Turán's theorem. We have seen this refinement as Theorem 7.2 and here it is again for easy reference:

THEOREM 11.1 (Erdős [118])   *Let r be an integer greater than 1. For every graph G with clique number less than r, there is a graph H such that*

(i) *G and H share their vertex-set V,*
(ii) *H is complete $(r-1)$-partite,*
(iii) *$d_G(v) \le d_H(v)$ for all v in V.*[a]

I was lucky to be in the audience. As I sat and listened, the beauty of this theorem took my breath away. Let me elaborate.

When G is a graph and r is a positive integer, the property that $\omega(G) \ge r$ can be certified by pointing out r pairwise adjacent vertices in G; finding such a certificate may be extremely difficult when r is large, but verifying it is straightforward and quick. By contrast, no easily verifiable certificate of the property that $\omega(G) < r$ is known.

To speculate about hypothetical certificates of $\omega(G) < r$, let us say that a graph G with $\omega(G) < r$ is *maximal* with respect to this property if adding any edge to G produces a graph H such that $\omega(H) \ge r$. The property that $\omega(G) < r$ could be certified by exhibiting a graph H that is maximal with respect to $\omega(H) < r$ and showing that G is a subgraph of H. However, this trick only shifts the burden of proof from all graphs with $\omega < r$ to all maximal graphs with $\omega < r$: how do we certify that $\omega(H) < r$? One way of doing this would be first to compile a catalogue of all maximal graphs with $\omega < r$ and then to simply point out H in this catalogue. Unfortunately, the catalogue is not only very large, but also very complex: some of its items H are wild in the sense that describing them is difficult. Yet the catalogue also includes items that are tame in the sense that describing them is easy: these tame items are the complete $(r-1)$-partite graphs.

---

[a] The fourth property of H in Erdős's theorem, *if $d_G(v) = d_H(v)$ for all v in V, then H = G,* is an icing on the cake irrelevant to the discussion that follows.

Theorem 11.1 shows that the tame items play a special role in the catalogue: for every wild item G there is a tame item H such that $d_G(v) \le d_H(v)$ for all vertices v of these two graphs. It was this beautiful surprise that took my breath away.

Aesthetics apart, Theorem 11.1 is useful in that it provides a sufficient condition for a graph to contain a large clique. Again, let me elaborate.

DEFINITION: When G is a graph with vertices $v_1, \ldots, v_n$, the sequence $d_G(v_1), \ldots, d_G(v_n)$ is called the *degree sequence* of G. (We do not insist on arranging its terms in any particular order, such as non-increasing or non-decreasing.)

THEOREM 11.2 (another form of Theorem 11.1)   *Let r be an integer greater than 1 and let $d_1, \ldots, d_n$ be an integer sequence. If there is no complete $(r-1)$-partite graph H with vertices $v_1, \ldots, v_n$ such that $d_H(v_i) \ge d_i$ for all i, then every graph G with degree sequence $d_1, \ldots, d_n$ has $\omega(G) \ge r$.*

Theorem 11.2 describes a large set of integer sequences such that all graphs G with these degree sequences have $\omega(G) \ge r$, but it does not describe all such sequences. For example, if $3 \le r \le n-1$, then there is a unique graph G that has

$\qquad r-3$ vertices   of degree $n-1$,
$\qquad\qquad 3$ vertices   of degree $r-1$,
$\qquad n-r$ vertices   of degree $r-3$,

and this graph has $\omega(G) = r$, but Theorem 11.2 does not guarantee that $\omega(G) \ge r$: there is a complete $(r-1)$-partite graph with

$\qquad r-3$ vertices   of degree $n-1$,
$\qquad\qquad 2$ vertices   of degree $n-2$,
$\qquad n-r+1$ vertices   of degree $r-1$.

Nevertheless, Theorem 11.2 is the best theorem of its kind. Let us explain in what sense it is the best.

DEFINITIONS: We say that a sequence $e_1, \ldots, e_n$ *tops* a sequence $d_1, \ldots, d_n$ if, and only if, $e_i \ge d_i$ for all i. A property of degree sequences is said to be *monotone* if, with every degree sequence d that has this property, all degree sequences that top d have the property, too.

Trivially, a monotone property of degree sequences forces $\omega(G) \ge r$ only if the degree sequence of G is not topped by a degree sequence of any complete $(r-1)$-partite graph. Theorem 11.2 asserts that $\omega(G) \ge r$ if the degree sequence of G is not topped by a degree sequence of any complete $(r-1)$-partite graph. It follows that all theorems asserting that some monotone property of degree sequences forces $\omega(G) \ge r$ are subsumed in Theorem 11.2.

Let us put this observation in different terms.

DEFINITIONS & NOTATION: Let $\mathcal{D}_n$ denote the set of all degree sequences of graphs of order n and let $\Omega_{n,r}$ denote the set of all d in $\mathcal{D}_n$ for which every G with degree sequence d has $\omega(G) \ge r$.

Let us say that a subset $\mathcal{U}$ of $\mathcal{D}_n$ is *upward closed* if

$$d \in \mathcal{U}, \ e \in \mathcal{D}_n, \ e \text{ tops } d \ \Rightarrow \ e \in \mathcal{U}.$$

Since the union of two upward closed sets is upward closed, the union of all upward closed subsets of any set $\mathcal{S}$ of degree sequences is upward closed; let $\mathcal{S}^{\uparrow}$ denote this largest upward closed subset of $\mathcal{S}$.

Theorem 11.2 decribes $\Omega^{\uparrow}_{n,r}$ with $n \geq r \geq 2$: if a degree sequence fails to satisfy its condition, then it is topped by a degree sequence outside $\Omega_{n,r}$, and so it does not belong to $\Omega^{\uparrow}_{n,r}$. The degree sequence in the preceding example,

$$r - 3 \text{ terms } n - 1, \quad 3 \text{ terms } r - 1, \quad n - r \text{ terms } r - 3,$$

belongs to $\Omega_{n,r} - \Omega^{\uparrow}_{n,r}$.

DEFINITION: A graph is called *hamiltonian* if it contains a *Hamilton cycle*, which means a cycle passing through all its vertices.

The property of being hamiltonian is like the property $\omega(G) \geq r$ in several respects. The claim that a graph G is hamiltonian can be certified by pointing out a Hamilton cycle in G; finding such a certificate may be extremely difficult, but verifying it is straightforward and quick. By contrast, no easily verifiable certificate of the claim that G is nonhamiltonian is known. Let us say that a nonhamiltonian graph is *maximal* with respect to this property if adding any edge to this graph produces a hamiltonian graph. The claim that G is nonhamiltonian could be certified by exhibiting a maximal nonhamiltonian graph H and showing that G is a subgraph of H. However, this trick only shifts the burden of proof from all nonhamiltonian graphs to all maximal nonhamiltonian graphs: how do we certify that H is nonhamiltonian? One way of doing this would be first to compile a catalogue of all maximal nonhamiltonian graphs and then to simply point out H in this catalogue. Unfortunately, the catalogue is not only very large, but also very complex: some of its items H are wild in the sense that describing them is difficult.

My doctoral advisor Crispin Nash-Williams (1932–2001) had coined in [303] the term *forcibly hamiltonian* for degree sequences d such that every G with degree sequence d is hamiltonian; by 1970, there was a progression of theorems ([100, Theorem 3], [316], [46]) that described larger and larger upward closed sets of forcibly hamiltonian sequences. Theorem 11.1 blazed a trail by displaying a new template for theorems that infer properties of graphs from properties of their degree sequences; having heard Erdős's lecture, I began to wonder what was a description of the largest upward closed set of forcibly hamiltonian sequences.

DEFINITIONS: Let us say that a degree sequence e *strictly tops* a degree sequence d if e tops d and $e \neq d$; let us say that a graph G is *degree-maximal* with respect to some property if G has the property and no graph with a degree sequence that strictly tops the degree sequence of G has the property.

Erdős's Theorem 11.1 amounts to a neat catalogue of all degree-maximal graphs G with $\omega(G) < r$; I was led to look for a catalogue of all degree-maximal nonhamiltonian graphs G. During the Christmas break, I compiled the catalogue of such graphs of orders $3, 4, 5, 6, 7$ and then I saw a pattern: just like in Erdős's prototype, graphs in the catalogue were tame in the sense of being easily described. In each of them, the vertex set could be split into three pairwise disjoint nonempty parts A, B, C such that $|A| = |B|$; every vertex in A was adjacent to all the remaining vertices, the vertices in C were pairwise adjacent, and there were no other edges. (To see that every such graph is nonhamiltonian, observe that the removal of A breaks it into $|A| + 1$ connected components, namely, the $|A|$ isolated vertices in B and the nonempty clique C. This could never happen if the graph contained a Hamilton cycle: the removal of A would break the cycle into at most $|A|$ segments and these segments would hold the rest of the graph in at most $|A|$ pieces.)

REMINDER: $G \oplus H$ denotes the *direct sum* of G and H, which is the graph consisting of a copy of G and a copy of H that have no vertices in common; $G - H$ denotes the *join* of G and H, which is $G \oplus H$ with additional edges that join every vertex in the copy of G to every vertex in the copy of H.

Each of the graphs in my catalogue could be specified as $K_k - (\overline{K_k} \oplus K_{n-2k})$ for some positive integer k less than $n/2$: the vertex set of the K_k is A, the vertex set of the $\overline{K_k}$ is B, and the vertex set of the K_{n-2k} is C. Now I knew what I had to prove:

THEOREM 11.3   *Let n be an integer at least 3. For every nonhamiltonian graph G of order n there is a graph H such that*
    (i) G and H share their vertex-set V,
    (ii) $d_G(v) \le d_H(v)$ for all v in V,
    (iii) $H = K_k - (\overline{K_k} \oplus K_{n-2k})$ for some positive integer k less than n/2.[b]

Theorem 11.3 can be paraphrased as follows:

> *Let n be an integer at least 3. If the degree sequence of a graph G of order n is not topped by the degree sequence of any $K_k - (\overline{K_k} \oplus K_{n-2k})$ with $1 \le k < n/2$, then G is hamiltonian.*

Since $K_k - (\overline{K_k} \oplus K_{n-2k})$ has
    k vertices   of degree k,
    $n - 2k$ vertices   of degree $n - k - 1$,
    k vertices   of degree $n - 1$,
the condition that the degree sequence of G is not topped by the degree sequence of any such graph with $1 \le k < n/2$ can be stated more directly:

THEOREM 11.4 (another form of Theorem 11.3)   *Let n be an integer at least 3. If G is a graph of order n that, for each positive integer k less than n/2, has fewer than k vertices of degree at most k or fewer than $n - k$ vertices of degree at most $n - k - 1$, then G is hamiltonian.*

I proved Theorem 11.4 in January 1971 and submitted the resulting paper [76] for publication on February 1.

---

[b] A fourth property of H, if $d_G(v) = d_H(v)$ for all v in V, then $H = G$, is trivial since this H does not share its degree sequence with any other graph.

By the way, this theorem describes a large set of forcibly hamiltonian sequences, but it does not describe all of them. For example, Nash-Williams [303] proved that, for every choice of positive integers k and n such that k is less than $n/2$ and even, every sequence of k terms equal to k and $n - k$ terms equal to $n - k - 1$ is forcibly hamiltonian. (In particular, every k-regular graph of order $2k + 1$ is hamiltonian.)

My proof of Theorem 11.4 was non-constructive: it consisted of showing that for every maximal nonhamiltonian graph G of order n there is a positive integer k less than $n/2$ such that G has at least k vertices of degree at most k and at least $n - k$ vertices of degree at most $n - k - 1$.

Three years and a few months after this, Adrian Bondy visited me in Montreal and complained about a student of his, who made no progress toward solving an easy problem that Adrian had proposed. The problem was to convert my proof into an efficient algorithm that, given a graph G satisfying the hypothesis of Theorem 11.4, returns a Hamilton cycle in G. I commiserated with Adrian but, as we talked about it, it began to dawn on us that the student may have not been all that weak and that the problem may have not been all that easy: we ourselves could not do it. Fortunately, this sorry state of affairs did not last long. Eventually we designed the algorithm that Adrian had wanted and then we wrote it up, along with a plethora of generalizations, in [48].

11.1.1 An Algorithmic Proof of Theorem 11.4

Our starting point (which I had also used in my proof of Theorem 11.4) was a proof of the following theorem by Øystein Ore (1899–1968):

THEOREM 11.5 (Ore [308])   *Let G be a graph of order n and let u, v be distinct nonadjacent vertices of G such that G with edge uv added is hamiltonian. If $d_G(u) + d_G(v) \geq n$, then G is hamiltonian.*

Proof   By assumption, G with edge uv added contains a cycle $u_1 u_2 \ldots u_n u_1$. If none of its n edges is uv, then this cycle is a Hamilton cycle of G and we are done; else we may assume, without loss of generality, that $uv = u_n u_1$. Write

$$S = \{i : u_1 \text{ is adjacent to } u_{i+1}\},$$
$$T = \{i : u_n \text{ is adjacent to } u_i\}$$

and note that S and T are subsets of $\{1, 2, \ldots, n - 1\}$. If $d_G(u) + d_G(v) \geq n$, then, since $|S| = d_G(u_1)$ and $|T| = d_G(u_n)$, we have $S \cap T \neq \emptyset$. With i standing for any subscript in $S \cap T \neq \emptyset$,

$$u_1 u_{i+1} u_{i+2} \ldots u_n, u_i u_{i-1} u_1$$

is a Hamilton cycle in G.   □

Now consider looking for a Hamilton cycle in a prescribed graph G. If the hypothesis of Theorem 11.5 is satisfied, then we may look instead for a Hamilton cycle in the graph H arising from G by adding edge uv: the proof of the theorem shows how a Hamilton cycle using this edge can be transformed into a Hamilton cycle not using it. Furthermore, if the new graph H and some other pair of distinct nonadjacent vertices x, y satisfy $d_H(x) + d_H(y) \geq n$, then we may in turn augment H by adding edge xy. Repeating this process as long as we can, we eventually arrive at a graph which we call the *closure of* G. (Here, the definite article is apt: it is easy to prove that the closure is unique, independent of the order in which new edges are added to the input graph. However, the argument is irrelevant to the discussion that follows and we will not let it distract us.) To be able to reverse the construction later on, we record with each new edge the time when it gets added to the emerging closure. Next, we propose

ALGORITHM 11.6 (Constructing the closure)

> $H = G$, `timestamp`$(e) = 0$ for all edges e of G, `maxtimestamp` $= 0$;
> **while** H has distinct nonadjacent vertices u, v
>   such that $d_H(u) + d_H(v) \geq n$
> **do**   add edge uv to H, `timestamp`$(uv) = $ `maxtimestamp` $+1$,
>   `maxtimestamp` $= $ `timestamp`(uv);
> **end**

to prove that, as long as the hypothesis of Theorem 11.4 is satisfied, the closure H of G is a complete graph: assuming that H is not complete, we will produce an integer k such that

   (i) $0 < k < n/2$,
   (ii) at least k vertices w of H have $d_H(w) \leq k$,
   (iii) at least $n - k$ vertices w of H have $d_H(w) \leq n - k - 1$.

Since $d_G(w) \leq d_H(w)$ for all w, properties (i), (ii), (iii) imply that the hypothesis of Theorem 11.4 is not satisfied.

In producing k, we may assume that H has no isolated vertices: else (i), (ii), (iii) are satisfied by setting $k = 1$. Under this assumption, let v denote a vertex maximizing $d_H(w)$ among all w such that $d_H(w) \leq n - 2$ (there are such vertices since $H \neq K_n$) and let u denote a vertex maximizing $d_H(w)$ among all w nonadjacent to v in H. We are going to show that (i), (ii), (iii) are satisfied by setting $k = d_H(u)$.

Since u is not an isolated vertex, we have $d_H(u) > 0$. The stopping condition of the **while** loop in Algorithm 11.6 guarantees that $d_H(u) + d_H(v) \leq n - 1$ and our choice of v guarantees that $d_H(v) \geq d_H(u)$; it follows that $d_H(u) < n/2$. Now (i) is verified. Vertex v is nonadjacent to $n - 1 - d_H(v)$ vertices other than itself; our choice of u guarantees that each of these vertices has degree at most $d_H(u)$; since $d_H(u) + d_H(v) \leq n - 1$, their number $n - 1 - d_H(v)$ is at least $d_H(u)$. Now (ii) is verified. Vertex u is nonadjacent to $n - 1 - d_H(u)$ vertices other than itself; our choice of v guarantees that each of these vertices has degree at most $d_H(v)$, which is at most $n - 1 - d_H(u)$; in addition to these $n - 1 - d_H(u)$ vertices, u also has degree at most $n - 1 - d_H(u)$. Now (iii) is verified.

Finally, any Hamilton cycle C in the closure of G can be transformed into a Hamilton cycle C in G by iterating the argument that proves Theorem 11.5:

ALGORITHM 11.7 (Recovering a Hamilton cycle)

> $uv =$ edge of C that maximizes `timestamp`(uv);
> **while** `timestamp`$(uv) > 0$
> **do**   list the vertices of C in their cyclic order as
> $u_1, u_2, \ldots, u_n, u_1$ with $uv = u_1 u_n$;
>   find a subscript i such that
>   `timestamp`$(u_1 u_{i+1}) <$ `timestamp`(uv) and
>   `timestamp`$(u_n u_i) <$ `timestamp`(uv);
> $C =$ the Hamilton cycle $u_1 u_{i+1} u_{i+2} \cdots u_n, u_i u_{i-1} u_1$;
> $uv =$ edge of C that maximizes `timestamp`(uv);
> **end**

With each iteration of the **while** loop in Algorithm 11.7, the largest `timestamp`(e) with e running through all the edges of C drops, and so the loop eventually terminates.

11.1.2   A Digression: Testing the Hypothesis of Theorem 11.2

It is obvious how to test the hypothesis of Theorem 11.4, but it may not be obvious how to test the hypothesis of Theorem 11.2. With $k_{\min}(d_1, \ldots, d_n)$ standing for the smallest k such that some complete k-partite graph H with vertices $v_1, \ldots, v_n$ has $d_H(v_i) \geq d_i$ for all i, the hypothesis of Theorem 11.2 can be stated as

$$r - 1 < k_{\min}(d_1, \ldots, d_n).$$

Owen Murphy [299] designed an efficient algorithm that, given an integer sequence $d_1, d_2, \ldots, d_n$ such that $0 \leq d_1 \leq \ldots \leq d_n$, returns the smallest k such that some graph G with vertices $v_1, \ldots, v_n$ consists of k pairwise vertex-disjoint cliques and has $d_G(v_i) \leq d_i$ for all i. Since a graph consists of k pairwise vertex-disjoint cliques if and only if its complement is complete k-partite, we have

$$k = k_{\min}(n - 1 - d_n, \ldots, n - 1 - d_1),$$

and so converting Murphy's algorithm into an algorithm for computing $k_{\min}(d_1, \ldots, d_n)$ is a matter of mechanical routine.

Let us begin our discussion of the converted version with the following lemma:

LEMMA 11.8   *For every integer sequence $d_1, \ldots, d_n$ such that $d_1 \leq \ldots \leq d_n \leq n - 1$ there is a complete k-partite graph H with vertices $v_1, \ldots, v_n$ such that*
(i) $k = k_{\min}(d_1, \ldots, d_n)$,
(ii) $d_H(v_i) \geq d_i$ for all $i = 1, \ldots, n$,
(iii)   *one of the k parts of H is $\{v_{a+1}, v_{a+2}, \ldots, v_n\}$, where $a = d_n$.*

Proof   Consider any complete k-partite graph H with properties (i), (ii) and let S denote the part of H that includes v_n. Since $d_H(v_n) = n - |S|$, we have $|S| \leq n - d_n$. If $|S| < n - d_n$, then transfer any $n - d_n - |S|$ vertices from the outside of S into S. This transformation maintains property (ii): all the vertices in S have degree d_n after the move and the degrees of all the other vertices remain unchanged or increase during the move. Since the transformation maintains property (ii) and does not increase the number of parts of H, it maintains property (i) as well. Now $|S| = n - d_n$. Finally, if there are subscripts i and j such that $1 \leq i < j \leq n$ and $v_i \in S$, $v_j \notin S$, then $d_H(v_i) = d_n \geq d_j$, $d_H(v_j) \geq d_j \geq d_i$, and so swapping the labels of these two vertices maintains property (ii); trivially, it maintains property (i) as well. Repeating this operation until no such pair v_i, v_j is present any more produces an H with all three properties (i), (ii), (iii). □

Lemma 11.8 points out the recursive Algorithm 11.9 that, given any integer sequence $d_1, d_2, \ldots, d_n$ such that $d_1 \leq \ldots d_n \leq n - 1$, returns $k_{min}(d_1, \ldots, d_n)$. Algorithm 11.10 is the same algorithm presented in an iterative form (chosen to

ALGORITHM 11.9 (Murphy's algorithm in recursive form)

> **if** $d_n \leq 0$
> **then return** 1;
> **else** $s = n - d_n$;
> > **return** $1 + k_{min}(d_1 - s, \ldots, d_{n-s} - s)$;
> **end**

resemble the iterative algorithm on page 209 of [299]). In each of its iterations, Lemma 11.8 is applied to $d_1 - t, d_2 - t, \ldots, d_{n-t} - t$ in place of $d_1, d_2, \ldots, d_n$.

ALGORITHM 11.10 (Murphy's algorithm in iterative form)

> $n_0 = n, k = 0$;
> **while** $n_k > 0$
> **do** $t = n - n_k$;
> > $n_{k+1} = d_{n-t} - t$;
> > $k = k + 1$;
> **end**
> **return** k;

For instance, given the sequence $1, 1, 1, 2, 3, 3, 5$, the algorithm proceeds as follows:

recursive form		iterative form	
$k_{min}(1, 1, 1, 2, 3, 3, 5)$	$[s = 2]$	$n_0 =$ 7, $k = 0$	
$= 1 + k_{min}(-1, -1, -1, 0, 1)$	$[s = 4]$	$t = 0,$ $n_1 =$ 5, $k = 1$	
$= 2 + k_{min}(-5)$		$t = 2,$ $n_2 =$ 1, $k = 2$	
$= 3$		$t = 6,$ $n_3 = -5,$ $k = 3$	

11.2   A Theorem That Involves Connectivity and Stability

DEFINITION: Let us say that a graph is *tough* if it does not contain a nonempty set A of vertices whose removal breaks the rest of the graph into more than $|A|$ connected components.

As we have observed, on page 178, the graphs $K_k - (\overline{K_k} \oplus K_{n-2k})$ with $1 \leq k < n/2$ that are featured in Theorem 11.3 are nonhamiltonian for the simple reason that they are not tough. Every hamiltonian graph is tough, but the converse is false: for instance, the graph with vertices $u_1, u_2, u_3, v_1, v_2, v_3, w$ and edges

$$u_1 u_2, \ u_1 u_3, \ u_2 u_3, \ u_1 v_1, \ u_2 v_2, \ u_3 v_3, \ u_1 w, \ u_2 w, \ u_3 w, \ v_1 w, \ v_2 w, \ v_3 w$$

is tough but nonhamiltonian.[c]

Having proved Theorem 11.3, I was drawn to musings on which graphs are tough and which graphs are not. Maximal non-tough graphs are easily described: each of them is the join of a complete graph whose order is some positive integer k and a graph that is the direct sum of $k + 1$ nonempty complete graphs. (These graphs include not only all the graphs of Theorem 11.3, but other graphs as well. For instance, $K_1 - (K_2 \oplus K_2)$ is a maximal non-tough graph, which does not appear in Theorem 11.3, since its degree sequence is strictly topped by the degree sequence of another maximal non-tough graph, $K_2 - (K_1 \oplus K_1 \oplus K_1)$.) The shape of these graphs suggests a relationship between toughness and two classic invariants of graphs, the stability number $\alpha(G)$ (defined on page 40 as the largest number of pairwise disjoint vertices in G) and connectivity.

DEFINITION: The *connectivity* $\kappa(G)$ of an incomplete graph G is defined as the smallest number of vertices whose removal breaks the rest of G into at least two connected components. (In particular, $\kappa(G) = 0$ means that G is disconnected.) Note that $\kappa(G) \leq n-2$ whenever G is an incomplete graph of order of n. The connectivity of complete graphs is defined by $\kappa(K_n) = n - 1$.

If G is not tough, then $\alpha(G) > \kappa(G)$: when the removal of a nonempty set A of vertices breaks the rest of the graph into more than $|A|$ connected components, we have $\kappa(G) \leq |A|$ and (as vertices in different components of G with A removed are nonadjacent) $\alpha(G) > |A|$. So now I had two different conditions implying that a graph G is tough: one was that G is hamiltonian and the other was that $\alpha(G) \leq \kappa(G)$. But I could not find any graph that would satisfy the latter condition without satisfying the former.

In the spring of 1971, I traveled through the West Coast of Canada and United States: I had hiring interviews in Victoria and at Stanford, I gave talks at the University of Washington in Seattle and at University of California at Los Angeles, I went to San Francisco just to see it and I visited the RAND Corporation in Santa Monica. All this time, I kept switching back and forth between two distinct modes of operation: when a university paid for a particular leg of the trip, I flew on airplanes and slept in three-star hotels; when I was on

---

[c] In [77], I conjectured that a weaker version of the converse might be true, though: I proposed to call a graph t-tough, with t a positive number, if it contains no nonempty set A of vertices whose removal breaks the rest of the graph into more than $t|A|$ connected components and I conjectured the existence of a positive number t_0 such that every t_0-tough graph is hamiltonian. This conjecture remains open; Doug Bauer, Hajo Broersma, and Hendrik Veldman [20] proved that every such t_0 is at least $9/4$. For more on the conjecture, see [274].

my own, I hitchhiked and stayed at the YMCA. I incorporated into this tour a number theory conference held in March at the Washington State University since Paul Erdős was among its participants.

When the conference ended, Richard Guy (1916–2020) and his wife Louise (1918–2010) were about to drive the PGOM to Calgary and offered to take me with them as far as I wanted to go. Just before we got into the car, I mentioned to Erdős my suspicion that the condition $\alpha(G) \leq \kappa(G)$ implied that G was hamiltonian. This stratagem earned me a place beside him in the back seat and his undivided attention for the duration of my ride with them. In just under two hours, we arrived in Spokane and by that time Erdős had explained to me a proof that my suspicion was justified. Then they dropped me off and my luck continued: within minutes I hitched a ride with a genial truck driver, who took me on I-90 all the way to Seattle.

When I wrote up the proof, I added a footnote that read

> This note was written in Professor Richard K. Guy's car on the way from Pullman to Spokane, Wash. The authors wish to express their gratitude to Mrs. Guy for smooth driving.

Erdős liked the footnote and I was glad.

THEOREM 11.11 ([81])   *If G is a graph of order at least 3 such that $\alpha(G) \leq \kappa(G)$, then G is hamiltonian.*

Proof   A *vertex cut* is a set of vertices whose removal leaves the rest of the graph disconnected. We will specify an efficient algorithm that, given any graph G of order at least 3, returns either a vertex cut K and a stable set A such that $|A| > |K|$ (which certifies that the hypothesis of the theorem is false) or else a Hamilton cycle in G (which certifies that the conclusion of the theorem is true).

In its simple preliminary phase, the algorithm produces either a vertex cut $\{w\}$ and a stable set $\{u, v\}$ (in which case it terminates) or else a cycle in G (in which case it proceeds to the subsequent main phase). The former outcome occurs when some vertex u has degree at most 1: vertex v is an arbitrary vertex nonadjacent to u and vertex w is either the unique neighbour of u or, if u is an isolated vertex, any vertex other than u and v. The latter outcome occurs when every vertex of G has degree at least 2. In this case, we build iteratively longer and longer paths $u_1 u_2 u_3 \ldots$. To initialize this process, take an arbitrary vertex u_2 and its distinct neighbours u_1, u_3. Once a path $u_1 u_2 \ldots u_k$ has been built, consider a neighbour v of u_k other than u_{k-1}. If $v = u_i$ for some i such that $1 \leq i \leq k - 2$, then the preliminary phase terminates with cycle $u_i u_{i+1} \ldots u_k u_i$; else we set $u_{k+1} = v$ and proceed to the next iteration.

The main phase of the algorithm is also iterative. Each of its iterations begins with a cycle in G (in particular, the first iteration begins with the cycle produced in the preliminary phase). Let C denote this cycle.

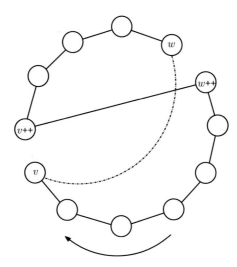

Figure 11.1 Switching to a longer cycle.

CASE 1: *C is not a Hamilton cycle.*
In this case, choose one of the two cyclic orientations of C and for each vertex v of C let $v++$ denote the immediate successor of v in this orientation. Let Q denote an arbitrary connected component of the graph arising from G by removing all vertices of C (and all edges incident with these vertices). Set

$$X = \{v \in C : v \text{ has a neighbour in } Q\},$$
$$Y = \{v++ : v \in X\}.$$

SUBCASE 1.1: $X \cap Y \neq \emptyset$.
In this subcase, there is a vertex v such that $v \in X$ and $v++ \in X$. Replace edge $vv++$ of C by a path from v to $v++$ that has length at least two and all interior vertices in Q; proceed to the next iteration with the longer cycle.

SUBCASE 1.2: $X \cap Y = \emptyset$ and Y is not a stable set.
In this subcase, there is an edge $v++w++$ such that $v, w \in X$ and $v \neq w++, w \neq v++$. As in Figure 11.1, replace edges $vv++$ and $ww++$ of C by edge $v++w++$ and a path from v to w that has length at least two and all interior vertices in Q; proceed to the next iteration with the longer cycle (consisting of the arc of C from $w++$ to v followed by the path from v to w followed by the arc of C with reversed orientation from w to $v++$ followed by edge $v++w++$).

SUBCASE 1.3: $X \cap Y = \emptyset$ and Y is a stable set.
In this subcase, return the vertex cut X and the stable set $Y \cup \{u\}$ with u an arbitrary vertex in Q.
CASE 2: *C is a Hamilton cycle.*
Return C. □

Which graphs satisfy the hypothesis of Theorem 11.11?

PROPOSITION 11.12   *With $m^\star(n)$ standing for the smallest number of edges in a graph G of order n such that $\alpha(G) \le \kappa(G)$, we have*

$$\frac{1}{2}n^{3/2} - \frac{1}{4}n \ < \ m^\star(n) \ \le \ \frac{1}{2}n^{3/2} + \frac{5}{2}n.$$

Proof   First, we will prove that every graph G with n vertices and m edges which is not complete has

$$\alpha(G) \ge n^2/(2m + n), \qquad\qquad (11.1)$$

$$\kappa(G) \le 2m/n. \qquad\qquad (11.2)$$

To prove (11.1), we appeal to the bound

$$\mathrm{ex}(K_r, n) \ \le \ \left(1 - \frac{1}{r-1}\right)\frac{n^2}{2} \qquad\qquad (11.3)$$

which follows directly from formula (7.1). If r is an integer such that $r \ge 2$ and if the number of edges of the complement $\overline{G}$ of G (which equals $\binom{n}{2} - m$) exceeds the right-hand side of (11.3), then $\overline{G}$ contains the complete graph K_r. This fact may be recorded as

$$\binom{n}{2} - m > \left(1 - \frac{1}{r-1}\right)\frac{n^2}{2} \ \Rightarrow \ \alpha(G) \ge r$$

or, after simplifications, as

$$r - 1 < n^2/(2m + n) \ \Rightarrow \ \alpha(G) \ge r,$$

which is logically equivalent to (11.1). To prove (11.2), let v denote a vertex of G that has the smallest degree. Since G is not complete, some vertex w is nonadjacent to v and distinct from it, and so the set K of all the neighbours of v is a vertex cut separating v from w. Since the average degree of a vertex in G is $2m/n$, we have $|K| \le 2m/n$.

Under the assumption that $\alpha(G) \le \kappa(G)$, bounds (11.1) and (11.2) imply that

$$4m^2 + 2mn \ \ge \ n^3;$$

since the left-hand side of this inequality is an increasing function of m and its value at $\frac{1}{2}n^{3/2} - \frac{1}{4}n$ equals $n^3 - \frac{1}{4}n^2$, the lower bound on $m^\star(n)$ follows.

To establish the upper bound on $m^\star(n)$, we will construct, for an arbitrary positive integer n, a graph G of order n with at most $\frac{1}{2}n^{3/2} + \frac{5}{2}n$ edges such that $\alpha(G) \le \kappa(G)$. For this purpose, set $t = \lfloor n^{1/2} \rfloor$. For the vertex set of G, we take the union of t pairwise disjoint sets $V_1, V_2, \ldots, V_t$ whose sizes differ from each other by at most 1: each $|V_i|$ is $\lfloor n/t \rfloor$ or $\lceil n/t \rceil$. Since $t \le n^{1/2}$, we have $n/t \ge t$, and so in each V_i we can choose pairwise distinct vertices $v_i^1, v_i^2, \ldots v_i^t$. For each choice of i and j such that $1 \le i \le t - 1$ and $1 \le j \le t$, we join vertex v_i^j to vertex v_{i+1}^j by an edge; for each i such that $1 \le i \le t$, we join every two vertices in V_i by an edge; apart from these, G has no other edges.

Since $n < (t+1)^2$, we have $n/t \le t + 2$, and so each $|V_i|$ is at most $t + 2$, and so the number m of edges of G satisfies $m \le t(t-1) + t\binom{t+2}{2} = \frac{1}{2}t^3 + \frac{5}{2}t^2 \le \frac{1}{2}n^{3/2} + \frac{5}{2}n$.

Since the vertex set of G is covered by the t cliques V_i, we have $\alpha(G) \le t$. To see that $\kappa(G) \ge t$, observe that no set K of $t-1$ vertices of G can be a vertex cut: since the t paths $v_1^j v_2^j \ldots v_t^j$ with $j = 1, 2, \ldots, t$ are pairwise vertex-disjoint, K cannot meet all of them. It follows that the graph $G - K$ arising from G by the deletion of all vertices in K (and all edges incident with these vertices) is connected: it contains at least one of the paths $v_1^j v_2^j \ldots v_t^j$ and, for each $i = 1, 2, \ldots, t$, all of its vertices in V_i are adjacent to v_i^j. $\qquad\square$

The upper bound in Proposition 11.12 shows that the hypothesis of Theorem 11.11 can be satisfied by relatively sparse graphs. Graphs that satisfy the hypothesis of Theorem 11.4, by contrast, must be dense:

PROPOSITION 11.13   *Let n be an integer at least 3. If G is a graph of order n that, for each positive integer k less than $n/2$, has fewer than k vertices of degree at most k or fewer than $n - k$ vertices of degree at most $n - k - 1$, then G has at least $n^2/8$ edges.*

Proof   To begin, set $k = \lceil n/2 \rceil - 1$ and note that

(i) $n - k \ge k + 1$.

Then arrange the degree sequence $d_1, d_2, \ldots, d_n$ of G in non-decreasing order,

(ii) $d_1 \le d_2 \le \ldots \le d_n$,

and note that $\sum_{i=1}^{n} d_i \ge \sum_{i=n-k}^{n} d_i \ge (k+1)d_{n-k}$. We will conclude the proof by showing that

(iii) $d_{n-k} \ge k + 1$,

which implies that $(k+1)d_{n-k} \ge (k+1)^2 \ge n^2/4$. For this purpose, let us distinguish between two cases. If $d_k \ge k+1$, then (iii) follows from (i) and (ii). If $d_k \le k$, then the sequence has at least k terms not exceeding k, and so it must have fewer than $n - k$ terms not exceeding $n - k - 1$, which means that $d_{n-k} \ge n - k$. This together with (i) implies (iii). $\qquad\square$

The lower bound in this proposition can be raised to $(3n^2 - 2n - 8)/16$ by more careful analysis [83, page 96]. (In the same article, its authors also prove that a graph of order n has complete closure only if it has at least $\lfloor (n+2)^2/8 \rfloor$ edges and that this bound cannot be improved.)

11.3   Hamilton Cycles in Random Graphs

Proposition 11.12 shows that in order to satisfy the hypothesis of Theorem 11.11, a graph of order n must have about $\frac{1}{2}n^{3/2}$ or more edges. As for graphs that are a little denser, the hypothesis is satisfied by an overwhelming majority of them:

PROPOSITION 11.14   *For every integer-valued function m of n such that*

$$n^{3/2}\sqrt{\ln n} \;\leq\; m(n) \;\leq\; \tbinom{n}{2} \text{ for all sufficiently large } n,$$

almost all graphs G with n vertices and m(n) edges have $\alpha(G) \leq \kappa(G)$.

To prove this proposition, we shall follow an argument used by Chvátal and Erdős [80, pp. 417–418]. Let us set a portion of it apart on its own:

LEMMA 11.15   *Let ε be a positive number and let m be an integer-valued function of n such that $1.05\varepsilon^{-1}n^{3/2} \leq m(n) \leq \binom{n}{2}$ for all sufficiently large n. Then almost all graphs with n vertices and m(n) edges have*

$$\kappa(G) > (1 - \varepsilon)\frac{2m(n)}{n}.$$

The coefficient $(1 - \varepsilon)$ in this lower bound on $\kappa(G)$ cannot be raised to 1: connectivity of a graph is at most the smallest degree of its vertex and $2m(n)/n$ is the average degree of a vertex in a graph with n vertices and $m(n)$ edges.

The lower bound on $m(n)$ featured in Lemma 11.15 can be reduced by a more laborious argument. (Such reductions cannot go too far: Theorem 10.1 shows that almost all graphs G with n vertices and $\frac{1}{2}n \ln n - f(n)$ edges such that $\lim_{n\to\infty} f(n)/n = +\infty$ have $\kappa(G) = 0$.) Nevertheless, for our purpose it is good enough as it stands.

REMINDER: $f(n) = o(g(n))$ means that $\lim_{n\to\infty} f(n)/g(n) = 0$.

Proof of Lemma 11.15   Let us write simply m for $m(n)$ and let us write

A_n for the set of all graphs with vertices $1, 2, \ldots, n$ and with m edges,
C_n for the set of all graphs in A_n that have a vertex cut of size at most $(1 - \varepsilon) \cdot 2m/n$,
D_n for the set of all graphs in A_n that have a vertex of degree at most $(1 - \varepsilon/2) \cdot 2m/n$.

We will prove the lemma by proving that

(i) $|C_n - D_n| = o(|A_n)|$ and
(ii) $|D_n| = o(|A_n)|$.

To prove (i), consider an arbitrary graph G in $C_n - D_n$. Since $G \in C_n$, its vertex set can be partitioned into pairwise disjoint sets A, B, C such that $|C| \leq (1 - \varepsilon) \cdot 2m/n$, the neighbours of every vertex in A come from $A \cup C$, the neighbours of every vertex in B come from $B \cup C$, and both A, B are nonempty. Since $G \notin D_n$, we have $|A \cup C| > (1 - \varepsilon/2) \cdot 2m/n$. It follows that $|A| > \varepsilon m/n$; similarly, $|B| > \varepsilon m/n$. Therefore all graphs in $C_n - D_n$ (as well as some graphs in D_n) can be manufactured by choosing first a partition of $\{1, 2, \ldots, n\}$ into pairwise disjoint sets A, B, C such that $|A| > \varepsilon m/n$, $|B| > \varepsilon m/n$, $|C| \leq (1 - \varepsilon) \cdot 2m/n$ and then a graph in A_n that has no edges with one endpoint in A and the other endpoint in B.

There are fewer than 3^n ways of choosing A, B, C, and for each of these choices there are precisely

$$\binom{\binom{n}{2} - |A| \cdot |B|}{m}$$

ways of choosing a graph in $\mathcal{A}_n$ so that no edge has one endpoint in A and the other endpoint in B. The latter number of choices can be bounded from above by bounding $|A| \cdot |B|$ from below: constraints $|A|+|B| \geq n-(1-\varepsilon)\cdot 2m/n$, $|A| > \varepsilon m/n$, $|B| > \varepsilon m/n$ imply

$$|A| \cdot |B| > \frac{\varepsilon m}{n}\left(n - (1 - \varepsilon)\frac{2m}{n} - \frac{\varepsilon m}{n}\right)$$

$$= \varepsilon m\left(1 - (2 - \varepsilon)\frac{m}{n^2}\right) > \varepsilon m\left(1 - \frac{2 - \varepsilon}{2}\right) = \frac{\varepsilon^2 m}{2}.$$

We conclude that

$$\frac{|\mathcal{C}_n - \mathcal{D}_n|}{|\mathcal{A}_n|} \leq 3^n \frac{\binom{\binom{n}{2} - \varepsilon^2 m/2}{m}}{\binom{\binom{n}{2}}{m}} \leq 3^n \left(1 - \frac{\varepsilon^2 m/2}{\binom{n}{2}}\right)^m$$

$$\leq 3^n \exp\left(-\left(\frac{\varepsilon m}{n}\right)^2\right) \leq \exp\left(n \ln 3 - 1.1025n\right) = o(1),$$

which proves (i).

To prove (ii), write $N = \binom{n}{2}$, $M = n - 1$, and $r = \lfloor(1 - \varepsilon/2)\cdot 2m/n\rfloor$. Every graph in $\mathcal{D}_n$ can be manufactured by choosing first a vertex v in its vertex set $\{1, 2, \ldots, n\}$, then an integer i such that $0 \leq i \leq r$, next a set S of i vertices other than v, and finally a set E of m edges such that $vw \in E$ if and only if $w \in S$. Since there are n ways of choosing v and $\binom{M}{i}$ ways of choosing S and, once S has been chosen, $\binom{N-M}{m-i}$ ways of choosing E, we have

$$\frac{|\mathcal{D}_n|}{|\mathcal{A}_n|} \leq n \sum_{i=0}^{r} \frac{\binom{M}{i}\binom{N-M}{m-i}}{\binom{N}{m}}$$

and so, by (A.23),

$$\frac{|\mathcal{D}_n|}{|\mathcal{A}_n|} \leq ne^{-pm}\left(\frac{epm}{r}\right)^r = n\left(\frac{1}{e}\left(\frac{epm}{r}\right)^{r/pm}\right)^{pm}$$

with $p = 2/n$. Since $(e/x)^x$ is an increasing function of x in the interval $(0, 1]$ and since $r/pm \leq 1 - \varepsilon/2$, we have $(epm/r)^{r/pm} \leq (1 - \delta)e$ for some positive δ, and so $|\mathcal{D}_n|/|\mathcal{A}_n| \leq n(1 - \delta)^{pm} = n(1 - \delta)^{2m/n} = o(1)$, which proves (ii). $\square$

Proof of Proposition 11.14   Let m be an integer-valued function of n such that

$$n^{3/2}\sqrt{\ln n} \leq m(n) \leq \binom{n}{2} \text{ for all sufficiently large } n.$$

By Lemma 9.22, almost all graphs G with n vertices and $m(n)$ edges have

$$\alpha(G) < \frac{4}{3} \cdot \frac{n^2}{m(n)} \ln n \le \frac{4}{3} \cdot n^{1/2} \sqrt{\ln n}.$$

By Lemma 11.15, almost all graphs G with n vertices and $m(n)$ edges have

$$\kappa(G) > \frac{2}{3} \cdot \frac{2m(n)}{n} \ge \frac{4}{3} \cdot n^{1/2} \sqrt{\ln n}. \qquad \square$$

The conjunction of Theorem 11.11 and Proposition 11.14 implies that almost all graphs G with n vertices and $\lceil n^{3/2}\sqrt{\ln n}\rceil$ edges are hamiltonian. Actually, the threshold function for the property of being hamiltonian is only $n \log n$, the same as the threshold function for the property of being connected. This breakthrough result follows from a theorem of Lajos Pósa [317]: Almost all graphs with n vertices and $m(n)$ edges are hamiltonian if $m(n) = (1 + o(1))cn \ln n$ with a sufficiently large constant c.

Independently and a little later, Aleksei Dmitrievich Korshunov (1936–2019) proved a stronger theorem [259, 260]: Almost all graphs with n vertices and $m(n)$ edges are hamiltonian if $m(n) = \frac{1}{2}n \ln \ln n + \frac{1}{2}n \ln \ln \ln n + c(n)n$ with $\lim_{n\to\infty} c(n) = \infty$. János Komlós and Endre Szemerédi [258] refined this result: With $h(n, m)$ standing for the probability that a random graph with n vertices and m edges is hamiltonian, we have

$$\lim_{n\to\infty} h\left(n, \frac{1}{2}n \ln n + \frac{1}{2}n \ln \ln n + cn + o(n)\right) = e^{-e^{-c}}. \qquad (11.4)$$

A special case of a classical result of Erdős and Rényi [150, Theorem 2 (with $r = 1$)] asserts that the probability of a random graph with n vertices and m edges having no vertex of degree less than 2 satisfies (11.4), too, in place of $h(n, m)$. This similarity between the property of having no vertex of degree less than 2 and the stronger property of being hamiltonian is highlighted by a striking theorem [258, Reformulation 2 on page 53], [37], [2]:

THEOREM 11.16 *Consider the random graph of order n that evolves step by step through adding edges one at a time. The probability that this graph becomes hamiltonian at the moment when all of its vertices have acquired at least two neighbours tends to 1 as n tends to infinity.* $\qquad \square$

Just like vertices of degree 0 are the last barrier to being connected (Theorem 10.6), vertices of degree 1 are the last barrier to having a Hamilton cycle.

UNIVERSITY OF COLORADO

BOULDER, COLORADO 80302

1971Ⅴ 10

DEPARTMENT OF MATHEMATICS

Dear Chvátal,

Many thanks for your letter and
our paper. It could certainly go to
Discrete Mathematics or the Monthly whichever
you prefer. Please send it away + give your
address as a return address in view of the
fact that my address changes all the time
   I will be in Montreal in June or July.
   It would be nice to prove that
almost all graphs of n^{++} n vertices and
$n^{1+\varepsilon}$ edges are Hamiltonian (perhaps it
remains true for c n log n if c is large)
   Kind regards to you + your boss +
all mathematicians   E. P.

Courtesy of Vašek Chvátal

Appendix A: A Few Tricks of the Trade

A.1 Inequalities

A.1.1 Two workhorses

Two simple and strikingly serviceable inequalities are

$$1 + x \le e^x \quad \text{for all real numbers } x, \tag{A.1}$$

$$k! > (k/e)^k \quad \text{for all positive integers } k. \tag{A.2}$$

Inequality (A.1) is an easy exercise in calculus and (A.2) follows from (A.1) by induction on k: the induction step consists of observing that

$$(k+1)! > (k+1)\left(\frac{k}{e}\right)^k = (k+1)^{k+1} \cdot \left(\frac{k}{k+1}\right)^k \cdot \frac{1}{e^k} > \left(\frac{k+1}{e}\right)^{k+1},$$

where the first inequality is the induction hypothesis and the second inequality is (A.1) with $1/k$ in place of x.

Inequality (A.1) can be written as $1/(1+x) \ge e^{-x}$. Substituting $-y/(1+y)$ with $y > -1$ for x in the latter inequality, we find that $1 + y \ge e^{y/(1+y)}$. Therefore

$$e^{x/(1+x)} \le 1 + x \le e^x \quad \text{whenever } x > -1.$$

We shall present refinements of the crude (A.2) in Section A.2. An immediate corollary of (A.2) is the inequality

$$\binom{n}{k} < \left(\frac{en}{k}\right)^k \quad \text{for all positive integers } k. \tag{A.3}$$

This bound is weak. In particular, when $k \ge 0.328n$, its right-hand side is greater than 2^n, which is a trivial upper bound on the left-hand side for all values of k. Nevertheless, (A.3) is not all bad: it has served satisfactorily in the proofs of Lemma 9.21 and Lemma 10.5. When $k = o(n^{1/2})$, its right-hand side is asymptotically equal to its left-hand side times $\sqrt{2\pi k}$ (this follows from the conjunction of (A.6) and Theorem A.3, where $j = 1$).

A.1.2 Cauchy–Bunyakovsky–Schwarz Inequality

In its finite form

$$\left(\sum_{i=1}^{n} a_i b_i\right)^2 \le \left(\sum_{i=1}^{n} a_i^2\right)\left(\sum_{i=1}^{n} b_i^2\right),$$

this inequality holds for all choices of real numbers a_i, b_i with i running through $\{1, \ldots, n\}$. It appeared in the second of the two notes on the theory of inequalities that constitute the last part of a textbook [66] published in 1821 by Augustin-Louis Cauchy (1789–1857). In 1859, Cauchy's former doctoral student Viktor Yakovlevich Bunyakovsky (1804–89) published a monograph [63], which includes an inequality just like Cauchy's except that its finite sums are replaced by the more general integrals. (Here, Bunyakovsky referred to Cauchy's original version as "well-known.") Apparently unaware of this work, Hermann Amandus Schwarz (1843–1921) rediscovered Bunyakovsky's integral inequality and came up with a remarkably elegant proof of it [338]. Restricted to the context of finite sums, Schwarz's proof consists of reflecting on the identity

$$\sum_{i=1}^{n}(a_ix + b_i)^2 \;=\; \left(\sum_{i=1}^{n} a_i^2\right)x^2 + \left(2\sum_{i=1}^{n} a_ib_i\right)x + \left(\sum_{i=1}^{n} b_i^2\right):$$

since its left-hand side is a sum of squares, both sides are nonnegative for all x, and so the discriminant $\left(2\sum_{i=1}^{n} a_ib_i\right)^2 - 4\left(\sum_{i=1}^{n} a_i^2\right)\left(\sum_{i=1}^{n} b_i^2\right)$ of the right-hand-side quadratic form cannot be positive. (For a wealth of information on the Cauchy–Bunyakovsky–Schwarz inequality, see [348].)

A.1.3 Jensen's Inequality

A real-valued function f defined on an interval I is called *convex* if

$$x, y \in I, \; 0 \le \lambda \le 1 \;\Rightarrow\; f(\lambda x + (1 - \lambda)y) \le \lambda f(x) + (1 - \lambda)f(y).$$

Straightforward induction on n shows that

$$f\left(\sum_{i=1}^{n} \lambda_i x_i\right) \;\le\; \sum_{i=1}^{n} \lambda_i f(x_i)$$

for every convex function f, for all numbers $x_1, \ldots, x_n$ in its domain, and for all choices of nonnegative numbers $\lambda_1, \ldots, \lambda_n$ that sum up to 1. This is the finite form of an inequality published in 1906 by Danish engineer Johan Ludwig William Valdemar Jensen (1859–1925), who, while successful in his career at the Copenhagen Telephone Company, worked on mathematics in his spare time [226]. Jensen's inequality subsumes a number of other classical inequalities. In particular, the Cauchy–Bunyakovsky–Schwarz inequality can be interpreted as a special case of Jensen's inequality: when no b_i equals 0 (which is not really a restrictive assumption), consider

$$f(x) = x^2, \;\; \lambda_i = \frac{b_i^2}{b_1^2 + b_2^2 + \ldots b_n^2}, \;\; x_i = \frac{a_i}{b_i}.$$

Jensen's Inequality and Binomial Coefficients

The following corollary of Jensen's inequality involves a generalization of the combinatorial notion $\binom{d}{k}$ to a real-valued function $\binom{x}{k}$ of a real variable x that reduces to $\binom{d}{k}$ when x assumes a nonnegative integer value d:

$$\binom{x}{k} \;=\; x(x-1)\ldots(x-k+1)/k!$$

LEMMA A.1   *If $d_1, \ldots, d_n$ are nonnegative integers and $\sum_{i=1}^{n} d_i \geq n(k-1)$, then*

$$\sum_{i=1}^{n} \binom{d_i}{k} \geq n \binom{\sum_{i=1}^{n} d_i/n}{k}.$$

Proof   Since the derivative of $\binom{x}{k}$ is an increasing function of x in the interval $(k - 1, \infty)$, function $\binom{x}{k}$ is convex in this interval; it follows that the function f defined on all reals by

$$f(x) = \begin{cases} 0 & \text{when } x \leq k - 1, \\ \binom{x}{k} & \text{when } x \geq k - 1 \end{cases}$$

is convex and so Jensen's inequality with $\lambda_i = 1/n$ for all i guarantees that

$$\sum_{i=1}^{n} \binom{d_i}{k} = \sum_{i=1}^{n} f(d_i) \geq nf\left(\sum_{i=1}^{n} d_i/n\right) = n\binom{\sum_{i=1}^{n} d_i/n}{k}. \qquad \square$$

A.2   Factorials and Stirling's Formula

The French-born mathematician Abraham De Moivre (1667–1754) proved that

$$n! = (1 + o(1))\lambda\sqrt{n}\left(\frac{n}{e}\right)^n \quad \text{for some constant } \lambda.$$

We will prove a more general theorem.

DEFINITION: A real-valued function f defined on an interval I is called *concave* if

$$x, y \in I, \ 0 \leq t \leq 1 \ \Rightarrow \ f(tx + (1 - t)y) \geq tf(x) + (1 - t)f(y).$$

(Should we point out the obvious, 'f is concave if and only if $-f$ is convex'? In any case, this observation is utterly useless to us.)

THEOREM A.2   *For every increasing concave function $f : [1, \infty) \to [1, \infty)$ there is a positive constant c such that*

$$\sum_{i=1}^{n} f(i) = \int_{1}^{n} f(x)dx + \frac{1}{2}f(n) + c + o(1).$$

To see that Theorem A.2 subsumes de Moivre's formula, set $f(x) = \ln x$, observe that $\ln x$ is a concave function of x (since its derivative is a decreasing function of x) and recall that $\int \ln x \, dx = x \ln x - x + \text{constant}$.

Proof of Theorem A.2   Define numbers $\Delta_1, \Delta_2, \Delta_3, \ldots$ by

$$\Delta_i = \int_{i}^{i+1} f(x)dx - \left(\frac{1}{2}f(i) + \frac{1}{2}f(i + 1)\right).$$

With this notation, we have

$$\int_1^n f(x)dx = \sum_{i=1}^{n-1} \int_i^{i+1} f(x)dx = \sum_{i=1}^{n-1} \left(\Delta_i + \tfrac{1}{2}f(i) + \tfrac{1}{2}f(i+1) \right)$$

$$= \sum_{i=1}^{n-1} \Delta_i + \sum_{i=1}^{n} f(i) - \left(\tfrac{1}{2}f(1) + \tfrac{1}{2}f(n) \right),$$

and so

$$\sum_{i=1}^{n} f(i) = \int_1^n f(x)dx + \tfrac{1}{2}f(n) + \left(\tfrac{1}{2}f(1) - \sum_{i=1}^{n-1} \Delta_i \right).$$

We will prove the theorem by proving that, as n tends to infinity, $\sum_{i=1}^{n-1} \Delta_i$ tends to a finite limit.

From the assumption that f is concave, we deduce that

$$\int_i^{i+1} f(x)dx \geq \tfrac{1}{2}f(i) + \tfrac{1}{2}f(i+1) \quad \text{whenever } i \geq 1, \tag{A.4}$$

$$\int_{i-1/2}^{i+1/2} f(x)dx \leq f(i) \qquad\qquad \text{whenever } i \geq 3/2: \tag{A.5}$$

inequality (A.4) follows from observing that

$$\int_i^{i+1} f(x)dx = \int_0^1 f(i+t)dt = \int_0^1 f(t(i+1) + (1-t)i)dt$$

$$\geq \int_0^1 (tf(i+1) + (1-t)f(i))dt = f(i+1) \int_0^1 t\,dt + f(i) \int_0^1 (1-t)dt$$

and inequality (A.5) follows from observing that

$$\int_{i-1/2}^{i+1/2} f(x)dx = \int_{i-1/2}^{i} (f(x) + f(2i-x))dx \leq \int_{i-1/2}^{i} 2f(i)dx = f(i).$$

Inequality (A.4) means that $\Delta_i \geq 0$ for all i; inequality (A.5) implies that

$$\int_{3/2}^{n+1/2} f(x)dx = \sum_{i=2}^{n} \int_{1/2}^{i+1/2} f(x)dx \leq \sum_{i=2}^{n} f(i),$$

and so

$$\sum_{i=1}^{n-1} \Delta_i = \int_1^n f(x)dx + \tfrac{1}{2}f(n) + \tfrac{1}{2}f(1) - \sum_{i=1}^{n} f(i)$$

$$= \int_{3/2}^{n+1/2} f(x)dx + \int_1^{3/2} f(x)dx - \int_n^{n+1/2} f(x)dx + \tfrac{1}{2}f(n) + \tfrac{1}{2}f(1) - \sum_{i=1}^{n} f(i)$$

$$\leq \int_1^{3/2} f(x)dx - \int_n^{n+1/2} f(x)dx + \tfrac{1}{2}f(n) - \tfrac{1}{2}f(1);$$

since f is an increasing function, it follows that

$$\sum_{i=1}^{n-1} \Delta_i \leq \int_1^{3/2} f(x)dx - \tfrac{1}{2}f(1).$$

To summarize, the sequence $\Delta_1, \Delta_1 + \Delta_2, \Delta_1 + \Delta_2 + \Delta_3, \ldots$ is nondecreasing and bounded from above by a constant; therefore it tends to a finite limit. $\qquad\square$

De Moivre's friend, the Scottish mathematician James Stirling (1692–1770), discovered that the constant λ in De Moivre's formula equals $\sqrt{2\pi}$. The resulting asymptotic equation,

$$n! = (1 + o(1))\sqrt{2\pi n} \left(\frac{n}{e} \right)^n,$$

is known as *Stirling's formula*.

Unfortunately, Stirling's formula is useless for dealing with any particular value of n. Fortunately, it can be refined as

$$1 < \frac{n!}{\sqrt{2\pi n}\left(\frac{n}{e}\right)^n} < \exp\frac{1}{12n} \quad \text{for all positive integers } n. \tag{A.6}$$

In 1955, the lower bound was strengthened [327] by the American mathematician Herbert Ellis Robbins (1915–2001):

$$\exp\frac{1}{12n+1} < \frac{n!}{\sqrt{2\pi n}\left(\frac{n}{e}\right)^n} < \exp\frac{1}{12n} \quad \text{for all positive integers } n.$$

A.3   Asymptotic Expressions for Binomial Coefficients

By definition,

$$\binom{n}{k} = \frac{n(n-1)\dots(n-k+1)}{k!} \leq \frac{n^k}{k!}.$$

When k is reasonably small relative to n, this crude upper bound on $\binom{n}{k}$ is asymptotically tight. In particular, when k is a constant, we have $n - i \sim n$ for all $i = 1, 2, \dots k - 1$, and so

$$\binom{n}{k} \sim \frac{n^k}{k!}.$$

Actually, this asymptotic formula remains correct even if k is allowed to increase with n, as long as it does not increase too fast: we are going to prove that

$$k(n) = o(n^{1/2}) \Rightarrow \binom{n}{k(n)} \sim \frac{n^{k(n)}}{k(n)!}.$$

This is just the first formula in a sequence that continues with

$$k(n) = o(n^{2/3}) \Rightarrow \binom{n}{k(n)} \sim \frac{n^{k(n)}}{k(n)!}\exp\left(-\frac{k(n)^2}{2n}\right),$$

$$k(n) = o(n^{3/4}) \Rightarrow \binom{n}{k(n)} \sim \frac{n^{k(n)}}{k(n)!}\exp\left(-\frac{k(n)^2}{2n} - \frac{k(n)^3}{6n^2}\right),$$

and so on as follows:

THEOREM A.3   *If $k : N \to N$ is a function such that $k(n) = o(n^{j/(j+1)})$ for some positive integer j, then*

$$\binom{n}{k(n)} \sim \frac{n^{k(n)}}{k(n)!}\exp\left(-\sum_{i=1}^{j-1}\frac{k(n)^{i+1}}{i(i+1)n^i}\right).$$

Proof Observing that

$$
\ln\left(\binom{n}{k}\frac{k!}{n^k}\right) = \ln\prod_{d=1}^{k-1}\left(\frac{n-d}{n}\right) = \sum_{d=1}^{k-1}\ln\left(1-\frac{d}{n}\right),
$$

we convert the desired asymptotic formula for $\binom{n}{k(n)}$ into its equivalent form

$$
\sum_{d=1}^{k(n)-1}\ln\left(1-\frac{d}{n}\right) = -\sum_{i=1}^{j-1}\frac{k(n)^{i+1}}{i(i+1)n^i} + o(1). \tag{A.7}
$$

The proof of (A.7) will rely on the fact that

$$
\lim_{x\to 0+}\frac{\ln(1-x)+\sum_{i=1}^{j}x^i/i}{x^j} = 0, \tag{A.8}
$$

which is guaranteed by Taylor's theorem taught as a rule in undergraduate calculus classes. Dyed-in-the-wool revolutionaries (cf. Chapter 8) may prove (A.8) by iterating l'Hôpital's rule: induction on t shows that

$$
\lim_{x\to 0+}\frac{\ln(1-x)+\sum_{i=1}^{j}x^i/i}{x^j} = \lim_{x\to 0+}\frac{-(t-1)!(1-x)^{-t}+\sum_{i=t}^{j}(i-1)!x^{i-t}/(i-t)!}{j!x^{j-t}/(j-t)!}
$$

for all $t = 1, 2, \ldots, j$.

Now write

$$
R_j(x) = \ln(1-x)+\sum_{i=1}^{j}x^i/i
$$

and observe that

$$
\sum_{d=1}^{k-1}\ln\left(1-\frac{d}{n}\right) = \sum_{d=1}^{k-1}\left(R_{j-1}\left(\frac{d}{n}\right)-\sum_{i=1}^{j-1}\frac{d^i}{in^i}\right)
$$

$$
= \sum_{i=1}^{j-1}\left(-\frac{1}{in^i}\sum_{d=1}^{k-1}d^i\right)+\sum_{d=1}^{k-1}R_{j-1}\left(\frac{d}{n}\right).
$$

We will complete the proof of (A.7) by proving that

$$
\sum_{d=1}^{k(n)-1}d^i = \frac{k(n)^{i+1}}{i+1} + o(n^i) \quad \text{for all } i = 1, 2, \ldots, j \tag{A.9}
$$

and that

$$
\sum_{d=1}^{k(n)-1}R_{j-1}\left(\frac{d}{n}\right) = o(1). \tag{A.10}
$$

To prove (A.9), note that

$$
\sum_{d=1}^{k-1}d^i \le \sum_{d=1}^{k-1}\int_d^{d+1}x^i dx = \int_1^k x^i dx \le \int_0^k x^i dx = k^{i+1}/(i+1),
$$

that $\sum_{d=1}^{k-1}d^i = \sum_{d=1}^{k}d^i - k^i$, and that

$$
\sum_{d=1}^{k}d^i \ge \sum_{d=1}^{k}\int_{d-1}^d x^i dx = \int_0^k x^i dx = k^{i+1}/(i+1).
$$

It follows that

$$k^{i+1}/(i+1) - k^i \le \sum_{d=1}^{k-1} d^i \le k^{i+1}/(i+1),$$

which implies (A.9) since $k(n) = o(n)$.

To prove (A.10), note first that (A.8) can be written in the form

$$\lim_{x \to 0+} \frac{R_{j-1}(x)}{x^j} = -\frac{1}{j}.$$

In particular, this implies the existence of a positive δ such that

$$0 < x < \delta \implies -2x^j/j < R_{j-1}(x) < 0;$$

since $k(n) = o(n)$, there is an integer n_0 such that

$$n \ge n_0 \implies k(n)/n < \delta;$$

it follows that

$$n \ge n_0, \ 1 \le d \le k(n) \implies |R_{j-1}(d/n)| < 2(d/n)^j/j \le 2(d/n)^j,$$

and so

$$n \ge n_0 \implies \left| \sum_{d=1}^{k(n)-1} R_{j-1}(d/n) \right| \le \sum_{d=1}^{k(n)-1} |R_{j-1}(d/n)| \le 2 \sum_{d=1}^{k(n)-1} (d/n)^j.$$

The proof of (A.10) is completed by observing that

$$\sum_{d=1}^{k(n)-1} (d/n)^j < \sum_{d=1}^{k(n)-1} (k(n)/n)^j < k(n)^{j+1}/n^j = o(1). \qquad \square$$

A.4   The Binomial Distribution

DEFINITION: When n is a nonnegative integer and p is a real number such that $0 \le p \le 1$, the probability distribution $B_{n,p} : \{0, 1, \ldots n\} \to [0, 1]$ defined by

$$B_{n,p}(k) = \binom{n}{k} p^k (1-p)^{n-k}$$

is called the *binomial distribution with parameters n and p.*

REMINDER: The *probability distribution of a random variable X* is defined as

$$p_X : \Omega_X \to [0, 1],$$

where Ω_X stands for the set of values of X and $p_X(i)$ is defined as $\mathsf{Prob}(X = i)$.

THEOREM A.4   *If the probability distribution of a random variable X is binomial with parameters n and p, then $E(X) = np$ and $Var(X) = np(1-p)$.*

Proof Given n and p, consider the probability space $(\{0,1\}^n, f)$ whose probability distribution f is defined by

$$f((a_1, a_2, \ldots a_n)) = p^{\sum_{i=1}^n a_i}(1-p)^{n-\sum_{i=1}^n a_i}.$$

In this probability space, define a random variable Y by

$$Y((a_1, a_2, \ldots a_n)) = \sum_{i=1}^n a_i$$

and note that Y has the binomial probability distribution with parameters n and p. Next, define random variables $Y_1, Y_2, \ldots Y_n$ by

$$Y_i((a_1, a_2, \ldots a_n)) = a_i \text{ for all } i;$$

observe that $Y = Y_1 + Y_2 + \ldots Y_n$, and so

$$Y^2 = \left(\sum_{i=0}^n Y_i\right) \cdot \left(\sum_{j=0}^n Y_j\right) = \sum_{i=0}^n \sum_{j=0}^n Y_i Y_j.$$

By linearity of expectation, we have

$$E(Y) = \sum_{i=0}^n E(Y_i) \text{ and } E(Y^2) = \sum_{i=0}^n \sum_{j=0}^n E(Y_i Y_j).$$

Note that $\mathsf{Prob}(Y_i = 1) = p$, and so

$$E(Y_i) = p \text{ for all } i.$$

Since Y_i is a zero-one random variable, we have $Y_i = Y_i^2$, and so

$$E(Y_i^2) = p \text{ for all } i.$$

Note that $\mathsf{Prob}(Y_i = Y_j = 1) = p^2$ whenever $i \neq j$, and so

$$E(Y_i Y_j) = p^2 \text{ whenever } i \neq j.$$

It follows that $E(Y) = np$ and $E(Y^2) = np + n(n-1)p^2$; since $\mathrm{Var}(Y) = E(Y^2) - E(Y)^2$ (see page 171), this finding can be recorded as

$$E(Y) = np \text{ and } \mathrm{Var}(Y) = np(1-p).$$

We have constructed a particular random variable Y that has the binomial probability distribution with parameters n and p and satisfies $E(Y) = np$, $\mathrm{Var}(Y) = np(1-p)$. Since the mean and variance of a random variable X are determined by its probability distribution p_X (see page 172), the theorem follows. $\square$

DEFINITION: A *mode* of a probability distribution $p : \Omega \to [0,1]$ is an element M of Ω such that

$$p(M) \geq p(i) \text{ for all } i \text{ in } \Omega.$$

A mode of a probability distribution may not be unique: if M and n are integers such that $1 \leq M \leq n$, then

$$\binom{n}{M}\left(\frac{M}{n+1}\right)^M \left(\frac{n-M+1}{n+1}\right)^{n-M} = \binom{n}{M-1}\left(\frac{M}{n+1}\right)^{M-1}\left(\frac{n-M+1}{n+1}\right)^{n-M+1},$$

and so both M and $M-1$ are modes of $B_{n,M/(n+1)}$. The following proposition implies that these are the only examples of a binomial probability distribution whose mode is not unique:

PROPOSITION A.5   *If M is a mode of the binomial distribution $B_{n,p}$, then*

$$np - (1-p) \leq M \leq np + p.$$

Proof   Since

$$\frac{B_{n,p}(i+1)}{B_{n,p}(i)} = \frac{n-i}{i+1} \cdot \frac{p}{1-p},$$

we have

$$i < np - (1-p) \implies B_{n,p}(i) < B_{n,p}(i+1),$$
$$i + 1 > np + p \implies B_{n,p}(i+1) < B_{n,p}(i). \qquad \square$$

COROLLARY A.6   *If a binomial probability distribution has an integer mean, then this mean is also its unique mode.*

Proof   Let M denote a mode of $B_{n,p}$. By definition, we have $M \in \{0, 1, \ldots n\}$; by Proposition A.5, we have $M \in [np + p - 1, np + p]$. By Theorem A.4, the mean of $B_{n,p}$ is np. If np is an integer, then the interval $[np + p - 1, np + p]$ contains another integer only if $p = 0$ or $p = 1$. If $p = 0$, then the other integer is -1; if $p = 1$, then the other integer is $n + 1$; neither of the two belongs to $\{0, 1, \ldots n\}$. $\qquad \square$

DEFINITION: When Ω is a set of real numbers, a *median* of a probability distribution $p : \Omega \to [0, 1]$ is an element m of Ω such that

$$\mathsf{Prob}(i \leq m) \geq 0.5 \text{ and } \mathsf{Prob}(i \geq m) \geq 0.5.$$

A median of a probability distribution may not be unique: for every positive integer k, both k and $k + 1$ are medians of $B_{2k+1,1/2}$. Nevertheless, Corollary A.6 has an analogue with 'median' in place of 'mode':

THEOREM A.7   *If a binomial probability distribution has an integer mean, then this mean is also its unique median.* $\qquad \square$

This theorem was proved first in 1966 by Peter Neumann [305, Satz 1] and independently, about a year later [228, Theorem 3.2], by Kumar Jogdeo and Stephen M. Samuels (1938–2012). Other proofs have been provided by Rob Kaas and Jan M. Buhrman in [231, Theorem 1] and by Knuth in [254, exercises MPR–23 and MPR–24].

Our (1.5) states that $\binom{2n}{n} \geq 4^n/2n$ whenever $n \geq 1$. Now we are going to raise this lower bound in a more general context:

THEOREM A.8   *For every choice of integers n, k such that $0 < k < n$ and a real number p such that $0 \leq p \leq 1$, we have*

$$\binom{n}{k} p^k (1-p)^{n-k} > \tfrac{2}{3} n^{-1/2} \left(\frac{pn}{k}\right)^k \left(\frac{(1-p)n}{n-k}\right)^{n-k}.$$

Proof Observe that

$$\binom{n}{k}p^k(1-p)^{n-k} = \frac{n!}{k!(n-k)!}p^k(1-p)^{n-k}$$

and that bounds (A.6) imply

$$\frac{n!}{k!(n-k)!}p^k(1-p)^{n-k} \geq \frac{1}{e^{1/12k}e^{1/12(n-k)}}\sqrt{\frac{n}{2\pi k(n-k)}}\left(\frac{pn}{k}\right)^k\left(\frac{(1-p)n}{n-k}\right)^{n-k}.$$

Since $1 \leq k \leq n-1$, we have $k(n-k) \geq n-1$, and so

$$\frac{1}{e^{1/12k}e^{1/12(n-k)}} = \exp\left(-\frac{1}{12k} - \frac{1}{12(n-k)}\right)$$

$$= \exp\left(-\frac{n}{12k(n-k)}\right) \geq \exp\left(-\frac{n}{12(n-1)}\right) \geq e^{-1/6};$$

since $k(n-k) = n^2/4 - (k-n/2)^2 \leq n^2/4$, we have

$$\sqrt{\frac{n}{2\pi k(n-k)}} \geq \sqrt{\frac{2}{\pi n}}.$$

To conclude, note that $e^{-1/6}(2/\pi)^{1/2} = 0.675394\ldots$. $\qquad\square$

A.5   Tail of the Binomial Distribution

Many counting arguments about randomly generated structures rely on the fact that, in a sense, most of these structures do not stray too far from the average. For example, the average number of heads in n flips of an unbiased coin equals $n/2$ and, as we are going to prove, large deviations of the actual number of heads from $n/2$ are unlikely. More precisely, with $\pi_n(i)$ standing for the probability that n flips of an unbiased coin produce precisely i heads, we will prove that, for every function d that maps positive integers into positive reals,

$$\lim_{n\to\infty} d(n)/\sqrt{n} = \infty \quad\Rightarrow\quad \sum_{|i-n/2|\geq d(n)} \pi_n(i) = o(1). \tag{A.11}$$

Actually, we will prove a finer result in a more general context.

To illustrate this more general context, let us consider a barrel of precisely N apples among which precisely M are rotten and let us write

$$p = M/N.$$

In this notation, an apple selected from the barrel at random with uniform probability is rotten with probability p. What is the probability that, in a sample of n apples selected from this barrel at random, precisely i apples will be rotten? The answer depends not only on n and i, but also on the procedure used to generate the sample; the sampling procedure that is simplest to analyze is *sampling with replacement*. Here, each apple is selected from the barrel at random with uniform probability and, once examined, it is thrown back into the barrel in such a way that all of the N apples in the barrel, including the one just examined, are again equally likely to be selected next. A model of this procedure is the uniform probability distribution on the sample space Ω that consists of the N^n sequences $a_1, a_2, \ldots, a_n$ where each a_i belongs to a fixed set of size N representing the barrel of apples. The number of sequences $a_1, a_2, \ldots, a_n$ in Ω with precisely i terms rotten equals $\binom{n}{i} M^i (N - M)^{n-i}$: to see this, observe that each of these sequences can be manufactured by first choosing a set R of i subscripts such that $j \in R$ if and only if apple a_j is rotten, then assigning one of the M rotten apples to each of the i subscripts in R (each rotten apple can be assigned to any number of subscripts), and finally assigning one of the $N - M$ good apples to each of the $n - i$ subscripts outside R (each good apple can be assigned to any number of subscripts, too). This quantity divided by $|\Omega|$ comes to $\binom{n}{i} p^i (1 - p)^{n-i}$, which is the probability that precisely i apples will be rotten in a sample of n generated by sampling with replacement.

We will prove that, for every real number p such that $0 \le p \le 1$,

$$\lim_{n \to \infty} d(n)/\sqrt{n} = \infty \quad \Rightarrow \quad \sum_{|i - pn| \ge d(n)} B_{n,p}(i) = o(1). \qquad (A.12)$$

This generalizes (A.11) since $\pi_n(i) = B_{n,1/2}(i)$ for all $i = 0, 1, \ldots, n$.

We will give two distinct proofs of (A.12). The first of these proofs relies on Chebyshev's inequality (which we proved in Section 10.4):

$$t > 0, \operatorname{Var}(X) > 0 \quad \Rightarrow \quad \operatorname{Prob}\left(|X - \mathrm{E}(X)| \ge t \cdot \operatorname{Var}(X)^{1/2} \right) \le \frac{1}{t^2}.$$

When X is a random variable with probability distribution $B_{n,p}$, we have $\mathrm{E}(X) = np$ and $\operatorname{Var}(X) = np(1 - p)$ by Theorem A.4, and so Chebyshev's inequality with this X states that

$$t > 0 \quad \Rightarrow \quad \sum_{|i - pn| \ge t \cdot \sqrt{np(1-p)}} B_{n,p}(i) \le \frac{1}{t^2}.$$

Substituting $d/\sqrt{np(1 - p)}$ for t, we find that

$$\sum_{|i - pn| \ge d} B_{n,p}(i) \le \frac{np(1 - p)}{d^2}, \qquad (A.13)$$

which implies (A.12).

Our second proof of (A.12) begins with a result of the Ukrainian mathematician Sergei Natanovich Bernstein (1880–1968):

THEOREM A.9 ([27], pp. 159–165 of [28])   *Let n, k be integers such that $0 < k < n$ and let p be a real number such that $0 \le p \le 1$. If $k \ge pn$, then*

$$\sum_{i=k}^{n} \binom{n}{i} p^i (1 - p)^{n-i} \le \left(\frac{pn}{k} \right)^k \left(\frac{(1 - p)n}{n - k} \right)^{n-k}. \qquad (A.14)$$

Proof If $x \geq 1$, then

$$\sum_{i=k}^{n} \binom{n}{i} p^i (1-p)^{n-i} \leq \sum_{i=k}^{n} x^{i-k} \binom{n}{i} p^i (1-p)^{n-i}$$

$$= x^{-k} \sum_{i=k}^{n} \binom{n}{i} (px)^i (1-p)^{n-i} \leq x^{-k} \sum_{i=0}^{n} \binom{n}{i} (px)^i (1-p)^{n-i}$$

$$= x^{-k} (px + 1 - p)^n.$$

In particular, if $k \geq pn$, then

$$\frac{(1-p)k}{p(n-k)} \geq 1,$$

and so

$$\sum_{i=k}^{n} \binom{n}{i} p^i (1-p)^{n-i} \leq \left(\frac{(1-p)k}{p(n-k)} \right)^{-k} \left(\frac{(1-p)k}{n-k} + 1 - p \right)^n$$

$$= \left(\frac{p(n-k)}{(1-p)k} \right)^k \left(\frac{(1-p)n}{n-k} \right)^n = \left(\frac{pn}{k} \right)^k \left(\frac{(1-p)n}{n-k} \right)^{n-k}. \quad \square$$

Theorem A.8 shows that Bernstein's bound (A.14) is close to best possible in the sense that already the first term of its left-hand-side sum is greater than $\frac{2}{3}n^{-1/2}$ times its right-hand-side upper bound.

Theorem A.9 is only a special case of a theorem proved by Bernstein in the more general framework of sums of independent random variables. This theorem had been mostly unknown outside the Russian-speaking world for decades: [91] and [369, pp. 204–205] are isolated attempts to bring it to the attention of English speakers. Eventually, the American applied mathematician, statistician, and physicist Herman Chernoff published another theorem[a] that subsumes Theorem A.9 as a special case [69, Example 3]. For this reason, bound (A.14) or its weaker form (A.17) is often referred to as *the Chernoff bound*.

DEFINITIONS: When $k \geq pn$, the quantity $\sum_{i=k}^{n} B_{n,p}(i)$ featured in Theorem A.9 is called *the right tail of the binomial distribution;* when $k \leq pn$, the quantity $\sum_{i=0}^{k} B_{n,p}(i)$ is called *the left tail of the binomial distribution.*

Since

$$\sum_{i=0}^{k} \binom{n}{i} p^i (1-p)^{n-i} = \sum_{i=0}^{k} \binom{n}{n-i} (1-p)^{n-i} p^i = \sum_{j=n-k}^{n} \binom{n}{j} (1-p)^j p^{n-j},$$

the left tail of the binomial distribution $B_{n,p}$ is the right tail of the binomial distribution $B_{n,1-p}$, and so every bound on one of the two tails provides a bound on the other tail. In particular, Theorem A.9 guarantees also that $k \leq pn$ implies

$$\sum_{i=0}^{k} \binom{n}{i} p^i (1-p)^{n-i} \leq \left(\frac{pn}{k} \right)^k \left(\frac{(1-p)n}{n-k} \right)^{n-k}. \tag{A.15}$$

---

[a] This theorem has been proved by Herman Rubin (1926–2018): see page 35 of H. Chernoff. A career in statistics, in: *Past, Present, and Future of Statistical Science* (X. Lin, C. Genest, D. L. Banks, G. Molenberghs, D. W. Scott, and J. L. Wang, eds.). pp. 29–40, CRC Press, 2014.

Relaxing the Bernstein Bound

In computations involving tails of binomial distributions, it is often useful to sacrifice precision for the sake of simplicity and to replace Bernstein's bound by estimates that are easier to manipulate. A particularly elegant estimate comes from Masashi Okamoto [307, Lemma 2(a)]:

$$0 < k < n, \ k = (p+t)n \ \Rightarrow \ \left(\frac{pn}{k}\right)^k \left(\frac{(1-p)n}{n-k}\right)^{n-k} \leq \exp(-2t^2 n). \qquad (A.16)$$

To prove (A.16), take a fixed p and define a function $f : (-p, 1-p) \to \mathbf{R}$ by

$$\left(\frac{pn}{k}\right)^k \left(\frac{(1-p)n}{n-k}\right)^{n-k} = \left(\left(\frac{p}{p+t}\right)^{p+t} \left(\frac{1-p}{1-p-t}\right)^{1-p-t}\right)^n = e^{-f(t)n}.$$

We have

$$f(t) = (p+t)(\ln(p+t) - \ln p) + (1-p-t)(\ln(1-p-t) - \ln(1-p)),$$

$$f'(t) = \ln(p+t) - \ln p - \ln(1-p-t) + \ln(1-p),$$

$$f''(t) = \frac{1}{p+t} + \frac{1}{1-p-t} = \frac{1}{(p+t)(1-p-t)} \geq 4;$$

since $f'(0) = 0$, it follows first that $f'(t) \geq 4t$ whenever $0 \leq t < 1 - p$ and that $f'(t) \leq 4t$ whenever $-p < t \leq 0$; since $f(0) = 0$, it follows in turn that $f(t) \geq 2t^2$ whenever $-p < t < 1 - p$.

Combining (A.14), (A.15), and (A.16), we conclude that

$$\sum_{|i-pn|\geq tn} B_{n,p}(i) \leq 2\exp(-2t^2 n);$$

this may be written as

$$\sum_{|i-pn|\geq d} B_{n,p}(i) \leq 2\exp(-2d^2/n), \qquad (A.17)$$

which again implies (A.12).

Both (A.13) and (A.17) imply (A.12), but the upper bound (A.17) is often better since it drops exponentially with d^2/n, whereas, for every fixed p, the upper bound (A.13) drops only as the reciprocal of d^2/n.

A simple alternative to (A.16),

$$0 < k < n \ \Rightarrow \ \left(\frac{pn}{k}\right)^k \left(\frac{(1-p)n}{n-k}\right)^{n-k} \leq e^{-pn}\left(\frac{epn}{k}\right)^k, \qquad (A.18)$$

can be verified by setting $y = (k-pn)/(n-k)$ in the inequality $1 + y \leq e^y$. When k is proportional to pn, bound (A.18) is better than bound (A.16) for all sufficiently small p. For instance,

if $k = 2pn$, then the right-hand side of (A.18) is $(e/4)^{pn}$,
which is less than $\exp(-2p^2 n)$, the right-hand side of (A.16),

whenever $p < \ln 2 - 0.5 \doteq 0.193$;
if $k = pn/2$, then the right-hand side of (A.18) is $(2/e)^{pn/2}$,
which is less than $\exp(-p^2 n/2)$, the right-hand side of (A.16),
whenever $p < 1 - \ln 2 \doteq 0.306$.

We will use inequality (A.18) in section A.6.

More on the tail of the binomial distribution can be found in [7, Appendix A] and in [225, Section 2.1].

Applications

To illustrate the use of (A.14), (A.16), and (A.18), we return to Chapter 9. At the end of its Section 9.3, we mentioned that in almost all graphs of order n, every subgraph with s vertices such that $s = \lceil \sqrt{8n} \rceil$ has fewer than

$$\frac{3}{4}\binom{s}{2}$$

edges. This follows from the combination of (A.14) and (A.16) with $\binom{s}{2}$ in place of n, $p = 1/2$, and $k = \lceil 3s(s-1)/8 \rceil$: these inequalities guarantee that

$$\sum_{j \geq 3s(s-1)/8} \binom{\binom{s}{2}}{j} 2^{-\binom{s}{2}} \leq e^{-s(s-1)/16}$$

and so, since $n \leq s^2$,

$$\binom{n}{s} \sum_{j \geq 3s(s-1)/8} \binom{\binom{s}{2}}{j} \cdot 2^{-\binom{s}{2}} \leq \binom{n}{s} e^{-s(s-1)/16}$$

$$\leq n^s e^{-s(s-1)/16} \leq s^{2s} e^{-s(s-1)/16} = o(1).$$

In Section 9.6, we mentioned that in almost all graphs of order n, every vertex has degree $(1 + o(1))n/2$. By (A.17) with $n - 1$ in place of n, with $p = 1/2$, and with $d = \sqrt{(n-1)\ln n}$, the probability that a particular prescribed vertex has a degree differing from $(n-1)/2$ by at least d is at most $2n^{-2}$, and so the probability that G has a vertex whose degree differs from $(n-1)/2$ by at least d is at most $2n^{-1}$.

A.6   Tail of the Hypergeometric Distribution

Let us return to our barrel of N apples, of which M are rotten; as in the preceding section, let p stand for the probability M/N that an apple selected from this barrel at random with uniform probability is rotten. This time, we will analyze *sampling without replacement*, where each of the n apples in the sample is selected from the barrel at random with uniform probability and, once examined, it is kept out of the barrel, so that it can never be selected again. A model of this procedure is the uniform probability distribution on the sample space Ψ of the $\binom{N}{n}$ subsets $\{a_1, a_2, \ldots, a_n\}$ of a fixed set of size N representing the barrel of apples. The number of these subsets with precisely i elements rotten equals $\binom{M}{i}\binom{N-M}{n-i}$: each of them can be manufactured by first choosing a set of its i rotten apples and then choosing a set of its $n - i$ good apples. This quantity divided by $|\Psi|$ is the probability that precisely i apples will be rotten in a sample of n apples generated by sampling without replacement.

DEFINITION: When n, M, N are nonnegative integers such that $0 \le M \le N$, the probability distribution $H_{n,M,N} : \{0, 1, \ldots n\} \to [0, 1]$ defined by

$$H_{n,M,N}(i) = \frac{\binom{M}{i}\binom{N-M}{n-i}}{\binom{N}{n}}$$

is called a *hypergeometric distribution with parameters* n, M, N.

The hypergeometric distribution $H_{n,M,N}$ and the binomial distribution defined by

$$B_{n,M/N}(i) = \binom{n}{i}\frac{M^i(N-M)^{n-i}}{N^n}$$

approximate each other: in particular, if n tends to infinity and M, N grow in such a way that

$$\lim_{n\to\infty} M/n^2 = \infty \quad \text{and} \quad \lim_{n\to\infty} (N - M)/n^2 = \infty$$

then Theorem A.3 with $j = 1$ guarantees that

$$\binom{M}{i} \sim M^i/i!\,,$$
$$\binom{N-M}{n-i} \sim (N - M)^{n-i}/(n - i)!\,,$$
$$\binom{N}{n} \sim N^n/n!\,,$$

and so

$$H_{n,M,N}(i) \sim B_{n,M/N}(i)$$

for all $i = 0, 1, \ldots, N$. Nevertheless, the two distributions are distinct: to take an extreme example, if $0 < M < N$, then $B_{N,M/N}(i)$ is strictly between 0 and 1 for all $i = 0, 1, \ldots, N$, but $H_{N,M,N}(i)$ equals 1 when $i = M$ and 0 otherwise.

Now write

$$p = M/N$$

as in the preceding section and consider the random variable

$$W : \Psi \to \{0, 1, \ldots, n\}$$

that counts the number $W(\psi)$ of rotten apples in each set ψ in our sample space Ψ. To see that $E(W) = np$, note that $W = W_1 + W_2 + \ldots + W_n$, where $W_i(\psi) = 1$ if the i-th apple in the sample ψ is rotten and $W_i(\psi) = 0$ otherwise: we have $E(W) = E(W_1)+E(W_2)+\ldots+E(W_n)$ by linearity of expectation and, since all of the N apples in the barrel are equally likely to show up as the i-th apple in the sample, $E(W_i) = p$ for all i.

DEFINITIONS: When $k \geq pn$, the quantity $\sum_{i=k}^{n} H_{n,M,N}(i)$ is called *the right tail of the hypergeometric distribution;* when $k \leq pn$, the quantity $\sum_{i=0}^{k} H_{n,M,N}(i)$ is called *the left tail of the hypergeometric distribution.*

The American statistician and probabilist Wassily Hoeffding (1914–91) proved a theorem [214, Theorem 4], whose special case states that Bernstein's bound on the tail of the binomial distribution is also a bound on the tail of the hypergeometric distribution:

THEOREM A.10 *Let n, k, M, N be integers such that $0 < k < n$ and $0 \leq M \leq N$; write $p = M/N$. If $k \geq pn$, then*

$$\sum_{i=k}^{n} \frac{\binom{M}{i}\binom{N-M}{n-i}}{\binom{N}{n}} \leq \left(\frac{pn}{k}\right)^{k} \left(\frac{(1-p)n}{n-k}\right)^{n-k}. \tag{A.19}$$

Proof We will show that

$$x \geq 1 \quad \Rightarrow \quad \sum_{i=0}^{n} \frac{\binom{M}{i}\binom{N-M}{n-i}}{\binom{N}{n}} x^{i} \leq \sum_{i=0}^{n} \binom{n}{i} p^{i}(1-p)^{n-i} x^{i}; \tag{A.20}$$

the theorem follows from (A.20) by setting $x = (1-p)k/p(n-k)$ as in the proof of Theorem A.9.

Inequality (A.20) cannot be proved by simply comparing the coefficients of the two polynomials in x: the two probability distributions, $H_{n,M,N}$ and $B_{n,M/N}$, are different and so, depending on the value of i, the left-hand-side coefficients are sometimes smaller and sometimes larger than their right-hand-side counterparts. The trick is to express both sides of (A.20) as polynomials in $x - 1$ and only then to compare coefficients. Since

$$\sum_{i=0}^{n} \frac{\binom{M}{i}\binom{N-M}{n-i}}{\binom{N}{n}} x^{i} = \sum_{i=0}^{n} \frac{\binom{M}{i}\binom{N-M}{n-i}}{\binom{N}{n}} (1 + (x-1))^{i}$$

$$= \sum_{i=0}^{n} \frac{\binom{M}{i}\binom{N-M}{n-i}}{\binom{N}{n}} \sum_{j=0}^{i} \binom{i}{j}(x-1)^{j} = \sum_{j=0}^{n} \sum_{i=j}^{n} \frac{\binom{M}{i}\binom{N-M}{n-i}}{\binom{N}{n}} \binom{i}{j}(x-1)^{j}$$

and

$$\sum_{i=0}^{n} \binom{n}{i} p^{i}(1-p)^{n-i} x^{i} = (px + 1 - p)^{n} = (1 + p(x-1))^{n}$$

$$= \sum_{j=0}^{n} \binom{n}{j} p^{j}(x-1)^{j},$$

this means proving (A.20) by proving that

$$\sum_{i=j}^{n} \frac{\binom{M}{i}\binom{N-M}{n-i}}{\binom{N}{n}} \binom{i}{j} \leq \binom{n}{j} p^{j} \tag{A.21}$$

for all $j = 0, 1, \ldots, n$. Since

$$\sum_{i=j}^{n} \binom{M}{i}\binom{i}{j}\binom{N-M}{n-i} = \sum_{i=j}^{n} \binom{M}{j}\binom{M-j}{i-j}\binom{N-M}{n-i}$$

$$= \binom{M}{j}\sum_{k=0}^{n-j}\binom{M-j}{k}\binom{N-M}{n-j-k} = \binom{M}{j}\binom{N-j}{n-j}$$

and

$$\binom{N}{n}\binom{n}{j} = \binom{N}{j}\binom{N-j}{n-j}$$

(A.21) can be written as $\binom{M}{j}/\binom{N}{j} \le (M/N)^j$, which is true since $M \le N$.   □

Since

$$\sum_{i=0}^{k} \frac{\binom{M}{i}\binom{N-M}{n-i}}{\binom{N}{n}} = \sum_{i=0}^{k} \frac{\binom{N-M}{n-i}\binom{M}{i}}{\binom{N}{n}} = \sum_{j=n-k}^{n} \frac{\binom{N-M}{j}\binom{M}{n-j}}{\binom{N}{n}},$$

the left tail of the hypergeometric distribution $H_{n,M,N}$ is the right tail of the hypergeometric distribution $H_{n,N-M,N}$, and so every bound on one of the two tails provides a bound on the other tail. In particular, Theorem A.10 guarantees also that $k \le pn$ implies

$$\sum_{i=0}^{k} \frac{\binom{M}{i}\binom{N-M}{n-i}}{\binom{N}{n}} \le \left(\frac{pn}{k}\right)^k \left(\frac{(1-p)n}{n-k}\right)^{n-k} \quad \text{with } p = M/N. \quad \text{(A.22)}$$

Combining inequalities (A.22), (A.18) and changing notation (replacing n by m and replacing k by r), we get

$$0 < r < m, \ r/m \le M/N \ \Rightarrow$$

$$\sum_{i=0}^{r} \frac{\binom{M}{i}\binom{N-M}{m-i}}{\binom{N}{m}} \le e^{-pm}\left(\frac{epm}{r}\right)^r \text{with } p = M/N; \quad \text{(A.23)}$$

combining inequalities (A.19), (A.18) and changing notation in the same way, we get

$$0 < r < m, \ r/m \ge M/N \ \Rightarrow$$

$$\sum_{i=r}^{m} \frac{\binom{M}{i}\binom{N-M}{m-i}}{\binom{N}{m}} \le e^{-pm}\left(\frac{epm}{r}\right)^r \text{with } p = M/N. \quad \text{(A.24)}$$

More on the tail of the hypergeometric distribution can be found in [225, Section 2.1].

Applications

To illustrate the use of (A.23) and (A.24), let us return again to Chapter 9. In the proof of its Lemma 9.13, we used ad hoc arguments to show that, when $m = \lfloor n^{1+3\delta} \rfloor$

and $s = \lfloor n^{1-\delta} \rfloor$ for some positive constant δ, we have

$$\sum_{i=0}^{n} \binom{\binom{s}{2}}{i}\binom{\binom{n}{2} - \binom{s}{2}}{m-i}\binom{\binom{n}{2}}{m}^{-1} \leq s^{2n} \exp\left(-0.99\, ms^2/n^2\right)$$

for all sufficiently large n. Inequality (A.23) with $M = \binom{s}{2}$, $N = \binom{n}{2}$, $r = n$ guarantees a stronger result:

$$\sum_{i=0}^{n} \binom{\binom{s}{2}}{i}\binom{\binom{n}{2} - \binom{s}{2}}{m-i}\binom{\binom{n}{2}}{m}^{-1} \leq \left(\frac{em}{n^3}\right)^n \cdot s^{2n} \exp\left(-0.99\, ms^2/n^2\right)$$

for all sufficiently large n. To see this, note that $0.99s^2/n^2 \leq p \leq s^2/n^2$ for all sufficiently large n.

In the proof of Lemma 9.21, we used ad hoc arguments to show that, when c is a positive integer and n is sufficiently large with respect to c and $s \leq \delta n$ with $\delta = 1/64e^5c^3$, we have

$$\sum_{j \geq 3s/2} \binom{n}{s}\binom{\binom{s}{2}}{j}\binom{\binom{n}{2} - \binom{s}{2}}{cn - j}\binom{\binom{n}{2}}{cn}^{-1} \leq \frac{4}{3}\left(\frac{1}{8}\right)^s.$$

Inequality (A.24) with $r = \lceil 3s/2 \rceil$, $m = cn$, $M = \binom{s}{2}$, and $N = \binom{n}{2}$ guarantees a stronger result:

$$\sum_{j \geq 3s/2} \binom{n}{s}\binom{\binom{s}{2}}{j}\binom{\binom{n}{2} - \binom{s}{2}}{cn - j}\binom{\binom{n}{2}}{cn}^{-1} \leq \left(\frac{1}{6\sqrt{6}}\right)^s.$$

To see this, note that

$$\left(\frac{epm}{r}\right)^r \leq \left(\frac{2ec}{3} \cdot \frac{s}{n}\right)^{3s/2},$$

and so

$$\binom{n}{s}\sum_{j \geq 3s/2} \binom{\binom{s}{2}}{j}\binom{\binom{n}{2} - \binom{s}{2}}{cn - j}\binom{\binom{n}{2}}{cn}^{-1} \leq \left(\frac{en}{s}\right)^s \left(\frac{2ec}{3} \cdot \frac{s}{n}\right)^{3s/2}$$

$$\leq \left(\frac{8e^5c^3}{27} \cdot \delta\right)^{s/2} = \left(\frac{1}{6\sqrt{6}}\right)^s.$$

A.7   Two Models of Random Graphs

NOTATION: In this section, we shall let Prob stand for the probability of events in a probability space which may change even within a single statement, but will be understood without ambiguity in each instance.

As suggested on page 151, the Erdős–Rényi random graphs with n vertices and m edges are elements of the probability space whose sample space is the set of all graphs with vertices $1, 2, \ldots, n$ and with m edges and whose probability distribution is uniform.

TERMINOLOGY & NOTATION: We shall refer to this probability space as the *uniform model of random graphs* and we shall let $\mathbb{G}(n, m)$ denote its generic element.

Another popular notion of random graphs was introduced by Edgar Gilbert (1923–2013) in [185]. It is modeled by probability spaces that are parametrized by positive integers n and real numbers p such that $0 \leq p \leq 1$. Here, each sample space consists of all graphs with vertices $1, 2, \ldots, n$ and the probability of each of these graphs is defined as

$$p^i(1 - p)^{N-i},$$

where $N = \binom{n}{2}$ and i is the number of edges of the graph. (This is the probability of the graph appearing when the N unordered pairs of distinct vertices are considered one by one and each of them made adjacent with probability p independently of all other adjacencies.)

TERMINOLOGY & NOTATION: We shall refer to this probability space as the *binomial model of random graphs* and we shall let $\mathbf{G}(n, p)$ denote its generic element.

Model $\mathbb{G}(n, m)$ is like model $\mathbf{G}(n, m\binom{n}{2}^{-1})$ in the sense that m is the mean number of edges as well as the most likely number of edges of $\mathbf{G}(n, m\binom{n}{2}^{-1})$: see Theorem A.4 and Corollary A.6. We shall point out that similarities between the two models go deeper.

Computations involving $\mathbb{G}(n, m)$ are often more convoluted than computations involving its counterpart $\mathbf{G}(n, m\binom{n}{2}^{-1})$. For illustration, consider the analogue of Lemma 9.13 in the binomial model:

> *For every constant δ such that $0 < \delta < 1/3$, the probability of $\mathbf{G}(n, \lfloor n^{1+3\delta} \rfloor \binom{n}{2}^{-1})$ having more than n edges in every subgraph induced by $\lfloor n^{1-\delta} \rfloor$ vertices is $1 - o(1)$.*

We are going to prove a stronger and more general statement:

LEMMA A.11   *Let n and s be positive integers such that*

$$3 \leq s \leq n \quad and \quad \tfrac{1}{2}\binom{s}{2} \geq n.$$

If p is a real number such that

$$\frac{12n \ln s}{s(s - 1)} \leq p \leq \frac{1}{2},$$

then the probability of $\mathbf{G}(n, p)$ having an induced subgraph with s vertices and at most n edges is at most 27^{-n}.

Proof Let $\mathcal{F}$ denote the set of all graphs F with s vertices, all of them coming from the set $\{1, 2, \ldots n\}$, and with at most n edges. Since $3 \leq s \leq n$, we have $\binom{n}{s} \leq n^s \leq s^n$; since $3 \leq n \leq \frac{1}{2}\binom{s}{2}$, it follows that

$$|\mathcal{F}| = \binom{n}{s} \sum_{i=0}^{n} \binom{\binom{s}{2}}{i} \leq s^n \sum_{i=0}^{n} \binom{\binom{s}{2}}{i} \leq s^n(n+1)\binom{\binom{s}{2}}{n} \leq s^n\binom{s}{2}^n \leq s^{3n}.$$

For each F in $\mathcal{F}$ define a random variable X_F by

$$X_F = \begin{cases} 1 & \text{when } F \text{ is an induced subgraph of } \mathbf{G}(n, p), \\ 0 & \text{otherwise.} \end{cases}$$

The lemma asserts that

$$\text{Prob}(\textstyle\sum_{F \in \mathcal{F}} X_F \geq 1) \leq 27^{-n}.$$

To prove this, consider first a single F in $\mathcal{F}$. With i standing for the number of its edges, we have

$$\mathrm{E}(X_F) = \text{Prob}(X_F = 1) = p^i(1-p)^{\binom{s}{2}-i} \leq (1-p)^{\binom{s}{2}} \leq \exp\left(-p\binom{s}{2}\right).$$

By linearity of expectation, it follows that

$$\mathrm{E}(\textstyle\sum_{F \in \mathcal{F}} X_F) = \textstyle\sum_{F \in \mathcal{F}} \mathrm{E}(X_F) \leq |\mathcal{F}| \exp\left(-p\binom{s}{2}\right)$$
$$\leq s^{3n}\exp\left(-p\binom{s}{2}\right) = \exp\left(3n \ln s - p\binom{s}{2}\right) \leq \exp\left(-\tfrac{1}{2}p\binom{s}{2}\right) \leq s^{-3n} \leq 27^{-n}.$$

By Markov's inequality or by common sense, whichever you prefer, we have

$$\text{Prob}\left(\textstyle\sum_{F \in \mathcal{F}} X_F \geq 1\right) \leq \mathrm{E}\left(\textstyle\sum_{F \in \mathcal{F}} X_F\right),$$

and this completes the proof. □

Now let δ be a constant such that $0 < \delta < 1/3$, let n be a positive integer, and let $\mathcal{X}$ denote the set of all graphs with vertices $1, 2, \ldots, n$ that have an induced subgraph with $\lfloor n^{1-\delta} \rfloor$ vertices and at most n edges. In this notation, Lemma 9.13 states that

$$\text{Prob}(\mathbb{G}(n, \lfloor n^{1+3\delta} \rfloor) \in \mathcal{X}) = o(1) \tag{A.25}$$

and Lemma A.11 guarantees that, for all n that are sufficiently large with respect to δ,

$$\text{Prob}(\mathbf{G}(n, \lfloor n^{1+3\delta} \rfloor \binom{n}{2}^{-1}) \in \mathcal{X}) \leq 27^{-n}. \tag{A.26}$$

The following theorem shows that (A.25) is a direct consequence of (A.26):

THEOREM A.12 *Let n and m be integers such that $n \geq 1$ and $0 \leq m \leq \binom{n}{2}$. If $\mathcal{X}$ is an arbitrary set of graphs with vertices $1, 2, \ldots, n$, then*

$$\text{Prob}(\mathbb{G}(n, m) \in \mathcal{X}) \leq 1.07n \cdot \text{Prob}(\mathbf{G}(n, m\binom{n}{2}^{-1}) \in \mathcal{X}).$$

Proof With N standing for $\binom{n}{2}$, we have

$$\mathrm{Prob}(\mathbf{G}(n, m\binom{n}{2}^{-1}) \in \mathcal{X}) = \sum_{i=0}^{N} \binom{N}{i} \left(\frac{m}{N}\right)^i \left(\frac{N-m}{N}\right)^{N-i} \mathrm{Prob}(\mathbb{G}(n, i) \in \mathcal{X})$$

$$\geq \binom{N}{m} \left(\frac{m}{N}\right)^m \left(\frac{N-m}{N}\right)^{N-m} \mathrm{Prob}(\mathbb{G}(n, m) \in \mathcal{X})$$

$$\geq \tfrac{2}{3} N^{-1/2} \cdot \mathrm{Prob}(\mathbb{G}(n, m) \in \mathcal{X})$$

$$\geq (8/9)^{1/2} n^{-1} \cdot \mathrm{Prob}(\mathbb{G}(n, m) \in \mathcal{X});$$

the first inequality is trivial, the second is guaranteed by Theorem A.8, and the third is trivial again. □

The multiplier $1.07n$ in the upper bound of Theorem A.12 cannot be reduced to $0.88n$. To see this, set $N = \binom{n}{2}$ and restrict n to multiples of 4. This restriction guarantees that N is even; with $\mathcal{Y}$ standing for the set of all graphs with vertices $1, 2, \ldots n$ that have precisely $\frac{1}{2}N$ edges, we have

$$\mathrm{Prob}(\mathbb{G}(n, \tfrac{1}{2}N) \in \mathcal{Y}) = 1.$$

By Stirling's formula, $\binom{N}{N/2}(\frac{1}{2})^N \sim (2/\pi N)^{1/2}$, and so

$$\mathrm{Prob}(\mathbf{G}(n, \tfrac{1}{2}) \in \mathcal{Y}) \sim 2/\pi^{1/2} n.$$

Nevertheless, a much better upper bound can be deduced when $\mathcal{X}$ is the set of graphs in our illustrative example or the set of all 3-colourable graphs with vertices $1, 2, \ldots n$ or the set of all hamiltonian graphs with vertices $1, 2, \ldots n$ or one of many other sets.

DEFINITION: A set $\mathcal{X}$ of graphs is said to be *downward closed* if

$$G \in \mathcal{X}, F \text{ is a subgraph of } G \Rightarrow F \in \mathcal{X};$$

it is said to be *upward closed* if

$$F \in \mathcal{X}, F \text{ is a subgraph of } G \Rightarrow G \in \mathcal{X}.$$

THEOREM A.13   *Let n and m be integers such that $n \geq 1$ and $0 \leq m \leq \binom{n}{2}$. If $\mathcal{X}$ is a downward closed or upward closed set of graphs with vertices $1, 2, \ldots, n$, then*

$$\mathrm{Prob}(\mathbb{G}(n, m) \in \mathcal{X}) \leq 2 \cdot \mathrm{Prob}(\mathbf{G}(n, m\binom{n}{2}^{-1}) \in \mathcal{X}).$$

Proof Let N stand for $\binom{n}{2}$.

CASE 1: $\mathcal{X}$ *is downward closed.*

Given an i such that $0 \leq i < N$, let $\mathbf{P}$ denote the set of pairs (F, G) such that F is a graph in $\mathcal{X}$ with i edges, G is a graph in $\mathcal{X}$ with $i + 1$ edges, and F is a subgraph of G. Since there are $\binom{N}{i}\mathrm{Prob}(\mathbb{G}(n, i) \in \mathcal{X})$ graphs in $\mathcal{X}$ with i edges and each of them appears as F in at most $N - i$ pairs (F, G) in $\mathbf{P}$, we have

$$|\mathbf{P}| \leq (N - i) \cdot \binom{N}{i} \mathrm{Prob}(\mathbb{G}(n, i) \in \mathcal{X}).$$

There are $\binom{N}{i+1}$Prob$(\mathbb{G}(n, i+1) \in \mathcal{X})$ graphs in $\mathcal{X}$ with $i+1$ edges; since $\mathcal{X}$ is downward closed, each of them appears as G in precisely $i+1$ pairs (F, G) in $\mathbf{P}$; it follows that

$$|\mathbf{P}| = (i+1)\binom{N}{i+1}\text{Prob}(\mathbb{G}(n, i+1) \in \mathcal{X}).$$

Comparing this formula with the upper bound on $|\mathbf{P}|$, we find that

$$\text{Prob}(\mathbb{G}(n, i) \in \mathcal{X}) \geq \text{Prob}(\mathbb{G}(n, i+1) \in \mathcal{X}) \text{ for all } i = 0, 1, \ldots N-1,$$

and so

$$\text{Prob}(\mathbb{G}(n, i) \in \mathcal{X}) \geq \text{Prob}(\mathbb{G}(n, m) \in \mathcal{X}) \text{ for all } i = 0, 1, \ldots m. \tag{A.27}$$

We conclude that

$$\begin{aligned}
\text{Prob}(\mathbb{G}(n, m\binom{n}{2}^{-1}) \in \mathcal{X}) &= \sum_{i=0}^{N} \binom{N}{i} \left(\frac{m}{N}\right)^i \left(\frac{N-m}{N}\right)^{N-i} \text{Prob}(\mathbb{G}(n, i) \in \mathcal{X}) \\
&\geq \sum_{i=0}^{m} \binom{N}{i} \left(\frac{m}{N}\right)^i \left(\frac{N-m}{N}\right)^{N-i} \text{Prob}(\mathbb{G}(n, i) \in \mathcal{X}) \\
&\geq \sum_{i=0}^{m} \binom{N}{i} \left(\frac{m}{N}\right)^i \left(\frac{N-m}{N}\right)^{N-i} \text{Prob}(\mathbb{G}(n, m) \in \mathcal{X}) \\
&\geq \frac{1}{2} \cdot \text{Prob}(\mathbb{G}(n, m) \in \mathcal{X}):
\end{aligned}$$

the first inequality is trivial, the second is guaranteed by (A.27), and the third by Theorem A.7.

CASE 2: $\mathcal{X}$ *is upward closed.*
The argument here is a mirror image of the argument used in the preceding case. We are going to spell out its details only to reassure the faint-hearted.

Given an i such that $0 \leq i < N$, let $\mathbf{P}$ denote the set of pairs (F, G) such that F is a graph in $\mathcal{X}$ with i edges, G is a graph in $\mathcal{X}$ with $i+1$ edges, and F is a subgraph of G. Since there are $\binom{N}{i+1}$Prob$(\mathbb{G}(n, i+1) \in \mathcal{X})$ graphs in $\mathcal{X}$ with $i+1$ edges and each of them appears as G in at most $i+1$ pairs (F, G) in $\mathbf{P}$, we have

$$|\mathbf{P}| \leq (i+1) \cdot \binom{N}{i+1}\text{Prob}(\mathbb{G}(n, i+1) \in \mathcal{X}).$$

There are $\binom{N}{i}$Prob$(\mathbb{G}(n, i) \in \mathcal{X})$ graphs in $\mathcal{X}$ with i edges; since $\mathcal{X}$ is upward closed, each of them appears as F in precisely $N - i$ pairs (F, G) in $\mathbf{P}$; it follows that

$$|\mathbf{P}| = (N - i)\binom{N}{i}\text{Prob}(\mathbb{G}(n, i) \in \mathcal{X}).$$

Comparing this formula with the upper bound on $|\mathbf{P}|$, we find that

$$\text{Prob}(\mathbb{G}(n, i+1) \in \mathcal{X}) \geq \text{Prob}(\mathbb{G}(n, i) \in \mathcal{X}) \text{ for all } i = 0, 1, \ldots N-1,$$

and so

$$\text{Prob}(\mathbb{G}(n,i) \in \mathcal{X}) \geq \text{Prob}(\mathbb{G}(n,m) \in \mathcal{X}) \quad \text{for all } i = m, m+1, \dots N. \quad \text{(A.28)}$$

We conclude that

$$
\begin{aligned}
\text{Prob}(\mathbf{G}(n, m\binom{n}{2}^{-1}) \in \mathcal{X}) &= \sum_{i=0}^{N} \binom{N}{i} \left(\frac{m}{N}\right)^i \left(\frac{N-m}{N}\right)^{N-i} \text{Prob}(\mathbb{G}(n,i) \in \mathcal{X}) \\
&\geq \sum_{i=m}^{N} \binom{N}{i} \left(\frac{m}{N}\right)^i \left(\frac{N-m}{N}\right)^{N-i} \text{Prob}(\mathbb{G}(n,i) \in \mathcal{X}) \\
&\geq \sum_{i=m}^{N} \binom{N}{i} \left(\frac{m}{N}\right)^i \left(\frac{N-m}{N}\right)^{N-i} \text{Prob}(\mathbb{G}(n,m) \in \mathcal{X}) \\
&\geq \frac{1}{2} \cdot \text{Prob}(\mathbb{G}(n,m) \in \mathcal{X}):
\end{aligned}
$$

the first inequality is trivial, the second is guaranteed by (A.28), and the third by Theorem A.7. □

Intricate theorems of Bollobás [40, Theorem 2.2] and Łuczak [282, Theorem 2] establish that, for most purposes, $\mathbb{G}(n,m)$ is asymptotically equivalent to $\mathbf{G}(n,p)$ as long as m stays close to $p\binom{n}{2}$ and both m and $\binom{n}{2} - m$ grow to infinity with n.

Appendix B: Definitions, Terminology, Notation

B.1   Graphs

By a *graph* we mean an ordered pair (V, E) such that V is a finite set and E is a set of two-point subsets of V. (There are also other kinds of graphs, namely, *multigraphs, infinite graphs,* and *directed graphs*; none of them appear in the present book. In the broader context, our kind of graphs are called *simple finite undirected graphs*.) Elements of V are called *vertices* (plural of *vertex*) and elements of E are called *edges* of the graph; the set of all vertices of a graph is its *vertex set* and the set of all edges of a graph is its *edge set*. The *order* of a graph is the number of its vertices.

We refer to an edge $\{u, v\}$ by writing simply uv; the two vertices u and v are the *endpoints* of this edge. Vertices u and v are said to be *adjacent* if uv is an edge; otherwise they are said to be *nonadjacent*.

In a *complete graph,* every two distinct vertices are adjacent. A complete graph with n vertices is denoted by K_n.

Adjacent vertices are called *neighbours;* $N_G(w)$ denotes the set of all neighbours of vertex w in G; we write $d_G(w) = |N_G(w)|$ and refer to $d_G(w)$ as the *degree* of w in G. When only one graph is involved in the discussion, we may write simply $N(w)$ for $N_G(w)$ and $d(w)$ for $d_G(w)$. A graph where all vertices have the same degree is called *regular.*

A *path of length $k - 1$ between vertices u and v* is a string $w_1 w_2 \ldots w_k$ of pairwise distinct vertices such that $w_1 = u$, $w_k = v$, and each w_i with $i = 1, 2, \ldots k - 1$ is adjacent to w_{i+1}. If, in addition, w_k is adjacent to w_1, then the string $w_1 w_2 \ldots w_k w_1$ is a *cycle of length k*. A path between vertices u and v is said to *join* u and v.

Sometimes we abuse this notation a little: The *path of length $k - 1$*, denoted by P_k, may also mean the graph with vertices $w_1, \ldots w_k$ and edges $w_1 w_2, \ldots w_{k-1} w_k$. Similarly, the *cycle of length k*, denoted by C_k, may also mean the graph with vertices $w_1, \ldots w_k$ and edges $w_1 w_2, \ldots w_{k-1} w_k, w_k w_1$.

A *subgraph* of a graph G is a graph whose vertex set is a subset of the vertex set of G and where two vertices are adjacent only if they are adjacent in G. (Two vertices may be adjacent in G and nonadjacent in a subgraph of G.) An *induced subgraph* of a graph G is a graph whose vertex set is a subset of the vertex set of G and where two vertices are adjacent if and only if they are adjacent in G. (Every graph is a subgraph of a complete graph, but only complete graphs are induced subgraphs of a complete graph.)

A graph is *connected* if, and only if, for every two of its vertices, u and v, there is a path (of any length) between u and v; otherwise the graph is *disconnected*. A *tree* is a connected graph which contains no cycle. A *star* is a tree where one vertex is adjacent to all the remaining vertices.

For every graph G, there are a positive integer k (possibly $k = 1$) and a partition of the vertex set of G into pairwise disjoint nonempty parts $V_1, V_2, \ldots, V_k$ such that each subgraph of G induced by one of the parts is connected and no edge of G has its two endpoints in two distinct parts. Each of the k subgraphs of G induced by one of the parts is called a *connected component of G* or simply a *component of G*.

A *complete k-partite graph* is a graph whose vertices can be distributed into k pairwise disjoint parts (not necessarily all of them empty) so that two vertices are adjacent if and only if they belong to different parts. When $k = 2$, the graph is *complete bipartite*.

A *clique* in a graph is a set of pairwise adjacent vertices; the *clique number* $\omega(G)$ of a graph G is the number of vertices in its largest clique. A *stable set* in a graph is a set of pairwise nonadjacent vertices; the *stability number* $\alpha(G)$ of a graph G is the number of vertices in its largest stable set. (Stable sets are often referred to as *independent sets*, in which case $\alpha(G)$ is called the *independence number* of G.)

The *chromatic number* $\chi(F)$ of a graph F is the smallest number of colours that can be assigned to the vertices of F in such a way that every two adjacent vertices receive distinct colours. Equivalently, $\chi(F)$ is the smallest r such that vertices of F can be distributed into r stable sets. Graphs F with $\chi(F) \leq 2$ are called *bipartite*.

The *complement* $\overline{G}$ of a graph G has the same vertices as G; two vertices are adjacent in $\overline{G}$ if and only if they are nonadjacent in G.

Two graphs are said to be *isomorphic* if some bijection between their sets of vertices maps pairs of adjacent vertices onto pairs of adjacent vertices and it maps pairs of nonadjacent vertices onto pairs of nonadjacent vertices.

$G \oplus H$ denotes the *direct sum* of G and H, which is the graph consisting of a copy of G and a copy of H that have no vertices in common; $G - H$ denotes the *join* of G and H, which is $G \oplus H$ with additional edges that join every vertex in the copy of G to every vertex in the copy of H. (This notation has been introduced by Knuth [253, page 26].)

B.2   Hypergraphs

A *hypergraph* is a set V along with a set E of subsets of V. Elements of V are the *vertices* of the hypergraph and members of E are its *hyperedges*. If, for some integer k, every hyperedge consists of k vertices, then the hypergraph is said to be *k-uniform*. In particular, a 2-uniform hypergraph is a graph.

The *chromatic number* $\chi(H)$ of a hypergraph H is the smallest r such that the vertices of H can be distributed into r sets, none of which contains a hyperedge. (This notion extends that of the chromatic number of a graph.)

B.3 Asymptotic Notation

When f, g, d are real-valued functions defined on positive integers, we write

- $f(n) \sim g(n)$ to mean that $\lim_{n \to \infty} f(n)/g(n) = 1$,
- $f(n) = o(g(n))$ to mean that $\lim_{n \to \infty} f(n)/g(n) = 0$,
- $f(n) = O(g(n))$ to mean that there are positive constants c and n_0
  such that $n \geq n_0 \Rightarrow |f(n)| \leq c|g(n)|$,
- $f(n) = \Omega(g(n))$ to mean that there are positive constants c and n_0
  such that $n \geq n_0 \Rightarrow |f(n)| \geq c|g(n)|$,
- $f(n) = \Theta(g(n))$ to mean that $f(n) = O(g(n))$ and $f(n) = \Omega(g(n))$,
- $f(n) = g(n) + O(d(n))$ to mean that $f(n) - g(n) = O(d(n))$,
- $f(n) = g(n) + o(d(n))$ to mean that $f(n) - g(n) = o(d(n))$.

O-notation was introduced by Paul Bachmann (1837–1920) in [13] and o-notation was introduced by Landau in [271]. Our definition of $f(n) = \Omega(g(n))$ was proposed by Knuth in [251] and subsequently embraced by computer scientists.[a] Knuth also introduced Θ-notation, following suggestions from Bob Tarjan and, independently, Mike Paterson [251, pp. 19–20].

The expression $5n^2 - 10 = \Theta(n^2)$ is not an equation: switching its two sides gives $\Theta(n^2) = 5n^2 - 10$, which is nonsense. As the symbol $\Theta(n^2)$ encompasses a set of functions, Ron Rivest [251, p. 20] suggested writing $5n^2 - 10 \in \Theta(n^2)$ instead. However, the usage $5n^2 - 10 = \Theta(n^2)$ is too firmly entrenched; four reasons for continuing with this abuse are listed in [193, pp. 446–447].

Additional discussion of the asymptotic notation appears in [89, Chapter 3] and [57, Chapter 3].

B.4 Sundry Notation

We let $\ln x$ stand for the natural logarithm $\log_e x$, we let $\lg x$ stand for the binary logarithm $\log_2 x$, and we let $\log x$ stand for the logarithm $\log_b x$ when its base b is immaterial (for instance, in expressions such as $O(n \log n)$).

We use the notation introduced in [222, page 12] by Ken Iverson (1920–2004): $\lfloor x \rfloor$ denotes the largest integer at most x (the 'floor' of x) and $\lceil x \rceil$ denotes the smallest integer at least x (the 'ceiling' of x).

---

[a] As Knuth pointed out, Hardy and John Edensor Littlewood (1885–1977) had defined $f(n) = \Omega(g(n))$ some sixty years previously in [209, page 225] to mean $f(n) \neq o(g(n))$. If $f(n) = \Omega(g(n))$ in the sense of Knuth, which we use, then evidently $f(n) \neq o(g(n))$. To see that the converse is false, set $f(n) = n$ for all odd n, $f(n) = 1$ for all even n, and $g(n) = n$ for all n.

Appendix C: More on Erdős

C.1   Selected Articles

- A. Ádám, K. Győry, and A. Sárközy. The life and mathematics of Paul Erdős (1913-1996), *Mathematica Japonica* **46** (1997), 517–526.
- I. H. Anellis. In memoriam: Paul Erdős (1913–1996), *Modern Logic* **7** (1997), 83–84.
- L. Babai. In and out of Hungary: Paul Erdős, his friends, and times. In: *Combinatorics, Paul Erdős is eighty* (D. Miklós, V. T. Sós, and T. Szőnyi, eds.), Volume 2 of Bolyai Society Mathematical Studies, pp. 7–95. J. Bolyai Mathematical Society, Budapest, 1996.
- L. Babai and J. Spencer. Paul Erdős (1913–1996), *Notices of the American Mathematical Society* **45** (1998), 64–73.
- L. Babai, C. Pomerance, and P. Vértesi. The mathematics of Paul Erdős, *Notices of the American Mathematical Society* **45** (1998), 19–31.
- A. Baker and B. Bollobás. Paul Erdős. 26 March 1913 – 20 September 1996, *Biographical Memoirs of Fellows of the Royal Society* **45** (1999), 147–164.
- J. E. Baumgartner. In memoriam: Paul Erdős, 1913–1996, *Bulletin of Symbolic Logic* **3** (1997), 70–72.
- B. Bollobás. Paul Erdős — Life and work. In: *The Mathematics of Paul Erdős I. Second Edition* (R. L. Graham, J. Nešetřil, and S. Butler, eds.), Springer, New York, Heidelberg, 2013. pp. 1–41.
- J. A. Bondy. Paul Erdős et la combinatoire, *La Gazette des mathématiciens* Société Mathématique de France **71** (1997), 25–30.
- R. Freud. Paul Erdős 80 – a personal account, *Periodica Mathematica Hungarica* **26** (1993), 87–93.
- A. Hajnal. Paul Erdős' set theory. In: *The Mathematics of Paul Erdős II. Second Edition* (R. L. Graham, J. Nešetřil, and S. Butler, eds.), Springer, New York, Heidelberg, 2013. pp. 379–425.
- M. Henriksen. Reminiscences of Paul Erdős (1913–1996), *Humanistic Mathematics Network Journal* Issue 15, Article 7 (1997). Available at: `http://scholarship.claremont.edu/hmnj/vol1/iss15/7`
- A. Ivić. Remembering Paul Erdős, *Nieuw Archief voor Wiskunde* **15** (1997), 79–90.

- G. O. H. Katona. Memories on shadows and shadows of memories. In: *The Mathematics of Paul Erdős II. Second Edition* (R. L. Graham, J. Nešetřil, and S. Butler, eds.), Springer, New York, Heidelberg, 2013. pp. 195–198.
- L. Lovász. Paul Erdős is 80. In: *Combinatorics, Paul Erdős is eighty* (D. Miklós, V. T. Sós. and T. Szőnyi, eds.), Volume 1 of Bolyai Society Mathematical Studies, pp. 9–11. J. Bolyai Mathematical Society, Budapest, 1993.
- J. Pach. Two places at once: A remembrance of Paul Erdős, *The Mathematical Intelligencer* **19** (1997), 38–48.
- R. Rado. Paul Erdős is seventy years old, *Combinatorica* **3** (1983), 243–244.
- I. Z. Ruzsa. Paul Erdős - from an epsilon's-eye view, *Periodica Mathematica Hungarica* **33** (1996), 73–81.
- A. Sárközy. Farewell, Paul, *Acta Arithmetica* **81** (1997), 299–300.
- A. Sárközy. Paul Erdős (1913–1996), *Acta Arithmetica* **81** (1997), 301–302.
- C. A. B. Smith. Did Erdős save western civilization? In: *The Mathematics of Paul Erdős I. Second Edition* (R. L. Graham, J. Nešetřil, and S. Butler, eds.), Springer, New York, Heidelberg, 2013. pp. 81–92.
- V. T. Sós. Paul Erdős, 1913–1996, *Aequationes Mathematicae* **54** (1997), 205–220.
- J. Spencer. Erdős magic, In: *The Mathematics of Paul Erdős I. Second Edition* (R. L. Graham, J. Nešetřil, and S. Butler, eds.), Springer, New York, Heidelberg, 2013. pp. 43–46.
- A. H. Stone. Encounters with Paul Erdős, In: *The Mathematics of Paul Erdős I. Second Edition* (R. L. Graham, J. Nešetřil, and S. Butler, eds.), Springer, New York, Heidelberg, 2013. pp. 93–98.
- E. G. Straus. Paul Erdős at 70, *Combinatorica* **3** (1983), 245–246.
- M. Svéd. Paul Erdős - Portrait of our new academician, *Gazette of the Australian Mathematical Society* **14** (1987), 59–62.
- M. Svéd. Old snapshots of the young, *Geombinatorics* **2** (1993), 47–52.
- G. Szekeres. Recollections. In: *Combinatorics, Paul Erdős is eighty* (D. Miklós, V. T. Sós, and T. Szőnyi, eds.), Volume 1 of Bolyai Society Mathematical Studies, pp. 15–17. J. Bolyai Mathematical Society, Budapest, 1993.
- G. Szekeres. Paul Erdős (1913–1996), *Gazette of the Australian Mathematical Society* **23** (1996), 189–191.
- A. Vázsonyi. Erdős stories. In: F. R. L. Chung and R. L. Graham, *Erdős on Graphs. His Legacy of Unsolved Problems*, A. K. Peters, Ltd., Wellesley, MA, 1998, pp. 119–138.

C.2 Selected Books

- Martin Aigner and Günter M. Ziegler. *Proofs from THE BOOK,* Springer, 2014.
- Noga Alon and Joel Spencer. *The Probabilistic Method. Third Edition,* John Wiley & Sons, Hoboken, New York, 2008 (Appendix B: Paul Erdős).

- Fan Chung and Ron Graham. *Erdős on Graphs. His Legacy of Unsolved Problems,* A. K. Peters, Ltd., Wellesley, MA, 1998.
- Bruce Schechter. *My Brain is Open: The Mathematical Journeys of Paul Erdős.* Simon & Schuster, 1998.

C.3   Films

- N is a Number. A Portrait of Paul Erdős
  `www.zalafilms.com/films/nisanumber.html`
- Erdős 100 Plus
  `www.zalafilms.com/films/erdos-plus.html`

C.4   Websites

- Institute of Mathematics, Budapest
  `http://web.cs.elte.hu/erdos/`
- Biography from MacTutor
  `www-history.mcs.st-andrews.ac.uk/history/Biographies/Erdos.html`
- Collected papers of Paul Erdős (up to 1989)
  `www.renyi.hu/en/kutatoknak/erdos-pal-cikkei?`
- Zentralblatt MATH
  `www.emis.de/classics/Erdos/index.htm`
- The Erdős Number Project
  `www.oakland.edu/enp/`

C.5   An FBI File

In an article published on July 21, 2015 on the collaborative news site

> `www.muckrock.com/news/archives/2015/jul/21/`,

Beryl C. D. Lipton of the non-profit MuckRock Foundation embedded 233 pages of a declassified FBI file on Erdős. Here are a few of its highlights:

> [from page 87] ...T-1 advised that subject had been described by a person who knew him reasonably well as the type of person who, if engaged in research work of a secret nature, would probably divulge the results of his research to a foreign power in the misguided belief that his action was serving the best interest of humanity as a whole...

[from page 115]...Erdos is quoted as stating in 1958 in a letter to an associate in the U.S., "I did not apply for U.S. citizenship since I am stateless by political conviction. I left (America) without a re-entry permit since I object to Iron Curtains - both Joe's (Stalin) and Sam's (U.S.)."...

[from page 123]...Bureau investigation disclosed that Erdos is one of world's outstanding pure or theoretical mathematicians, specializing in numbers and analysis. ...

[from page 173]...He has repelled at being asked the question by any government as to whether he is or has been a member of the Communist-Party (CP). He reacts violently to such questions, and refuses to comply with the various immigration requirements of certain governments. ...

[from page 185]...ERDOS is an outgoing, extroverted individual who is quite unconventional and has a wide range of knowledge on many subjects. His political views most closely coincide with those of the British Labor Party and he delights in political discussions where he might take different positions just for the sake of argument. ...

A part of the file has been reproduced by JPat Brown, B. C. D. Lipton, and Michael Morisy in *Scientists under surveillance*, MIT Press, Cambridge, MA, 2019.

C.6   A Photo Album

Erdős's public transport ID card

With László Fejes Tóth[a]   With Sophie Pach[b]

All four photos courtesy of János Pach

---

[a] Along with Harold Scott MacDonald Coxeter (1915–2005) and Paul Erdős, one of the founders of discrete geometry.

[b] Now a physician in London.

With co-authors

Photos by George Csicsery for his film "N is a Number" (1993).

All four photos by George Csicsery for his film "N is a Number" (1993).

Photo by George Csicsery for his film "N is a Number" (1993).

Photo by George Csicsery for his film "N is a Number" (1993).

Bibliography

[1] M. Ajtai, J. Komlós, and E. Szemerédi. A note on Ramsey numbers, *Journal of Combinatorial Theory. Series A* **29** (1980), 354–360.

[2] M. Ajtai, J. Komlós, and E. Szemerédi. First occurrence of Hamilton cycles in random graphs, *North-Holland Mathematics Studies* **115** (1985), 173–178.

[3] N. Alon. Hypergraphs with high chromatic number number, *Graphs and Combinatorics* **1** (1985), 387–389.

[4] N. Alon, S. Hoory, and N. Linial. The Moore bound for irregular graphs, *Graphs and Combinatorics* **18** (2002), 53–57.

[5] N. Alon, M. Krivelevich, and B. Sudakov. Turán numbers of bipartite graphs and related Ramsey-type questions, *Combinatorics, Probability and Computing* **12** (2003), 477–494.

[6] N. Alon, L. Rónyai, and T. Szabó. Norm-graphs: variations and applications, *Journal of Combinatorial Theory. Series B* **76** (1999), 280–290.

[7] N. Alon and J. H. Spencer. *The Probabilistic Method. Third Edition,* John Wiley & Sons, Hoboken, New York, 2008.

[8] R. Alweiss, S. Lovett, K. Wu, and J. Zhang. Improved bounds for the sunflower lemma, in: *Proceedings of the 52nd Annual ACM SIGACT Symposium on Theory of Computing, 2020,* pp. 624–630.

[9] B. Andrásfai, P. Erdős, and V. T. Sós. On the connection between chromatic number, maximal clique and minimal degree of a graph, *Discrete Mathematics* **8** (1974), 205–218

[10] V. Angeltveit and B. D. McKay. $R(5,5) \leq 48$, *Journal of Graph Theory* **89** (2018), 5–13.

[11] K. Appel and W. Haken. Every planar map is four colorable. I. Discharging, *Illinois Journal of Mathematics* **21** (1977), 429–490.

[12] K. Appel, W. Haken, and J. Koch. Every planar map is four colorable. II. Reducibility, *Illinois Journal of Mathematics* **21** (1977), 491–567.

[13] P. Bachmann. *Die analytische Zahlentheorie,* Teubner, Leipzig, 1894.

[14] R. Baer. Polarities infinite projective planes. *Bulletin of the American Mathematical Society* **52** (1946), 77–93.

[15] R. C. Baker, G. Harman, and J. Pintz. The difference between consecutive primes, II, *Proceedings of the London Mathematical Society (3)* **83**, (2001) 532–562.

[16] B. Barak, A. Rao, R. Shaltiel, and A. Wigderson. 2-source dispersers for $n^{o(1)}$ entropy, and Ramsey graphs beating the Frankl-Wilson construction, *Annals of Mathematics* **176** (2012), 1483–1543.

[17] I. Bárány. A short proof of Kneser's conjecture, *Journal of Combinatorial Theory. Series A* **25** (1978), 325–326.

[18] R. Barrington Leigh and A. Liu, eds. *Hungarian Problem Book IV*, Mathematical Association of America, 2011.

[19] J. G. Basterfield and L. M. Kelly. A characterization of sets of n points which determine n hyperplanes, *Mathematical Proceedings of the Cambridge Philosophical Society* **64** (1968), 585–588.

[20] D. Bauer, H. J. Broersma, and H. J. Veldman. Not every 2-tough graph is Hamiltonian, *Discrete Applied Mathematics* **99** (2000), 317–321.

[21] M. D. Beeler and P. E. O'Neil. Some new van der Waerden numbers, *Discrete Mathematics* **28** (1979), 135–146.

[22] J. Beck. On a combinatorial problem of P. Erdős and L. Lovász, *Discrete Mathematics* **17** (1977), 127–131.

[23] J. Beck. On 3-chromatic hypergraphs, *Discrete Mathematics* **24** (1978), 127–137.

[24] L. Bellmann and C. Reiher. Turán's theorem for the Fano plane, *Combinatorica* **39** (2019), 961–982.

[25] C. T. Benson. Minimal regular graphs of girths eight and twelve, *Canadian Journal of Mathematics* **18** (1966), 1091–1094.

[26] E. R. Berlekamp. A construction for partitions which avoid long arithmetic progressions, *Canadian Mathematical Bulletin* **11** (1968), 409–414.

[27] S. Bernstein. On a modification of Chebyshev's inequality and of the error formula of Laplace, *Section Mathématique des Annales Scientifiques des Institutions Savantes de l'Ukraine*, **1** (1924), 38–49. (Russian)

[28] S. Bernstein. *Theory of Probability,* Moscow, 1927.

[29] J. Bertrand. Mémoire sur le nombre de valeurs que peut prendre une fonction quand on y permute les lettres qu'elle renferme, *Journal de l'École Polytechnique* **18** (1848), 123–140.

[30] N. L. Biggs, E. K. Lloyd, and R. J. Wilson. *Graph Theory: 1736–1936*, Clarendon Press, Oxford, 1976.

[31] T. Blankenship, J. Cummings, and V. Taranchuk. A new lower bound for van der Waerden numbers, *European Journal of Combinatorics* **69** (2018), 163–168.

[32] T. F. Bloom and O. Sisask. Breaking the logarithmic barrier in Roth's theorem on arithmetic progressions, `arXiv:2007.03528 [math.NT]`

[33] T. Bohman and P. Keevash. Dynamic concentration of the triangle-free process, in: *The Seventh European Conference on Combinatorics, Graph Theory and Applications* (J. Nešetřil and M. Pellegrini, eds.), Edizioni della Normale, Pisa, 2013, pp. 489–495.

[34] B. Bollobás. On generalized graphs, *Acta Mathematica Hungarica* **16** (1965), 447–452.

[35] B. Bollobás. *Extremal Graph Theory,* London Mathematical Society Monographs, 11. Academic Press, Inc. [Harcourt Brace Jovanovich, Publishers], London-New York, 1978.

[36] B. Bollobás. The evolution of random graphs, *Transactions of the American Mathematical Society* **286** (1984), 257–274.

[37] B. Bollobás. The evolution of sparse graphs, in: *Graph Theory and Combinatorics: Proceedings of the Cambridge Combinatorial Conference in Honour of Paul Erdős*, (B. Bollobás, ed.), Academic Press, London, 1984, pp. 35–57.

[38] B. Bollobás. The chromatic number of random graphs, *Combinatorica* **8** (1988), 49–55.

[39] B. Bollobás. Extremal graph theory, in: *Handbook of Combinatorics* (R. L. Graham, M. Grötschel, and L. Lovász, eds.), Elsevier Science B.V., Amsterdam; MIT Press, Cambridge, MA, 1995, pp. 1231–1292.

[40] B. Bollobás. *Random Graphs. Second Edition,* Cambridge University Press, Cambridge, 2001.

[41] B. Bollobás. *Modern Graph Theory,* Springer, 2013.

[42] B. Bollobás, P. A. Catlin, and P. Erdős. Hadwiger's conjecture is true for almost every graph, *European Journal of Combinatorics* **1** (1980), 195–199.

[43] B. Bollobás and P. Erdős. On the structure of edge graphs, *Bulletin of the London Mathematical Society* **5** (1973), 317–321.

[44] B. Bollobás and P. Erdős. Cliques in random graphs, *Mathematical Proceedings of the Cambridge Philosophical Society* **80** (1976), 419–427.

[45] B. Bollobás and A. Thomason. Random graphs of small order, *North-Holland Mathematics Studies* **118** (1985) 47–97.

[46] J. A. Bondy. Properties of graphs with constraints on degrees, *Studia Scientiarum Mathematicarum Hungarica* **4** (1969), 473–475.

[47] J. A. Bondy, Extremal problems of Paul Erdős on circuits in graphs, in: *Paul Erdős and his mathematics II* (G. Halász et al., eds.), Bolyai Society Mathematical Studies **11**, János Bolyai Mathematical Society, Budapest, 2002, pp. 135–156.

[48] J. A. Bondy and V. Chvátal. A method in graph theory, *Discrete Mathematics* **15** (1976), 111–135.

[49] J. A. Bondy and M. Simonovits. Cycles of even length in graphs, *Journal of Combinatorial Theory. Series B* **16** (1974), 97–105.

[50] C. E. Bonferroni. Teoria statistica delle classi e calcolo delle probabilita, *Pubblicazioni del Reale Istituto Superiore di Scienze Economiche e Commerciali di Firenze,* **8**, Libreria Internazionale Seeber, Firenze, 62 pp. (1936).

[51] K. Borsuk. Drei Sätze über die *n*-dimensionale euklidische Sphäre, *Fundamenta Mathematicae* **20** (1933), 177–190.

[52] P. Borwein and W. O. J. Moser. A survey of Sylvester's problem and its generalizations, *Aequationes Mathematicae* **40** (1990), 111–135.

[53] R. C. Bose. Strongly regular graphs, partial geometries and partially balanced designs, *Pacific Journal of Mathematics* **13** (1963), 389–419.

[54] C. W. Bostwick. Elementary Problem E 1321, *American Mathematical Monthly* **65** (1958), 446. Solutions in *American Mathematical Monthly* **66** (1959), 141–142.

[55] J. M. Boyer and W. J. Myrvold. On the cutting edge: simplified $O(n)$ planarity by edge addition. *Journal of Graph Algorithms and Applications* **8** (2004), 241–273.

[56] P. Brass, W. O. J. Moser, and J. Pach. *Research Problems in Discrete Geometry,* Springer Science & Business Media, 2006.

[57] G. Brassard and P. Bratley. *Fundamentals of Algorithmics,* Prentice Hall, 1996.

[58] A. E. Brouwer and J. H. van Lint. Strongly regular graphs and partial geometries, in: *Enumeration and Design* (D. M. Jackson and S. A. Vanstone, eds.), Academic Press, Toronto, 1984, pp. 85–122.

[59] W. G. Brown. On graphs that do not contain a Thomsen graph, *Canadian Mathematical Bulletin* **9** (1966), 281–285.

[60] W. G. Brown. On an open problem of Paul Turán concerning 3-graphs, in: *Studies in pure mathematics* (P. Erdős et al., eds.), pp. 91–93. Birkhäuser, Basel, 1983.

[61] T. Brown, B. M. Landman, and A. Robertson. Bounds on van der Waerden numbers and some related functions, arXiv:0706.4420 [math.CO] (2007).

[62] R. H. Bruck and H. J. Ryser. The nonexistence of certain finite projective planes, *Canadian Journal of Mathematics* **1** (1949), 88–93.

[63] V. Y. Bunyakovski. Sur quelques inégalités concernant les intégrales ordinaires et les intégrales aux differences finies, *Mémoires de l'Académie Impériale des Sciences de St.-Pétersbourg,* VII ^{e} Série, Tome I, N ^{o} 9 (1859), 18 pp.

[64] L. E. Bush. The William Lowell Putnam Mathematical Competition, *The American Mathematical Monthly* **60** (1953), 539–542.

[65] P. A. Catlin. Hajós' graph-coloring conjecture: variations and counterexamples, *Journal of Combinatorial Theory. Series B* **26** (1979), 268–274.

[66] A. L. Cauchy. *Cours d'Analyse de l'École Royale Polytechnique, Ire Partie, Analyse Algébrique,* Paris, 1821.

[67] P. L. Chebyshev. Mémoire sur les nombres premiers, *Journal de Mathématiques Pures et Appliquées* **17** (1852), 366–390. Also in: *Oeuvres de P.L. Tchebychef* (A. Markoff and N. Sonin, eds.), St. Petersburg, 1899–1907, Vol. 1, pp. 49–70. (Reprint Chelsea, New York, 1952)

[68] D. D. Cherkashin and J. Kozik. A note on random greedy coloring of uniform hypergraphs, *Random Structures & Algorithms* **47** (2015), 407–413.

[69] H. Chernoff. A measure of asymptotic efficiency of tests of a hypothesis based on the sum of observations, *Annals of Mathematical Statistics* **23** (1952), 493–507.

[70] F. R. K. Chung and R. L. Graham. *Erdős on Graphs. His Legacy of Unsolved Problems,* A. K. Peters, Ltd., Wellesley, MA, 1998.

[71] F. R. K. Chung and R. L. Graham. Forced convex *n*-gons in the plane, *Discrete & Computational Geometry* **19** (1998), 367–371.

[72] F. R. K. Chung and C. M. Grinstead. A survey of bounds for classical Ramsey numbers, *Journal of Graph Theory* **7** (1983), 25–37.

[73] V. Chvátal. On finite Δ-systems of Erdős and Rado. *Acta Mathematica Academiae Scientiarum Hungaricae* **21** (1970) 341–355.

[74] V. Chvátal. Some unknown van der Waerden numbers, in: *Combinatorial Structures and their Applications. Proceedings of the Calgary International Conference on Combinatorial Structures and Their Applications* (R. Guy et al., eds.), Gordon and Breach, New York, 1970, pp. 31–33.

[75] V. Chvátal. Hypergraphs and Ramseyian theorems, *Proceedings of the American Mathematical Society* **27** (1971), 434–440.

[76] V. Chvátal. On Hamilton's ideals, *Journal of Combinatorial Theory. Series B* **12** (1972), 163–168.

[77] V. Chvátal. Tough graphs and Hamiltonian circuits, *Discrete Mathematics* **5** (1973), 215–228.

[78] V. Chvátal. The minimality of the Mycielski graph, in: *Graphs and Combinatorics. Proceedings of the Capital Conference on Graph Theory and Combinatorics at the George Washington University* (R. A. Bari and F. Harary, eds.), Lecture Notes in Mathematics **406**, Springer-Verlag, 1974, pp. 243–246.

[79] V. Chvátal. Cutting planes in combinatorics, *European Journal of Combinatorics* **6** (1985), 217–226.

[80] V. Chvátal. Hamiltonian cycles, in: *The Traveling Salesman Problem* (E. L. Lawler et al., eds.), John Wiley, 1985, pp. 403–429.

[81] V. Chvátal and P. Erdős. A note on Hamiltonian circuits, *Discrete Mathematics* **2** (1972), 111–113.

[82] V. Chvátal, P. Erdős, and Z. Hedrlín. Ramsey's theorem and self-complementary graphs, *Discrete Mathematics* **3** (1972), 301–304.

[83] L. Clark, R. C. Entringer, and D. E. Jackson. Minimum graphs with complete k-closure, *Discrete Mathematics* **30** (1980), 95–101.

[84] G. Cohen. Towards optimal two-source extractors and Ramsey graphs, in: *Proceedings of 49th Annual ACM SIGACT Symposium on the Theory of Computing,* 2017.

[85] C. J. Colbourn and J. D. Dinitz. *Handbook of Combinatorial Designs,* 2nd edition, Chapman and Hall/CRC, 2006.

[86] M. J. Collison. The Unique Factorization Theorem: From Euclid to Gauss, *Mathematics Magazine* **53** (1980), 96–100.

[87] W. S. Connor and W. H. Clatworthy. Some theorems for partially balanced designs, *Annals of Mathematical Statistics* **25** (1954), 100–112.

[88] J. H. Conway and V. Pless. On primes dividing the group order of a doubly-even (72, 36, 16) code and the group order of a quaternary (24, 12, 10) code, *Discrete Mathematics* **38** (1982), 143–156.

[89] T. H. Cormen, C. E. Leiserson, R. L. Rivest, and C. Stein. *Introduction to Algorithms, Third Edition,* MIT Press, 2009.

[90] H. S. M. Coxeter. A problem of collinear points, *American Mathematical Monthly* **55** (1948), 26–28.

[91] C. C. Craig. On the Tchebychef inequality of Bernstein, *Annals of Mathematical Statistics* **4** (1933), 94–102.

[92] G. Csicsery. *N is a Number: A Portrait of Paul Erdős* [DVD (NTSC)], Zala Films, 1993. ISBN: 0-933621-62-0 https://vimeo.com/ondemand/nisanumber.

[93] N. G. de Bruijn and P. Erdős. On a combinatorial problem, *Indagationes Mathematicae* **10** (1948), 421–423.

[94] N. G. de Bruijn and P. Erdős. A colour problem for infinite graphs and a problem in the theory of relations, *Indagationes Mathematicae* **13** (1951), 369–373.

[95] B. Descartes. A three colour problem, *Eureka* **9** (1947), 21. Solution in *Eureka* **10** (1948), 24–25.

[96] B. Descartes. Solution of Advanced Problem 4526, *American Mathematical Monthly* **61** (1954), 352–353.

[97] M. Deza. Une propriété extrémale des plans projectifs finis dans une classe de codes équidistants, *Discrete Mathematics* **6** (1973), 343–352.

[98] M. Deza. Solution d'un problème de Erdős-Lovász, *Journal of Combinatorial Theory. Series B* **16** (1974), 166–167.

[99] G. A. Dirac. A property of 4-chromatic graphs and some remarks on critical graphs, *Journal of the London Mathematical Society* **27**, (1952), 85–92.

[100] G. A. Dirac. Some theorems on abstract graphs, *Proceedings of the London Mathematical Society (3)* **2** (1952), 69–81.

[101] G. A. Dirac. Chromatic number and topological complete subgraphs, *Canadian Mathematical Bulletin* **8** (1965), 711–715.

[102] P. Dodos, V. Kanellopoulos, and K. Tyros. A simple proof of the density Hales-Jewett theorem, *International Mathematics Research Notices,* Issue 12 **2014** (2014), 3340–3352.

[103] N. Eaton and G. Tiner. On the Erdős-Sós conjecture and graphs with large minimum degree, *Ars Combinatoria* **95** (2010), 373–382.

[104] J. Edmonds. Matroids and the greedy algorithm, *Mathematical Programming* **1** (1971), 127–136.

[105] P. Erdős. A magasabb rendű számtani sorokról, *Középiskolai Matematikai és Fizikai Lapok* **36** (1929), 187–189.

[106] P. Erdős. Beweis eines Satzes von Tschebyschef, *Acta Litterarum ac Scientiarum Szeged* **5** (1932), 194–198.

[107] P. Erdős. Three point collinearity, *American Mathematical Monthly* **50** (1943), Problem 4065, p. 65. Solutions in Vol. **51** (1944), 169–171.

[108] P. Erdős. Some remarks on the theory of graphs, *Bulletin of the American Mathematical Society* **53** (1947), 292–294.

[109] P. Erdős. Some unsolved problems, *Michigan Mathematical Journal* **4** (1957), 291–300.

[110] P. Erdős. Graph theory and probability, *Canadian Journal of Mathematics* **11** (1959), 34–38.

[111] P. Erdős. On circuits and subgraphs of chromatic graphs, *Mathematika* **9** (1962), 170–175.

[112] P. Erdős. On a combinatorial problem, *Nordisk Matematisk Tidskrift* **11** (1963). 5–10.

[113] P. Erdős. Extremal problems in graph theory, in: *Theory of Graphs and its Applications. Proceedings of the Symposium held in Smolenice in June 1963* (M. Fiedler, ed.), Publishing House of the Czechoslovak Academy of Sciences, Prague, 1964, pp. 29–36.

[114] P. Erdős. On a combinatorial problem. II, *Acta Mathematica Academiae Scientiarum Hungaricae* **15** (1964), 445–447.

[115] P. Erdős. A problem on independent *r*-tuples, *Annales Universitatis Scientiarum Budapestinensis de Rolando Eötvös Nominatae. Sectio Mathematica* **8** (1965), 93–95.

[116] P. Erdős. Extremal problems in graph theory, in: *A Seminar on Graph Theory* (F. Harary, ed.), pp. 54–59, Holt, Rinehart and Winston, New York, 1967.

[117] P. Erdős. Some recent results on extremal problems in graph theory, in: *Theory of Graphs. International Symposium held at the International Computation Center in Rome, July 1966* (P. Rosenstiehl, ed.), Gordon and Breach, New York; Dunod, Paris, 1967, pp. 117–123 (English), 124–130 (French).

[118] P. Erdős. Turán Pál gráf tételéről (On the graph theorem of Turán, in Hungarian), *Középiskolai Matematikai és Fizikai Lapok* **21** (1970), 249–251 (1971).

[119] P. Erdős. Extremal problems on graphs and hypergraphs, in: *Hypergraph Seminar. Ohio State University, 1972* (C. Berge and D. Ray-Chaudhuri, eds.), Lecture Notes in Mathematics **411**, Springer, Berlin, Heidelberg, 1974, pp. 75–84.

[120] P. Erdős. Remarks on some problems in number theory, *Mathematica Balkanica* **4** (1974), 197–202.

[121] P. Erdős. Problems and results in combinatorial number theory, in: *Journées Arithmétiques de Bordeaux. Astérisque*, Nos. 24–25, pp. 295–310, Société Mathématique de France, Paris, 1975.

[122] P. Erdős. Problems and results on finite and infinite combinatorial analysis, in: *Infinite and finite sets: to Paul Erdős on his 60th birthday* (A. Hajnal, R. Rado, and V. T. Sós, eds.), Colloquia mathematica Societatis János Bolyai **10**, North-Holland, Amsterdam, 1975, Vol. I, pp. 403–424.

[123] P. Erdős. Some recent progress on extremal problems in graph theory, *Congressus Numerantium* **14** (1975), 3–14.

[124] P. Erdős. Problems and results on combinatorial number theory, II., *Journal of the Indian Mathematical Society (N.S.)* **40** (1976), 285–298.

[125] P. Erdős. Some extremal problems on families of graphs and related problems, in: *Combinatorial Mathematics,* pp. 13–21, Springer, Berlin, Heidelberg, 1978.

[126] P. Erdős. Problems and results in graph theory and combinatorial analysis, in: *Graph theory and related topics. Proceedings of the conference held in honour of Professor W.T. Tutte on the occasion of his sixtieth birthday, University of Waterloo, July 5–9, 1977* (J. A. Bondy and U. S. R. Murty, eds.), *Academic Press, New York-London,* 1979, pp. 153–163.

[127] P. Erdős. On the combinatorial problems which I would most like to see solved, *Combinatorica* **1** (1981), 25–42.

[128] P. Erdős. Personal reminiscences and remarks on the mathematical work of Tibor Gallai, *Combinatorica* **2** (1982), 207–212.

[129] P. Erdős. Extremal problems in number theory, combinatorics and geometry, *Proceedings of the International Congress of Mathematicians (Warsaw, 1983),* pp. 51–70, PWN, Warsaw, 1984.

[130] P. Erdős. Two problems in extremal graph theory, *Graphs and Combinatorics* **2** (1986), 189–190.

[131] P. Erdős. Ramanujan and I. *Number Theory, Madras 1987*, Lecture Notes in Mathematics 1395, Springer, Berlin, 1989, pp. 1–20.

[132] P. Erdős. Problems and results on graphs and hypergraphs: similarities and differences, in: *Mathematics of Ramsey Theory* (J. Nešetřil and V. Rödl, eds.), Springer, Berlin, 1990, pp. 12–28.

[133] P. Erdős. Some of my favourite unsolved problems, in: *A tribute to Paul Erdős* (A. Baker, B. Bollobás, A. Hajnal, eds.), Cambridge University Press, 1990, pp. 467–469.

[134] P. Erdős. Some of my favorite problems and results, in: *The Mathematics of Paul Erdős, Vol. I* (R. L. Graham and J. Nešetřil, eds.), Springer-Verlag, Berlin, 1997, pp. 47–67.

[135] P. Erdős and S. Fajtlowicz. On the conjecture of Hajós, *Combinatorica* 1 (1981), 141–143.

[136] P. Erdős, R. J. Faudree, J. Pach, and J. H. Spencer. How to make a graph bipartite, *Journal of Combinatorial Theory. Series B* **45** (1988), 86–98.

[137] P. Erdős and T. Gallai. On maximal paths and circuits of graphs. *Acta Mathematica Academiae Scientiarum Hungaricae* **10** (1959), 337–356.

[138] P. Erdős and A. Hajnal. On a property of families of sets, *Acta Mathematica Academiae Scientiarum Hungaricae* **12** (1961), 87–123.

[139] P. Erdős and A. Hajnal. Some remarks on set theory. IX. Combinatorial problems in measure theory and set theory, *Michigan Mathematical Journal* **11** (1964), 107–127.

[140] P. Erdős and A. Hajnal. On chromatic number of graphs and set-systems. *Acta Mathematica Academiae Scientiarum Hungaricae* **17** (1966), 61–99.

[141] P. Erdős and H. Hanani. On a limit theorem in combinatorical analysis, *Publicationes Mathematicae* **10** (1963), 10–13.

[142] P. Erdős, C. Ko, and R. Rado. Intersection theorems for systems of finite sets, *The Quarterly Journal of Mathematics. Oxford Second Series* **12** (1961), 313–320.

[143] P. Erdős and L. Lovász. Problems and results on 3-chromatic hypergraphs and some related questions, in: *Infinite and finite sets: to Paul Erdős on his 60th birthday* (A. Hajnal, R. Rado, and V. T. Sós, eds.), Colloquia mathematica Societatis János Bolyai **10**, North-Holland, Amsterdam, 1975, Vol. II, pp. 609–627.

[144] P. Erdős and L. Pósa. On the maximal number of disjoint circuits of a graph, *Publicationes Mathematicae* **9** (1962), 3–12.

[145] P. Erdős and R. Rado. Combinatorial theorems on classifications of subsets of a given set, *Proceedings of the London Mathematical Society(3)* **2** (1952), 417–439.

[146] P. Erdős and R. Rado. Intersection theorems for systems of sets, *Journal of London Mathematical Society* **35** (1960) 85–90.

[147] P. Erdős and A. Rényi. On random graphs. I. *Publicationes Mathematicae* **6** (1959), 290–297.

[148] P. Erdős and A. Rényi. On the evolution of random graphs, *A Magyar Tudomanyos Akademia. Matematikai es Fizikai Tudomanyok Osztalyanak Közlemenyei* **5** (1960), 17–61.

[149] P. Erdős and A. Rényi. On the evolution of random graphs, *Bulletin de l'Institut International de Statistique.* **38** (1961), 343–347.

[150] P. Erdős and A. Rényi. On the strength of connectedness of a random graph, *Acta mathematica Academiae Scientiarum Hungaricae* **12** (1961), 261–267.

[151] P. Erdős, A. Rényi, and V. T. Sós. On a problem of graph theory, *Studia Scientiarum Mathematicarum Hungarica* **1** (1966), 215–235.

[152] P. Erdős and M. Simonovits. A limit theorem in graph theory, *Studia Scientiarum Mathematicarum Hungarica* **1** (1966), 51–57.

[153] P. Erdős and J. Spencer. *Probabilistic methods in combinatorics*, Academic Press, New York, 1974.

[154] P. Erdős and A. H. Stone. On the structure of linear graphs, *Bulletin of the American Mathematical Society* **52**, (1946), 1087–1091.

[155] P. Erdős and J. Surányi. *Válogatott fejezetek a számelméletből* [Selected Chapters from Number Theory], Tankönyvkiadó Vállalat, Budapest, 1960.

[156] P. Erdős and J. Surányi. *Topics in the Theory of Numbers* (B. Guiduli, translator), Springer Science+Business Media, 2003.

[157] P. Erdős and G. Szekeres. A combinatorial problem in geometry, *Compositio Mathematica* **2** (1935), 463–470.

[158] P. Erdős and G. Szekeres. On some extremum problems in elementary geometry, *Annales Universitatis Scientiarum Budapestinensis de Rolando Eötvös Nominatae. Sectio Mathematica* **3–4** (1960/1961), 53–62.

[159] P. Erdős and P. Turán. On some sequences of integers, *Journal of the London Mathematical Society* **1** (1936), 261–264.

[160] European Mathematical Society & FIZ Karlsruhe & Springer-Verlag, Publications of (and about) Paul Erdős, www.emis.de/classics/Erdos/zbl.htm, 1998.

[161] G. Exoo. A lower bound for $R(5,5)$, *Journal of Graph Theory* **13** (1989), 97–98.

[162] G. Fan, Y. Hong, and Q. Liu. The Erdős-Sós conjecture for spiders, arXiv:1804.06567 [math.CO], 18 April 2018

[163] G. Fano. Sui postulati fondamentali della geometria proiettiva in uno spazio lineare a un numero qualunque di dimensioni, *Giornale di Matematiche* **30** (1892), 106–132.

[164] I. Fáry. On straight line representation of planar graphs, *Acta Scientarum Mathematicarum (Szeged)* **11** (1948), 229–233.

[165] G. Fiz Pontiveros, S. Griffiths, and R. Morris. The triangle-free process and the Ramsey number $R(3, k)$. *Memoirs of the American Mathematical Society* **263** (2020), No. 1274.

[166] D. G. Fon-Der-Flaass. A method for constructing $(3, 4)$-graphs. (Russian), *Matematicheskie Zametki* **44** (1988), 546–550. Translation in *Mathematical notes of the Academy of Sciences of the USSR* **44** (1988), 781–783.

[167] K. Ford, B. Green, S. Konyagin, and T. Tao, Large gaps between consecutive prime numbers, *Annals of Mathematics* **183** (2016), 935–974.

[168] J. Fox, J. Pach, and A. Suk. Bounded VC-dimension implies the Schur-Erdős conjecture, `arXiv:1912.02342 [math.CO]`

[169] P. Frankl. A constructive lower bound for some Ramsey numbers, *Ars Combinatoria* **3** (1977), 297–302.

[170] P. Frankl. Asymptotic solution of a Turán-type problem, *Graphs and Combinatorics* **6** (1990), 223–227.

[171] P. Frankl and Z. Füredi. A new generalization of the Erdős-Ko-Rado theorem, *Combinatorica* **3** (1983), 3–349.

[172] P. Frankl and R. M. Wilson. Intersection theorems with geometric consequences, *Combinatorica* **1** (1981), 357–368.

[173] H. de Fraysseix, P. O. de Mendez, and P. Rosenstiehl. Trémaux trees and planarity, *International Journal of Foundations of Computer Science* **17** (2006), 1017–1029.

[174] A. M. Frieze. On the independence number of random graphs, *Discrete Mathematics* **81** (1990), 171–175.

[175] A. Frieze and M. Karoński. *Introduction to Random Graphs,* Cambridge University Press, 2015.

[176] Z. Füredi. Turán type problems, in: *Surveys in Combinatorics,* Cambridge University Press, 1991, pp. 253–300.

[177] Z. Füredi. An upper bound on Zarankiewicz'problem, *Combinatorics, Probability and Computing* **5** (1996), 29–33.

[178] Z. Füredi. New asymptotics for bipartite Turán numbers, *Journal of Combinatorial Theory. Series A* **75** (1996), 141–144.

[179] Z. Füredi and M. Simonovits. Triple systems not containing a Fano configuration, *Combinatorics, Probability and Computing* **14** (2005), 467–484.

[180] Z. Füredi and M. Simonovits. The history of degenerate (bipartite) extremal graph problems, in: *Erdős Centennial* (L. Lovász, I. Z. Ruzsa, and V. T. Sós, eds.), pp. 169–264. Springer, Berlin, Heidelberg, 2013.

[181] H. Furstenberg. Ergodic behavior of diagonal measures and a theorem of Szemerédi on arithmetic progressions. *Journal d'Analyse Mathématique* **31** (1977), 204–256.

[182] H. Furstenberg and Y. Katznelson. A density version of the Hales-Jewett theorem, *Journal d'Analyse Mathématique* **57** (1991), 64–119.

[183] D. Gale. Neighboring vertices on a convex polyhedron, in: *Linear inequalities and related systems, Annals of Mathematics Studies* **38** (H. W. Kuhn and A. W. Tucker, eds.), pp. 255–263, Princeton University Press, 1956.

[184] C. F. Gauss. *Disquisitiones arithmeticae,* Vol. 157. Yale University Press, 1966.

[185] E. N. Gilbert. Random graphs, *Annals of Mathematical Statistics* **30** (1959), 1141–1144.

[186] C. Godsil and G. Royle. *Algebraic Graph Theory*, Springer-Verlag, New York, 2001.

[187] Journal of Goedgebeur. On minimal triangle-free 6-chromatic graphs, *Journal of Graph Theory* **93** (2020), 34–48.

[188] P. Gorroochurn. Some laws and problems of classical probability and how Cardano anticipated them, *Chance* **25** (2012), 13–20.

[189] R. Gould. *Graph Theory,* Dover Publications, 2013.

[190] T. Gowers. A new proof of Szemerédi's theorem, *GAFA, Geometric and Functional Analysis* **11** (2001), 465–588. Erratum in the same volume, p. 869.

[191] R. L. Graham. Some of my favorite problems in Ramsey theory, *Integers: Electronic Journal of Combinatorial Number Theory* 7(2) (2007) #A15.

[192] R. L. Graham and S. Butler. *Rudiments of Ramsey Theory.* American Mathematical Society, 2015.

[193] R. L. Graham, D. E. Knuth, and O. Patashnik. *Concrete Mathematics,* Second edition, Addison-Wesley, 1994.

[194] R. L. Graham and B. L. Rothschild, Ramsey's theorem for n-parameter sets, *Transactions of American Mathematical Society* **159** (1971), 257–292.

[195] R. L. Graham and B. L. Rothschild. A short proof of van der Waerden's theorem on arithmetic progressions, *Proceedings of the American Mathematical Society* **42** (1974), 385–386.

[196] R. L. Graham, B. L. Rothschild, and J. H. Spencer. *Ramsey Theory,* 2nd ed. Wiley-Interscience Series Vol. 20, John Wiley & Sons. 1990.

[197] J. E. Graver and J. Yackel. Some graph theoretic results associated with Ramsey's theorem, *Journal of Combinatorial Theory* **4** (1968), 125–175.

[198] B. Green and T. Tao. New bounds for Szemerédi's theorem, II: A new bound for $r_4(N)$, arXiv:math/0610604 [math.NT] (2006).

[199] B. Green and T. Tao. The primes contain arbitrarily long arithmetic progressions, *Annals of Mathematics* **167** (2008), 481–547.

[200] R. E. Greenwood and A. M. Gleason. Combinatorial relations and chromatic graphs, *Canadian Journal of Mathematics* **7** (1955), 1–7.

[201] C. Grinstead and S. Roberts. On the Ramsey Numbers $R(3, 8)$ and $R(3, 9)$, *Journal of CombinatorialTheory. Series B* **33** (1982), 27–51.

[202] J. Grossman. List of publications of Paul Erdős, January 2013, in: *The Mathematics of Paul Erdős II. Second Edition* (R. L. Graham, J. Nešetřil, and S. Butler, eds.), Springer, New York, Heidelberg, 2013. pp. 497–603.

[203] H. Hadwiger. Über eine Klassifikation der Streckenkomplexe, *Vierteljahrsschrift der Naturforschenden Gesellschaft in Zürich,* **88** (1943), 133–143.

[204] A. W. Hales and R. I. Jewett. Regularity and positional games, *Transactions of American Mathematical Society* **106** (1963), 222–229.

[205] H. Hanani. On the number of straight lines determined by n points, *Riveon Lematematika* **5** (1951), 10–11.

[206] H. Hanani. On the number of lines and planes determined by d points, *Scientific Publications Technion, Israel Institute of Technology* **6** (1954), 58–63.

[207] H. Hanani. On quadruple systems, *Canadian Journal of Mathematics* **12** (1960), 145–157.

[208] H. Hanani. The existence and construction of balanced incomplete block designs, *The Annals of Mathematical Statistics* **32** (1961), 361–386.

[209] G. H. Hardy and J. E. Littlewood. Some problems of Diophantine approximation. I. The fractional part of $n^k\theta$, *Acta Mathematica* **37** (1914), 155–191.

[210] G. H. Hardy and J. E. Littlewood. Some problems of 'partitio numerorum'. III. On the expression of a number as a sum of primes, *Acta Mathematica* **44** (1923), 1–70.

[211] G. H. Hardy and E. M. Wright. *An Introduction to the Theory of Numbers,* Oxford University Press, Oxford, 1938.

[212] T. L. Heath, ed. *Thirteen Books of Euclid's Elements,* Courier Corporation, 1956.

[213] D. Hilbert and S. Cohn-Vossen. *Anschauliche Geometrie,* Springer, 1932; English translation: *Geometry and the Imagination,* AMS Chelsea Publishing, 1999.

[214] W. Hoeffding, Probability inequalities for sums of bounded random variables, *Journal of the American Statistical Association* **58** (1963), 13–30.

[215] A. J. Hoffman and R. R. Singleton. On Moore graphs with diameters 2 and 3, *IBM Journal of Research and Development* **4** (1960), 497–504.

[216] G. Hoheisel. Primzahlprobleme in der Analysis, *Sitzungsberichte der Preußischen Akademie der Wissenschaften, Physikalisch-Mathematische Klasse* **33** (1930), 3–11.

[217] A. F. Holmsen, H. N. Mojarrad, J. Pach, and G. Tardos. Two extensions of the Erdős-Szekeres problem. `arXiv:1710.11415` `[math.CO]` (2017).

[218] D. J. Houck and M. E. Paul. On a theorem of de Bruijn and Erdős. *Linear Algebra and its Applications* **23** (1979), 157–165.

[219] J. Hopcroft and R. Tarjan. Efficient planarity testing, *Journal of the Association for Computing Machinery* **21** (1974), 549–568.

[220] J. R. Isbell. $N(4, 4; 3) \geq 13$, *Journal of Combinatorial Theory* **6** (1969), 210–210.

[221] Y. Ishigami. Proof of a conjecture of Bollobás and Kohayakawa on the Erdős-Stone theorem, *Journal of Combinatorial Theory. Series B* **85** (2002), 222–254.

[222] K. E. Iverson. *A Programming Language,* Wiley, 1962.

[223] F. Jaeger and C. Payan. Determination du nombre maximum d'arêtes d'un hypergraphe τ-critique de rang h, *Comptes Rendus Hebdomadaires des Séances de l'Académie des Sciences, Paris* **273** (1971), 221–223.

[224] S. Janson, D. E. Knuth, T. Łuczak, and B. Pittel. The birth of the giant component, *Random Structures & Algorithms* **4** (1993), 233–358.

[225] S. Janson, T. Łuczak, and A. Ruciński. *Random Graphs,* Wiley-Interscience, New York, 2000.

[226] J. L. W. V. Jensen. Sur les fonctions convexes et les inégalités entre les valeurs moyennes, *Acta Mathematica* **30** (1906), 175–193.

[227] T. Jensen and G. F. Royle. Small graphs with chromatic number 5: A computer search, *Journal of Graph Theory* **19** (1995), 107–116.

[228] K. Jogdeo and S. M. Samuels. Monotone convergence of binomial probabilities and a generalization of Ramanujan's equation, *The Annals of Mathematical Statistics* **39** (1968), 1191–1195.

[229] S. Johnson. A new proof of the Erdős-Szekeres convex k-gon result, *Journal of Combinatorial Theory. Series A* **42** (1986), 318–319.

[230] K. Jordán. A valószínűségszámítás alapfogalmai, *Középiskolai Matematikai és Fizikai Lapok* **34** (1927), 109–136.

[231] R. Kaas and J. M. Buhrman. Mean, median and mode in binomial distributions, *Statistica Neerlandica* **34** (1980), 13–18.

[232] J. Kahn and P. D. Seymour. A fractional version of the Erdős-Faber-Lovász conjecture, *Combinatorica* **12** (1992), 155–160.

[233] J. D. Kalbfleisch, J. G. Kalbfleisch, and R. G. Stanton. A combinatorial problem on convex *n*-gons, in: *Proceedings of the Louisiana Conference on Combinatorics, Graph Theory, and Computing: Louisiana State University, Baton Rouge, March, 1–5, 1970, Congressus Numerantium I* (R. C. Mullin, K. B. Reid, and D. P. Roselle, eds.), Utilitas Mathematica, Winnipeg, Manitoba, 1970, pp. 180–188.

[234] J. G. Kalbfleisch. Construction of special edge-chromatic graphs, *Canadian Mathematical Bulletin* **8** (1965), 575–584.

[235] J. G. Kalbfleisch. *Chromatic graphs and Ramsey's theorem,* Ph.D. thesis, University of Waterloo, January 1966.

[236] G. O. H. Katona. A simple proof of the Erdős - Chao Ko - Rado theorem, *Journal of Combinatorial Theory. Series B* **13** (1972), 183–184.

[237] G. O. H. Katona. Solution of a problem of A. Ehrenfeucht and J. Mycielski, *Journal of Combinatorial Theory. Series A* **17** (1974), 265–266.

[238] G. Katona, T. Nemetz, and M. Simonovits, On a problem of Turán in the theory of graphs (Hungarian), *Középiskolai Matematikai és Fizikai Lapok* **15** (1964), 228–238.

[239] P. Keevash. Hypergraph Turán problems, in: *Surveys in combinatorics 2011* (R. Chapman, ed.), London Mathematical Society Lecture Note Series **392**, Cambridge Univ. Press, Cambridge, 2011, pp. 83–139.

[240] P. Keevash. The existence of designs, `arXiv:1401.3665` `[math.CO]`, 2019.

[241] P. Keevash and B. Sudakov, On a hypergraph Turán problem of Frankl, *Combinatorica* **25** (2005), 673–706.

[242] P. Keevash and B. Sudakov. The Turán number of the Fano plane, *Combinatorica* **25** (2005), 561–574.

[243] J. B. Kelly and L. M. Kelly. Paths and circuits in critical graphs, *American Journal of Mathematics* **76** (1954), 786–792.

[244] L. M. Kelly and W. O. J. Moser. On the number of ordinary lines determined by *n* points, *Canadian Journal of Mathematics* **10** (1958), 210–219.

[245] G. Kéry. On a theorem of Ramsey (in Hungarian), *Középiskolai Matematikai és Fizikai Lapok* **15** (1964), 204–224.

[246] H. A. Kierstead, E. Szemerédi, and W. T. Trotter, Jr. On coloring graphs with locally small chromatic number, *Combinatorica* **4** (1984), 183–185.

[247] J. H. Kim. The Ramsey number $R(3,t)$ has order of magnitude $t^2/\log t$, *Random Structures & Algorithms* **7** (1995), 173–207.

[248] T. P. Kirkman. On a problem in combinatorics, *The Cambridge and Dublin Mathematical Journal* **2** (1847), 191–204.

[249] L. Kirousis, J. Livieratos, and K. I. Psaromiligkos. Directed Lovász local lemma and Shearer's lemma, *Annals of Mathematics and Artificial Intelligence* **88** (2020), 133–155.

[250] M. Kneser. Aufgabe 300, *Jahresbericht der Deutschen Mathematiker-Vereinigung* **58** (1955).

[251] D. E. Knuth. Big omicron and big omega and big theta, *ACM Sigact News* **8** (1976), 18–24.

[252] D. E. Knuth. Mathematics and computer science: coping with finiteness, *Science* **194** (1976), 1235–1242.

[253] D. E. Knuth. *The Art of Computer Programming, Volume 4A: Combinatorial Algorithms, Part 1,* Addison–Wesley, 2011.

[254] D. E. Knuth. *The Art of Computer Programming, Volume 4, Fascicle 5: Mathematical Preliminaries Redux; Introduction to Backtracking; Dancing Links,* Addison–Wesley, 2020.

[255] V. F. Kolchin. On the behavior of a random graph near a critical point (Russian). *Teoriya Veroyatnosteĭ i ee Primeneniya* **31** (1986), 503–515. English translation in *Theory of Probability & Its Applications* **31** (1987), 439–451.

[256] J. Kollár, L. Rónyai, and T. Szabó. Norm-graphs and bipartite Turán numbers, *Combinatorica* **16** (1996), 399–406.

[257] J. Komlós and M. Simonovits, Szemerédi's regularity lemma and its applications in graph theory, in: *Combinatorics, Paul Erd?os is Eighty* (D. Miklós et al., eds.), Bolyai Society Mathematical Studies **2**, János Bolyai Mathematical Society, Budapest, 1996, pp. 295–352.

[258] J. Komlós and E. Szemerédi. Limit distribution for the existence of Hamiltonian cycles in a random graph, *Discrete Mathematics* **43** (1983), 55–63.

[259] A. D. Koršunov. Solution of a problem of Erdős and Rényi on Hamiltonian cycles in nonoriented graphs (in Russian), *Doklady Akademii Nauk SSSR* **228** (1976), pp. 529–532. English translation in *Soviet Mathematics. Doklady.* **17** (1976), 760–764.

[260] A. D. Koršunov. A solution of a problem of P. Erdős and A. Rényi about Hamiltonian cycles in undirected graphs (in Russian), *Metody Diskretnogo Analiza* **31** (1977), 17–56.

[261] A. V. Kostochka. A class of constructions for Turán's $(3, 4)$-problem. *Combinatorica* **2** (1982), 187–192.

[262] A. V. Kostochka. The minimum Hadwiger number for graphs with a given mean degree of vertices, *Metody Diskretnogo Analiza* **38** (1982), 37–58 [in Russian].

[263] A. V. Kostochka. A lower bound for the Hadwiger number of graphs by their average degree, *Combinatorica* **4** (1984), 307–316.

[264] M. Kouril. Computing the van der Waerden number $W(3, 4) = 293$, *Integers* **12** (2012), A46.

[265] M. Kouril and J. L. Paul. The Van der Waerden Number $W(2, 6)$ is 1132, *Experimental Mathematics* **17** (2008), 53–61.

[266] T. Kővári, V. T. Sós, and P. Turán. On a problem of K. Zarankiewicz. *Colloquium Mathematicum* **3** (1954), 50–57.

[267] J. Kozik and D. Shabanov. Improved algorithms for colorings of simple hypergraphs and applications, *Journal of Combinatorial Theory. Series B* **116** (2016), 312–332.

[268] K. Kuratowski. Sur le problème des courbes gauches en topologie, *Fundamenta Mathematicae* **15** (1930), 271–283.

[269] C. W. H. Lam. The search for a finite projective plane of order 10, *American Mathematical Monthly* **98** (1991), 305–318.

[270] C. W. H. Lam, L. Thiel, and S. Swiercz. The nonexistence of finite projective planes of order 10, *Canadian Journal of Mathematics* **41** (1989), 1117–1123.

[271] E. Landau. *Handbuch der Lehre von der Verteilung der Primzahlen,* B. G. Teubner, Leipzig, Berlin, 1909. Reprinted (with an Appendix by P. T. Bateman) by Chelsea Publishing Co., New York, 1953.

[272] F. Lazebnik. V. A. Ustimenko, and A. J. Woldar. A new series of dense graphs of high girth, *Bulletin of the American Mathematical Society* **32** (1995), 73–79.

[273] A.-M. Legendre. *Essai sur la Théorie des Nombres,* Courcier, Paris, 1808.

[274] L. Lesniak. Chvátal's t_0-tough conjecture, in: *Graph Theory: Favorite Conjectures and Open Problems - 1* (G. Ralucca, S. Hedetniemi, and C. Larson, eds), pp. 135–147, Springer, 2016.

[275] M. Lewin. A new proof of a theorem of Erdős and Szekeres, *The Mathematical Gazette* **60** (1976), 136–138.

[276] J. Q. Longyear and T. D. Parsons. The friendship theorem, *Indagationes Mathematicae* **34** (1972), 257–262.

[277] L. Lovász. On chromatic number of finite set-systems, *Acta Mathematica Academiae Scientiarum Hungaricae* **19** (1968), 59–67.

[278] L. Lovász. Kneser's conjecture, chromatic number, and homotopy, *Journal of Combinatorial Theory. Series A* **25** (1978), 319–324.

[279] L. Lovász. *Combinatorial Problems and Exercises*, Second edition, North-Holland Publishing Co., Amsterdam, 1993.

[280] D. Lubell. A short proof of Sperner's theorem, *Journal of Combinatorial Theory* **1** (1966), 299.

[281] T. Łuczak. Component behavior near the critical point of the random graph process, *Random Structures & Algorithms* **1** (1990), 287–310.

[282] T. Łuczak. On the equivalence of two basic models of random graphs, in: *Proceedings of Random Graphs'87* (M. Karoński, J. Jaworski, A. Ruciński, eds.), pp. 151–157, Wiley, 1990.

[283] T. Łuczak. B. Pittel, and J. C. Wierman. The structure of a random graph at the point of the phase transition, *Transactions of American Mathematical Society* **341** (1994), 721–748.

[284] C. McDiarmid and A. Steger. Tidier examples for lower bounds on diagonal Ramsey numbers, *Journal of Combinatorial Theory. Series A* **74** (1996), 147–152.

[285] B. D. McKay and S. P. Radziszowski. The first classical Ramsey number for hypergraphs is computed, *Proceedings of the second annual ACM-SIAM symposium on Discrete algorithms (SODA '91)*, 304–308.

[286] B. D. McKay and S. P. Radziszowski. $R(4, 5) = 25$, *Journal of Graph Theory* **19** (1995), 309–322.

[287] B. D. McKay and S. P. Radziszowski. Subgraph counting identities and Ramsey numbers, *Journal of Combinatorial Theory* Series B, **69** (1997), 193–209.

[288] B. D. McKay and Zhang Ke Min. The value of the Ramsey number $R(3, 8)$, *Journal of Graph Theory* **16** (1992), 99–105.

[289] K. N. Majumdar. On some theorems in combinatorics relating to incomplete block designs, *Annals of Mathematical Statistics* **24** (1953), 377–389.

[290] W. Mantel. Problem 28. Solution by H. Gouwentak, W. Mantel, J. Teixeira de Mattes, F. Schuh, and W. A. Wythoff. *Wiskundige Opgaven* **10** (1907), 60–61.

[291] J. Maynard. Large gaps between primes, *Annals of Mathematics* **183** (2016), 915–933.

[292] E. Melchior. Über Vielseite der Projektive Ebene, *Deutsche Mathematik* **5** (1940), 461–475.

[293] L. D. Meshalkin. Generalization of Sperner's theorem on the number of subsets of a finite set, *Theory of Probability and Its Applications* **8** (1963), 203–204.

[294] E. W. Miller. On the property of families of sets, *Comptes Rendus des Séances de la Société des Sciences et des Lettres de Varsovie. Classe III* **30** (1937), 31–38.

[295] M. Molloy and B. Reed. *Graph Colouring and the Probabilistic Method*, Springer-Verlag, Berlin, 2002.

[296] L. Moser. Notes on number theory. II. On a theorem of van der Waerden. *Canadian Mathematical Buletin* **3** (1960), 23–25.

[297] W. O. J. Moser. On the relative widths of coverings by convex bodies, *Canadian Mathematical Bulletin* **1** (1958), 154–154.

[298] Th. Motzkin. The lines and planes connecting the points of a finite set, *Transactions of American Mathematical Society* **70** (1951), 451–464.

[299] O. Murphy. Lower bounds on the stability number of graphs computed in terms of degrees, *Discrete Mathematics* **90** (1991), 207–211.

[300] J. Mycielski. Sur le coloriage des graphes, *Colloquium Mathematicum* **3** (1955), 161–162.

[301] A. Nachmias, Y. Peres. The critical random graph, with martingales, *Israel Journal of Mathematics* **176** (2010), 29–41.

[302] B. Nagle, V. Rödl, and M. Schacht. The counting lemma for regular k-uniform hypergraphs, *Random Structures & Algorithms* **28** (2006), 113–179.

[303] C. St J. A. Nash-Williams. Hamiltonian arcs and circuits, in: *Recent Trends in Graph Theory. Proceedings of the First New York City Graph Theory Conference, June 11 – 13, 1970* (M. Capobianco, J. B. Frechen, and M. Krolik, eds.), Lecture Notes in Mathematics **186**, Springer, Berlin, pp. 197–210.

[304] J. Nešetřil and V. Rödl. A short proof of the existence of highly chromatic hypergraphs without short cycles, *Journal of Combinatorial Theory. Series B* **27** (1979), 225–227.

[305] P. Neumann. Über den Median der Binomial- and Poissonverteilung, *Wissenschaftliche Zeitschrift der Technischen Universität Dresden* **15** (1966), 229–233.

[306] K. O'Bryant. Sets of integers that do not contain long arithmetic progressions, *The Electronic Journal of Combinatorics* **18** (2011), #P59.

[307] M. Okamoto. Some inequalities relating to the partial sum of binomial probabilities, *Annals of the Institute of Statistical Mathematics,* 10 (1958), 29–35.

[308] O. Ore. Note on Hamilton circuits, *American Mathematical Monthly* **67** (1960), 55.

[309] P. R. J. Östergård. On the minimum size of 4-uniform hypergraphs without property B, *Discrete Applied Mathematics* **163** (2014), 199–204.

[310] J. Pach and P. K. Agarwal. *Combinatorial Geometry.* John Wiley & Sons, 2011.

[311] E. M. Palmer. *Graphical Evolution. An Introduction to the Theory of Random Graphs,* Wiley-Interscience, John Wiley & Sons, Chichester, 1985.

[312] O. Pikhurko. A note on the Turán function of even cycles, *Proceedings of the American Mathematical Society* **140** (2012). 3687–3692.

[313] B. Pittel. On the largest component of the random graph at a nearcritical stage, *Journal of Combinatorial Theory, Series B* **82** (2001), 237–269.

[314] D. H. J. Polymath. A new proof of the density Hales-Jewett theorem, *Annals of Mathematics* **175** (2012), 1283–1327.

[315] D. H. J. Polymath. Variants of the Selberg sieve, and bounded intervals containing many primes, *Research in the Mathematical Sciences* **1** (2014), Article 12, 83 pages.

[316] L. Pósa. A theorem concerning Hamilton lines, *A Magyar Tudományos Akadémia Matematikai Kutató Intézetének Közleményei* **7** (1962), 225–226.

[317] L. Pósa. Hamiltonian circuits in random graphs, *Discrete Mathematics* **14** (1976), 359–364.

[318] J. Radhakrishnan and S. Shannigrahi. Streaming algorithms for 2-coloring uniform hypergraphs, in: *Algorithms and Data Structures,* Lecture Notes in Computer Science **6844** (2011), pp. 667–678.

[319] J. Radhakrishnan and A. Srinivasan. Improved bounds and algorithms for hypergraph 2-coloring, *Random Structures & Algorithms* **16** (2000), 4–32. Also in *Proceedings of 39th Annual Symposium on Foundations of Computer Science, 1998,* pp. 684–693.

[320] S. P. Radziszowski. Small Ramsey numbers, *The Electronic Journal of Combinatorics* (2017), revision #15.

[321] A. M. Raĭgorodskiĭ and D. A. Shabanov. The Erdős-Hajnal problem of hypergraph colorings, its generalizations, and related problems (Russian), *Uspekhi Matematicheskikh Nauk [N. S.]* **66** (2011), 109–182; translation in *Russian Mathematics Surveys* **66** (2011), 933–1002.

[322] S. Ramanujan. A proof of Bertrand's postulate, *Journal of the Indian Mathematical Society* **11** (1919), 181–182. Also in *Collected Papers of Srinivasa Ramanujan* (G. H. Hardy, P. V. Seshu Aiyar, B. M. Wilson, eds.), AMS/Chelsea Publication, 2000, pp. 208–209.

[323] F. P. Ramsey. On a problem of formal logic, *Proceedings of the London Mathematical Society* **30** (1930), 361–376.

[324] R. A. Rankin. The difference between consecutive primes, *Journal of the London Mathematical Society* **13** (1938), 242–247.

[325] A. Rao. Coding for sunflowers, *Discrete Analysis* 2020:2, 8 pp.

[326] G. Ringel. Extremal problems in the theory of graphs, in: *Theory of Graphs and its Applications. Proceedings of the Symposium held in Smolenice in June 1963* (M. Fiedler, ed.), Publishing House of the Czechoslovak Academy of Sciences, Prague, 1964, pp. 85–90.

[327] H. Robbins. A remark on Stirling's formula, *American Mathematical Monthly* **62** (1955), 26–29.

[328] N. Robertson, D. Sanders, P. Seymour, and R. Thomas. The four-colour theorem. *Journal of Combinatorial Theory. Series B* **70** (1997), 2–44.

[329] N. Robertson, P. Seymour, and R. Thomas. Hadwiger's conjecture for K_6-free graphs, *Combinatorica* **13** (1993), 279–361.

[330] V. Rödl, On a packing and covering problem, *European Journal of Combinatorics* **5** (1985), 69–78.

[331] V. Rödl and E. Šiňajová. Note on Ramsey numbers and self-complementary graphs, *Mathematica Slovaca* **45** (1995), 243–249.

[332] V. Rödl and J. Skokan. Regularity lemma for k-uniform hypergraphs, *Random Structures & Algorithms* **25** (2004), 1–42.

[333] V. Rödl and J. Skokan. Applications of the regularity Lemma for uniform hypergraphs, *Random Structures & Algorithms* **28** (2006), 180–194.

[334] K. F. Roth. On certain sets of integers, *Journal of the London Mathematical Society* **28** (1953), 104–109.

[335] H. J. Ryser. *Combinatorial Mathematics*, The Carus Mathematical Monographs, No. 14. Published by The Mathematical Association of America; distributed by John Wiley and Sons, Inc., New York, 1963.

[336] H. J. Ryser. An extension of a theorem of de Bruijn and Erdős on combinatorial designs, *Journal of Algebra* **10** (1968), 246–261.

[337] A. Schrijver. Vertex-critical subgraphs of Kneser graphs, *Nieuw Archief voor Wiskunde. Derde Serie* **26** (1978), 454–461.

[338] H. A. Schwarz. Über ein die Flächen Kleinsten Flächeninhalts betreffendes Problem der Variationsrechnung, *Acta Societatis Scientiarum Fennicae* **15** (1885), 315–362.

Reprinted in *Gesammelte Mathematische Abhandlungen, Vol. 1,* New York: Chelsea, pp. 224–269, 1972.

[339] P. Seymour. A note on a combinatorial problem of Erdős and Hajnal, *Bulletin of the London Mathematical Society* **8** (1974), 681–682.

[340] P. Seymour. Hadwiger's conjecture, in: *Open Problems in Mathematics* (J. F. Nash, Jr. and M. T. Rassis, eds.), Springer, 2016, pp. 417–437.

[341] J. B. Shearer. A note on the independence number of triangle-free graphs, *Discrete Mathematics* **46** (1983), 83–87.

[342] A. Sidorenko. Upper bounds for Turán numbers, *Journal of Combinatorial Theory. Series A* **77** (1997), 134–147.

[343] M. Simonovits. Paul Erdős' influence on extremal graph theory, in: *The Mathematics of Paul Erdős, II* (R. L. Graham and J. Nešetřil, eds.), Algorithms and Combinatorics, 14, Springer, Berlin, 1997, pp. 148–192.

[344] C. A. B. Smith and S. Abbott. The story of Blanche Descartes, *The Mathematical Gazette* **87** (2003), 23–33.

[345] V. T. Sós. Remarks on the connection of graph theory, finite geometry and block designs, in: *Colloquio Internazionale sulle Teorie Combinatorie, Roma, 3–15 settembre 1973,* Atti dei Convegni Lincei, No. 17, Accademia nazionale dei Lincei, Rome, 1976, Tomo II, pp. 223–233.

[346] J. Spencer. Ramsey's theorem — A new lower bound, *Journal of Combinatorial Theory. Series A* **18** (1975), 108–115.

[347] E. Sperner. Ein Satz über Untermengen einer endliche Menge, *Mathematische Zeitschrift* **27** (1928), 544–548.

[348] M. J. Steele. The Cauchy-Schwarz master class. An introduction to the art of mathematical inequalities. MAA Problem Books Series. *Mathematical Association of America, Washington, DC; Cambridge University Press, Cambridge,* 2004

[349] R. S. Stevens and R. Shantaram. Computer-generated van der Waerden partitions, *Mathematics of Computation* **32** (1978), 635–636.

[350] A. Suk. On the Erdős-Szekeres convex polygon problem, *Journal of the American Mathematical Society* **30** (2017), 1047–1053.

[351] J. J. Sylvester. Mathematical Question 11851, *Educational Times* **59** (1893), p. 98.

[352] G. Szekeres. A combinatorial problem in geometry: Reminiscences, *P. Erdős: The Art of Counting. Selected Writings* (J. Spencer, ed.), Mathematicians of Our Time, Vol. 5. The MIT Press, Cambridge, MA; London, 1973, pp. xix–xxii.

[353] G. Szekeres and L. Peters. Computer solution to the 17-point Erdős-Szekeres problem, *The ANZIAM Journal* **48** (2006), 151–164.

[354] G. Szekeres and H. S. Wilf. An inequality for the chromatic number of a graph, *Journal of Combinatorial Theory* **4** (1968), 1–3.

[355] E. Szemerédi. On sets of integers containing no four elements in arithmetic progression, *Acta Mathematica Academiae Scientiarum Hungaricae* **20** (1969), 89–104.

[356] E. Szemerédi. On sets of integers containing no k elements in arithmetic progression, *Acta Arithmetica* **27** (1975), 199–245.

[357] T. Tao. A variant of the hypergraph removal lemma, *Journal of Combinatorial Theory, Series A* **113** (2006), 1257–1280.

[358] T. Tao. A correspondence principle between (hyper)graph theory and probability theory, and the (hyper)graph removal lemma, *Journal d'Analyse Mathématique* **103** (2007), 1–45.

[359] A. Thomason. An extremal function for contractions of graphs, *Mathematical Proceedings of the Cambridge Philosophical Society* **95** (1984), 261–265.

[360] G. Tiner. On the Erdős-Sós Conjecture and double-brooms, *Journal of Combinatorial Mathematics and Combinatorial Computing* **93** (2015), 291–296.

[361] B. Toft. On colour-critical hypergraphs, in: *Infinite and Finite Sets: To Paul Erdős on his 60th Birthday* (A. Hajnal, R. Rado, V. T. Sós, eds.), North Holland Publishing Co., 1975, pp. 1445–1457.

[362] P. Turán. Egy gráfelméleti szélsőértékfeladatról (On an extremal problem in graph theory, in Hungarian), *Középiskolai Matematikai és Fizikai Lapok* **48** (1941), 436–452.

[363] P. Turán. On the theory of graphs, *Colloquium Mathematicum* **3** (1954), 19–30.

[364] P. Turán. Research problems, *A Magyar Tudományos Akadémia Matematikai Kutató Intézetének Közleményei* **6** (1961), 417–423.

[365] Zs. Tuza. Applications of the set-pair method in extremal hypergraph theory, in: *Extremal Problems for Finite Sets* (P. Frankl et al., eds.), Bolyai Society Mathematical Studies **3**, János Bolyai Mathematical Society, Budapest, 1994, pp. 479–514.

[366] Zs. Tuza. Applications of the set-pair method in extremal problems, II, in: *Combinatorics, Paul Erdős is Eighty* (D. Miklós, V. T. Sós, T. Szőnyi, eds.), Bolyai Society Mathematical Studies **2**, János Bolyai Mathematical Society, Budapest, 1996, pp. 459–490.

[367] P. Ungar. Advanced Problem 4526, *American Mathematical Monthly* **60** (1953), 123 and 336.

[368] P. Ungar. E-mail message to V.C. on 31 October 2009.

[369] J. V. Uspensky. *Introduction to Mathematical Probability,* McGraw-Hill, New York and London, 1937.

[370] O. Veblen and W. H. Bussey. Finite projective geometries, *Transactions of the American Mathematical Society,* **7** (1906), 241–259.

[371] B. L. van der Waerden. Beweis einer Baudetschen Vermutung, *Nieuw Archief voor Wiskunde* **15** (1927), 212–216.

[372] B. L. van der Waerden. How the proof of Baudet's conjecture was found, in: *Studies in Pure Mathematics, Papers Presented to Richard Rado on the Occasion of His Sixty-Fifth Birthday* (L. Mirsky, ed.), Academic Press, London and New York, 1971, pp. 251–260.

[373] K. Wagner. Über eine Eigenschaft der ebenen Komplexe, *Mathematische Annalen* **114** (1937), 570–590.

[374] K. Wagner. Beweis einer Abschwächung der Hadwiger-Vermutung, *Mathematische Annalen* **153** (1964), 139–141.

[375] H. S. Wilf. The friendship theorem, in: *Combinatorial Mathematics and Its Applications* (D. J. A. Welsh, ed.), Academic Press, London, 1971, pp. 307–309.

[376] K. Yamamoto. Logarithmic order of free distributive lattice, *Journal of the Mathematical Society of Japan* **6** (1954), 343–353.

[377] Y. Zhang. Bounded gaps between primes, *Annals of Mathematics* **179** (2014), 1121–1174.

[378] A. A. Zykov. On some properties of linear complexes (in Russian), *Matematicheskiĭ Sbornik. Novaya Seriya* **24** (1949), 163–188; translated in *American Mathematical Society Translations Series 1* **7** (1952), 418–449.

Index

Printed in the United States
by Baker & Taylor Publisher Services